666·767 : 539·4·011·25 : 539·219·2

Fracture Mechanics of Ceramics

Volume 4
Crack Growth and Microstructure

Volume 1   Concepts, Flaws, and Fractography
Volume 2   Microstructure, Materials, and Applications
Volume 3   Flaws and Testing
Volume 4   Crack Growth and Microstructure

Fracture Mechanics of Ceramics

Volume 4
Crack Growth and Microstructure

Edited by R. C. Bradt

Department of Materials Science and Engineering
Ceramic Science and Engineering Section
Pennsylvania State University
University Park, Pennsylvania

D. P. H. Hasselman

Department of Materials Engineering
Virginia Polytechnic Institute and State University
Blacksburg, Virginia

and F. F. Lange

Science Center
Rockwell International
Thousand Oaks, California

PLENUM PRESS · NEW YORK-LONDON

Library of Congress Cataloging in Publication Data

Symposium on the Fracture Mechanics of Ceramics, Pennsylvania State University, 1973.
    Fracture mechanics of ceramics.

    Includes bibliographical references.
    CONTENTS: v. 1. Concepts, flaws, and fractography. – v. 2. Microstructure, materials,
and applications. – v. 3. Flaws and testing. – v. 4. Crack growth and microstructure.
    1. Ceramics – Fracture – Congresses. I. Bradt, Richard Carl, 1938-   ed. II. Hassel-
man, D. P. H., 1931-   ed. III. Lange, F. F., 1939-   ed. IV. Pennsylvania. State
University. V. Symposium on the Fracture Mechanics of Ceramics, Pennsylvania State University. 1977. VI. Title.
TA430.S97 1973   620.1'4   73-20399
ISBN 0-306-37594-X (v. 4)

© 1978 Plenum Press, New York
A Division of Plenum Publishing Corporation
227 West 17th Street, New York, N.Y. 10011

Printed in the United States of America

PREFACE

These volumes, 3 and 4, of Fracture Mechanics of Ceramics constitute the proceedings of an international symposium on the fracture mechanics of ceramics held at the Pennsylvania State University, University Park, PA on July 27, 28, and 29, 1977. Volumes 1 and 2 were published previously as the proceedings of a symposium of the same name held July 11, 12, and 13, 1973, also at Penn State.

All four volumes published to date concentrate on the fracture aspects of the mechanical behavior of brittle ceramics in terms of the characteristics of cracks.

The program chairmen gratefully acknowledge the financial assistance for the symposium provided by the Office of Naval Research, the Energy Research and Development Administration, and the Army Research Office. Without their support the quality and magnitude of this conference simply would not have been possible.

Numerous individuals contributed to the success of the conference, but unfortunately they cannot all be listed here. However the program chairmen would especially like to recognize the contributions of Penn State Conference Coordinator, Mr. Ronald Avillion, whose expertise in planning and organization was indispensable; Dr. Fred R. Matson for his interesting after dinner speech; and Drs. A. M. Diness, J. C. Hurt, and D. W. Readey for their encouragement and valuable suggestions regarding the program.

Finally, we wish to also thank our joint secretaries for the patience and help in bringing these proceedings to press.

University Park, PA	R. C. Bradt
Blacksburg, VA	D. P. H. Hassleman
Thousand Oaks, CA	F. F. Lange

July, 1977

v

CONTENTS OF VOLUME 4

Contents of Volume 3   xi

Lattice Theories of Fracture   507
    E. R. Fuller, Jr., and R. M. Thomson

Mechanisms of Subcritical Crack Growth in Glass   549
    S. M. Wiederhorn

Stress Corrosion Mechanisms in E- Glass Fiber   581
    C. L. McKinnis

A Multibarrier Rate Process Approach to Subcritical
          Crack Growth   597
    S. D. Brown

Stress Rupture Evaluations of High Temperature
          Structural Materials   623
    R. J. Charles

Growth of Cracks Partly Filled with Water   639
    T. A. Michalske, J. R. Varner, V. D. Frechette

Electrolytic Degradation of Lithia–Stabilized
          Polycrystalline β"- Alumina   651
    D. K. Shetty, A. V. Virkar, and R. S. Gordon

Engineering Design and Fatigue Failure of Brittle
          Materials   667
    J. E. Ritter, Jr.

Subcritical Crack Growth in PZT   687
    J. G. Bruce, W. W. Gerberich, and B. G. Koepke

Subcritical Crack Growth in Electrical Porcelains   711
    M. Matsui, T. Soma, and I. Oda

Fracture Mechanics of Alumina in a Simulated
        Biological Environment   725
    E. M. Rockar and B. J. Pletka

Modified Double Torsion Method for Measuring Crack
        Velocity in NC-132 (Si$_3$N$_4$)   737
    C. G. Annis and J. S. Cargill

Subcritical Crack Growth in Glass Ceramics   745
    B. J. Pletka and S. M. Wiederhorn

Dynamic Fatigue of Foamed Glass   761
    P. H. Conley, H. C. Chandan, and R. C. Bradt

Static Fatigue Behavior in Chemically Strengthened
        Glass   773
    C. E. Olsen

Prediction of the Self-Fatigue of Surface Compression –
        Strengthened Glass Plates   787
    M. Bakioglu, F. Erdogan, and D. P. H. Hassleman

Fracture Mechanics and Microstructural Design   799
    F. F. Lange

Microcracking in a Process Zone and Its Relation to
        Continuum Fracture Mechanics   821
    R. F. Pabst, J. Steeb, and N. Claussen

A Structure Sensitive K_{Ic}-Value and Its Dependence
        on Grain Size Distribution, Density, and
        Microcrack Formation   835
    F. E. Buresch

Microstructural Dependence of Fracture Mechanics
        Parameters in Ceramics   849
    R. W. Rice, S. W. Freiman, R. C. Pohanka,
    J. J. Mecholsky, Jr., and C. Cm. Wu

Role of Stress Induced Phase Transformation in Enhancing
        Strength and Toughness of Zirconia
        Ceramics   877
    T. K. Gupta

Toughness and Fractography of TiC and WC   891
    J. L. Chermant, A. Deschanvres, and F. Osterstock

Mechanical Properties of Al$_2$O$_3$-HfO$_2$ Eutectic Micro-
        structures   903
    C. O. Hulse

Fracture Resistance and Temperatures in Metal
        Infiltrated Porous Ceramics   913
  R. A. Queeney and N. Rupert

Reaction Sintered Si_3N_4: Development of Mechanical
        Properties Relative to Microstructure and
        the Nitriding Environment   921
  M. W. Lindley, K. C. Pitman, and B. F. Jones

Mechanical Properties of Porous PNZT Poly-
        crystalline Ceramics   933
  D. R. Biswas and R. M. Fulrath

Fracture of Brittle Particulate Composites   945
  D. J. Green and P. S. Nicholson

Some Effects of Dispersed Phases on the Fracture
        Behavior of Glass   961
  J. S. Nadeau and R. C. Bennett

Fracture Toughness of Reinforced Glasses   973
  J. C. Swearengen, E. K. Beauchamp, and R. J. Eagen

Contributors   990

Index   995

CONTENTS OF VOLUME 3

Contents of Volume 4   x

Fundamentals of the Statistical Theory of Fracture   1
  S. B. Batdorf

A General Approach to the Statistical Analysis of
        Fracture   31
  A. G. Evans

Application of the Four Function Weibull Equation in
        the Design of Brittle Components   51
  P. Stanley, A. D. Sivill, and H. Fessler

Multiple Flaw Fracture Mechanics Model for Ceramics . . .   67
  H. A. Nied and K. Arin

Analysis of Microvoids in Si_3N_4 Ceramics by Small
        Angle Neutron Scattering   85
  P. Pizzi

Crack Blunting in Sintered SiC   99
  C. A. Johnson

Alteration of Flaw Sizes and K_{Ic}'s of Si_3N_4 Ceramics
        by Molten Salt Exposure   113
  W. C. Bourne and R. E. Tressler

Weibull Parameters and the Strength of Long
        Glass Fibers   125
  W. D. Scott and A. Gaddipati

Surface Flaws and the Mechanical Behavior of Glass
        Optical Fibers   143
  W. E. Snowden

Mechanical Behavior of Optical Fibers   161
  B. K. Tariyal and D. Kalish

Spin Testing of Ceramic Materials   177
   G. G. Trantina and C. A. Johnson

Effects of Specimen Size on Ceramic Strengths   189
   G. K. Bansal and W. H. Duckworth

Indentation Fracture and Strength Degradation in
           Ceramics   205
   B. R. Lawn and D. B. Marshall

Indentation Induced Strength Degradation and Stress
           Corrosion of Tempered Glasses   231
   M. V. Swain, J. T. Hagan, and J. E. Field

Compressive Microfracture and Indentation Damage
           in Al_2O_3   245
   J. Lankford

Microcracking Associated with the Scratching of
           Brittle Solids   257
   M. V. Swain

Crack Formation During Scratching of Brittle Materials . .   273
   J. D. B. Veldkamp, N. Hattu, and V. A. C. Snijders

Impact Damage in Ceramics   303
   A. G. Evans

Particle Impact Regimes in Single Crystals   333
   S. V. Hooker and W. F. Adler

A High-Speed Photographic Investigation of the Impact
           Damage in Soda-Lime and Borosilicate Glasses
           by Small Glass and Steel Spheres   349
   M. M. Chaudhri and S. M. Walley

Localized Impact Damage in a Viscous Medium (Glass) . . .   365
   H. P. Kirchner and R. M. Gruver

Erosion of Brittle Materials by Solid Particle Impact . .   379
   B. J. Hockey, S. M. Wiederhorn, and H. Johnson

Compression Testing of Ceramics   403
   G. Sines and M. Adams

The Strength of Ceramics Under Biaxial Stresses   435
   G. Tappin, R. W. Davidge, and J. R. McLaren

A Fracture Specimen for High Temperature Testing   451
    A. S. Kobayashi, L. I. Staley, A. F. Emery,
    and W. J. Love

Application of Fracture Mechanics to the Adherence of
            Thick Films and Braze Joints   463
    P. F. Becher, W. L. Newell, and S. A. Halen

A Notched Ring Fracture Toughness Test for Ceramics . . .   473
    D. J. Rowcliffe, R. L. Jones, and J. K. Gran

Short Rod K_{Ic} Measurements of Al_2O_3   483
    L. M. Barker

Impact Fracture of Ceramics at High Temperature   495
    S. T. Gonczy and D. L. Johnson

Contributors   xv

Index   xxi

LATTICE THEORIES OF FRACTURE

E. R. Fuller, Jr. and R. M. Thomson*

Institute for Materials Research
National Bureau of Standards
Washington, D. C.   20234

ABSTRACT

This paper briefly reviews our current understanding of the role of physics in elementary crack-tip processes. Particular attention is given to the influence of atomic discreteness on brittle fracture and crack propagation. One-dimensional lattice models are used to illustrate the phenomenon of "lattice trapping", whereby a crack is stably trapped by the atomic structure. These models are also used to discuss the role of thermodynamic surface energy in the brittle fracture process and in a microscopic Griffith criterion for fracture. Finally, two atomic mechanisms of subcritical crack growth are discussed with reference to a crack that is trapped by a lattice. One mechanism of crack growth results from thermal activation over the lattice-trapping energy barriers; and the other mechanism results from quantum tunneling through the barrier.

1.   INTRODUCTION

In the years following Griffith's original works [1], the field of fracture physics remained relatively inactive, broken only by conjectures from Elliott [2] and Orowan [3] in the late 1940's and from Barenblatt [4] in the late 1950's. Recently, however, progress in this field has been quite rapid. The work not only has been growing in sophistication, but also has been raising a number of important fundamental questions about fracture. Current theories of fracture provide models that allow us to

---

*Supported in part by ARPA through the ARPA Materials Research Council of the University of Michigan.

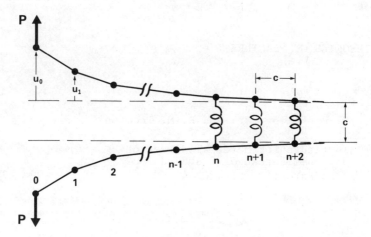

Figure 1.   Quasi-one-dimensional lattice model of a crack
            subjected to an opening mode of loading (Mode I).

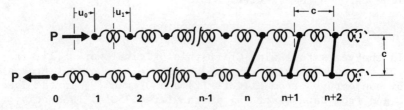

Figure 2.   Quasi-one-dimensional lattice model of a crack
            subjected to a sliding shear mode of loading
            (Mode II).

understand complicated phenomena in a predictive, but qualitative
manner. These theories display the general qualitative aspects
of fracture so that basic causes of behavior are explicitly
demonstrated, and so that qualitative interrelationships are
displayed between such things as crystalline structure, surface
energy, and interatomic forces for different classes of matter.

The simplest models of fracture and crack propagation are
one-dimensional lattice models, which are analogous to the
Frenkel-Kontorova model of a dislocation [5]. Two such quasi-
one-dimensional lattice models of a crack have been presented in
the recent literature. One, a model proposed by Thomson, Hsieh,
and Rana [6], simulates an opening mode of fracture, or mode I
fracture. As illustrated in Fig. 1, this model is composed of
two semi-infinite chains of atoms that interact through two types
of interactions. The interatomic forces in each chain are such
as to resist flexure of that chain, and the cohesive forces,
modeled as stretchable spring elements, are such as to bond the
chains cohesively to each other. Opening forces are applied to
the atoms at one end of the chains in such a way as to split the
chains apart. The crack corresponds to a finite number of broken
stretchable springs. A recent variation of this model [7], and
the one to be discussed in this paper, replaces the linear "bond-
snapping" cohesive force of the earlier model by an arbitrary
nonlinear spring element across the crack plane. The other one-
dimensional lattice model was advanced by Smith [8]. As illustrated in Fig. 2, it corresponds to a sliding shear mode of
fracture, or mode II fracture, but otherwise shows similar attributes to those of the mode I model. Smith also extended his
model to include a variety of nonlinear force laws that resist
shear of the crack faces [9, 10].

Both of these lattice models provide valuable insight into
the elementary physics of crack-tip processes. Their main disadvantage is that they do not lend themselves directly to quantitative
descriptions of real solids. This disadvantage is offset to some
extent, however, by the fact that these one-dimensional theories
are usually analytical so that interrelationships between parameters
in the models can be explicitly displayed.

The first studies of fracture in a discrete lattice were
conducted by computer simulation of a crack in a two-dimensional
atomic lattice. Beginning with the calculations of Goodier and
Kanninen [11], in which they considered nonlinear interatomic
potentials, these computer studies have been directed towards
more quantitative descriptions of fracture in real solids.
Subsequent computer calculations for equilibrium crack configurations have been extended to more realistic interatomic force laws
for iron by Kanninen, Gehlen, and coworkers [12-16] and for
silicon by Sinclair and Lawn [17-19]. The main difficulty with these

computer simulations has been the specification of the external boundary conditions. Limitations of computer memory have required that solutions for a discrete lattice either be truncated at a finite, and relatively small, number of atomic lattice points, or be matched to a continuum elasticity solution at the boundary. The nature of this boundary specification has been a continuing source of discrepancy between various computer computations.

This difficulty has been circumvented by the crack solutions of Hsieh and Thomson [20] which are obtained for a discrete two-dimensional lattice of infinite extent. Their stratagem was to modify known lattice-statics solutions for a perfect square lattice by the annihilation of a finite number of atomic bonds across the crack plane to form the crack. Although their calculations were limited to a simple linear bond-snapping force law, their solutions had the advantage of being partially analytical. Recently, their results for mode I fracture were extended by Esterling [21] to three dimensions (i.e., solutions for kink stability) and to a wide class of force laws. Esterling's work forms an intermediate stage between the simple two-dimensional lattice models and the full scale material simulation studies. Expanding the stratagem of Hsieh and Thomson, Esterling showed how to rigorously fit the linear elastic solution from the region outside of the vicinity of the crack tip to the nonlinear, and possibly nonelastic, region surrounding the crack tip. Following Esterling's procedure, limitations of computer memory no longer need to require that computer solutions be terminated at a small number of lattice points. Instead, the computer memory can be utilized to handle the interatomic interactions in the crack-tip region, and these solutions can be matched to the lattice-statics solutions for an infinite lattice. In addition to these lattice calculations, Smith [22] considered a variant of Elliott's model to obtain a crack solution that is appropriate for a two-dimensional lattice subjected to an antiplanar tearing mode of fracture, or mode III fracture.

Dynamics of crack propagation have been studied in computer simulation by Sanders [23], by Weiner and Pear [24], and by Ashurst and Hoover [25]. The dynamical studies by Ashurst and Hoover demonstrated that atomic discreteness can influence dynamic crack propagation. Their computer calculations showed that the lowest limiting velocity depended upon the assumed force laws, and that supersonic crack velocities (i.e., velocities greater than the longitudinal sound velocity) were, in principle, possible when the stress was large enough. Also, at the lowest limiting velocity the energy absorbed by the surface in the dynamic case was only approximately half of the total energy input, even under constant velocity conditions, indicating that the instabilities experienced at the tip of a crack when a bond snaps make a considerable energy contribution by emitting phonons.

Weiner and Pear showed that spontaneous generation of dislocations is possible when a crack is continuously accelerated.
In this case, the energy losses due to dislocation generation can
stop crack acceleration and cause the crack to achieve a steady
state velocity, which may be subsonic under appropriate conditions.
Both Weiner and Pear, and Ashurst and Hoover included thermal
effects in their calculations, but neither found any significant
influence on the dynamical properties.

The theoretical observation of spontaneous dislocation
generation is of particular interest, because of the importance of
dislocations in an understanding of ductile versus brittle behavior
in solids. Kelly, Tyson and Cottrell [26] first proposed a condition
that separated the ductile/brittle character of solids by comparing
the theoretical shear strength, σ_s, of a material with its theoretical tensile strength, σ_t. They reasoned that the stress field
surrounding an atomically sharp crack possessed very large shear
stress components, and that if these stresses exceeded the theoretical shear strength of the material, then crack blunting would
occur with concomitant ductile behavior. Since the cohesive bonds
in the vicinity of the crack tip must sustain the theoretical
tensile strength, the stress field in this region is normalized by
σ_t, and accordingly, the condition for stability of an atomically
sharp crack can be stated in terms of the ratio, σ_t/σ_s. Specifically,

$$(\sigma_t/\sigma_s) < (\sigma_{tensile}/\sigma_{shear})_{\text{maximum crack-}} \tag{1.1}$$
$$\text{tip stresses.}$$

Another condition for a ductile/brittle characterization was
proposed by Rice and Thomson [27]. They postulated that in order
for blunting to occur, a dislocation must be formed, and that if
dislocations are spontaneously generated from an atomically sharp
crack (i.e., they have a negative energy of formation), then crack
blunting will occur. If, on the other hand, the energy of formation is large, the crystal can sustain the sharp crack. Their
condition can be expressed approximately as

$$\mu b/\gamma \geq 10, \text{ for brittle solids,} \tag{1.2}$$

where μ is the elastic shear modulus, b is the Burgers vector of
the dislocation, and γ is the thermodynamic surface energy for the
fracture plane. This expression is formally quite different from
Eqn. (1.1). Iron turns out to be a special intermediate case for
this criterion, and indeed the computer solutions of Kanninen and
Gehlen indicate dislocation-like distortions in the highly strained
cohesive region of their computer simulations. Also, as mentioned
above, Weiner and Pear find force law parameters which give rise
to spontaneous dislocation generation.

The predictions of Eqns. (1.1) and (1.2) have recently been investigated in more detail from computer simulations by Tyson [28]. He reports cases where the numerical predictions of Eqn. (1.1) and Eqn. (1.2) are significantly different, and notes that the computer simulation generally favors Eqn. (1.1).

A further point of interest arises when one considers that a dislocation generated near the tip of a crack can be trapped by the Peierls lattice interaction [29]. Accordingly, the energy of formation for such a dislocation would be much less than that computed on a continuum basis by Rice and Thomson. Hence, the possibility for spontaneous generation of dislocations by a static crack, and the role of crack acceleration in aiding such spontaneous dislocation generation are still unsolved fundamental problems of prime importance.

In the remainder of the paper, we shall focus our attention on the phenomenon of lattice trapping. Lattice effects are ultimately responsible for subcritical crack growth in brittle materials, whether *in vacuo* or in the presence of a corrosive environment. Therefore, intrinsic lattice-trapping effects must be understood before the more complex chemical effects which are caused by external atmospheres can be understood. Our analysis of lattice trapping also examines a disturbing question recently raised by Esterling [21], namely that the Griffith stress lies outside the lattice-trapping region. In the next section, we shall present two one-dimensional lattice models of a crack, and shall use them to illustrate the phenomenon of lattice trapping, to investigate Esterling's proposition in a simple and analytic fashion, and to clarify the difference between global versus local conditions for fracture. In the final section, we investigate the possibility that quantum fluctuations, or quantum tunneling, can be an important factor in the slow creep of a crack pass atomic trapping barriers. An interesting proposal along this line was made by Gilman and Tong [30], and has been used by Doremus [31] and by Wiederhorn [32] in plotting experimental results for ceramics.

For other comprehensive discussions of a general nature, the reader is referred to a number of excellent reviews that have appeared in the recent literature [9, 33-36]. These papers deal with various atomistic aspects of fracture and crack propagation in a discrete lattice.

2.   ONE-DIMENSIONAL LATTICE THEORIES

In this section details of the two one-dimensional lattice models are presented, and some of the physical insight obtained from these models is discussed. The phenomenon of lattice trapping is seen to result from requirements for mechanical stability of a lattice. When expressed in terms of a "one-dimensional stress intensity factor," these expressions for mechanical stability are found to have a similar mathematical form for both lattice models. Although Griffith's fracture criterion is seen to be a necessary condition for crack growth, spontaneous rapid fracture of a crack trapped by the lattice is shown to occur only when a larger value of stress intensity factor is attained. Finally, the section is concluded with a discussion on the subcritical crack growth that can occur as a lattice-trapped crack is thermally activated over the lattice energy barriers.

2.1 Mode I Lattice Model

This model consists of two semi-infinite chains of atoms that are bonded with two types of interatomic force interactions, as illustrated in Fig. 1. These force interactions are modeled as flexural spring elements that resist flexure of each chain of atoms, and as stretchable spring elements that cohesively bond the two chains to one another. The stretchable elements up to the nth atoms are considered to be stretched beyond their finite range of interaction, or "broken", thus forming a crack of finite length, a = nc. The free ends of this crack at the zeroth atoms are subjected to equal and opposite transverse opening forces P. All displacements are constrained to be transverse to the undistorted chain with the displacement of the jth atom from its equilibrium separation, c, being denoted by u_j.

The total potential energy of this system is composed of three contributions:

(i) *Potential energy of external loading system.* As the external forces do work against the lattice, their potential energy diminishes by an equivalent amount. For an outward displacement of the zeroth atoms, this change in potential energy of the external loading system is given by

$$U_{external} = -2Pu_0 \; , \tag{2.1}$$

which is equal to the negative of the work done by the external forces.

(ii) *Strain energy of the flexural bonds.* The interaction of the flexural spring elements is modeled as a second-neighbor interaction between the atoms at j-1 and j+1 that resists flexure about their common nearest neighbor at j. The strain energy for this interaction about atom j is given by

$$\tfrac{1}{2} \beta [(u_{j-1}-u_j) - (u_j-u_{j+1})]^2 = \tfrac{1}{2} \beta [u_{j+1} - 2u_j + u_{j-1}]^2, \tag{2.2}$$

where β is the spring constant for this interaction. The strain energy for each chain of atoms is the summation of these contributions for atoms $j=1, 2, \cdots$. Counting one contribution for each side of the crack, the total flexural strain energy is twice this summation, or

$$U_{flexural} = \beta \sum_{j=1}^{\infty} (u_{j+1} - 2u_j + u_{j-1})^2. \tag{2.3}$$

(iii) Surface energy of the cohesive bonds. The cohesive bonds across the crack surface are modeled as nonlinear spring elements, that are allowed to have an arbitrary restoring force, $f_I(u)$, when the spring has been extended a distance $2u$ from its equilibrium length c. Such an idealized nonlinear atomic force law is illustrated schematically in Fig. 3. The linear elastic stiffness of these spring elements is defined by a spring constant, α, according to

$$\alpha = [df_I(u)/d(2u)]_{u=0}. \tag{2.4}$$

In addition, a finite range of interaction is assumed for the nonlinear cohesive forces so that elements which are stretched beyond a critical separation, $c + 2u_c$, are taken to be "broken". The energy contained in the jth stretchable spring element across the crack plane can be written as

$$2\gamma_I(u_j)c = 2 \int_0^{u_j} f_I(u)du, \tag{2.5}$$

where $\gamma_I(u_j)$ is defined as the surface energy density, or the surface energy per unit length (factor of c) assigned to each surface (factor of 2) of the chain of atoms [37]. Since the cohesive force is assumed to have a finite range, the cohesive energy of a bond is given by

$$U_b \equiv 2\gamma_{Io}c = 2 \int_0^{u_c} f_I(u)du, \tag{2.6}$$

where γ_{Io} is the macroscopic surface energy density as defined by one-half the work of separation per unit length of surface formed. The surface energy for the cohesive bonds is the summation of the contributions, Eqn. (2.5), for bonds $j = 0, 1, 2, \cdots$. Since the first n bonds (j=0 to n-1) are assumed to be broken, their contribution to the surface energy is $2\gamma_{Io}nc = 2\gamma_{Io}a$. Thus, the total cohesive surface energy is

$$U_{cohesive} = 2\gamma_{Io}nc + 2c\sum_{j=n}^{\infty} \gamma_I(u_j). \tag{2.7}$$

The inclusion in Eqn. (2.7) of an infinite number of stretchable spring elements, that have a nonlinear elastic interaction, makes the resulting solution intractable. It is reasonable to assume

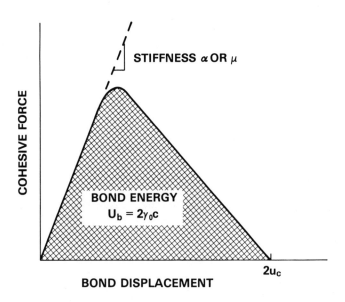

Figure 3.   Idealized interatomic cohesive force law. The linear
elastic stiffness, α or μ, is determined by the initial
slope and the cohesive bond energy is determined by the
area under the curve.

that a finite distance away from the crack tip that the cohesive
bonds are only stretched into their linear elastic region, where

$$f_I(u) = \alpha(2u),$$

and

$$2\gamma_I(u)c = \frac{1}{2}\alpha(2u)^2. \qquad (2.8)$$

Several general nonlinear spring elements can easily be included
with the aid of a computer, but for the present paper we will
assume that only one cohesive bond is nonlinear *elastic*, and that
all stretchable elements beyond j=n are *assumed* to be linear
elastic. Thus, Eqn. (2.7) for the total cohesive energy becomes

$$U_{\text{cohesive}} = 2\gamma_{Io}nc + 2\gamma_I(u_n)c + 2\alpha\Sigma_{j=n+1}^{\infty} u_j^2. \qquad (2.9)$$

However, this assumption of elastic linearity for the stretchable
elements beyond the crack tip should be verified from the solutions
for displacement, in order to assure self-consistency of the model
(see discussion in Section 2.4).

Combining these potential energy terms, Eqns. (2.1), (2.3), and (2.9), the total potential energy of the system is given by

$$U_I = U_{external} + U_{flexural} + U_{cohesive}$$

$$= -2Pu_0 + \beta\Sigma_{j=1}^{\infty} \left(u_{j+1} - 2u_j + u_{j-1}\right)^2$$

$$+2\gamma_{Io}nc + 2\gamma_I(u_n)c + 2\alpha\Sigma_{j=n+1}^{\infty} u_j^2. \tag{2.10}$$

For a given applied force P and crack length a=nc, necessary conditions for equilibrium configurations of the crack are

$$(\partial U_I/\partial u_j) = 0, \text{ for } j=0,1,2,\cdots. \tag{2.11}$$

These equations of stability give the following set of equations to be solved for the displacements u_j:

$$\beta(u_2 - 2u_1 + u_0) = P; \tag{2.12}$$

$$\beta(u_3 - 4u_2 + 5u_1 - 2u_0) = 0; \tag{2.13}$$

$$\beta(\delta^4 u_j) = 0, \text{ for } j=2,3,\cdots, n-1; \tag{2.14}$$

$$\beta(\delta^4 u_n) + f_I(u_n) = 0; \tag{2.15}$$

$$\beta(\delta^4 u_j) + 2\alpha u_j = 0, \text{ for } j=n+1, n+2, \cdots; \tag{2.16}$$

where the difference operator δ^4 is given by

$$(\delta^4 u_j) = u_{j+2} - 4u_{j+1} + 6u_j - 4u_{j-1} + u_{j-2}. \tag{2.17}$$

Two of these equations, Eqn. (2.14) and Eqn. (2.16), are fourth-order difference equations which are to be solved for the displacements; whereas the remaining equations are in effect boundary conditions on the general solution, or coupling equations for the two general solutions.

The general solution of Eqn. (2.14) is

$$u_j^A = A_0 + A_1 j + A_2 j^2 + A_3 j^3. \tag{2.18}$$

The boundary conditions for this solution are the end conditions that describe the application of the external forces, Eqns. (2.12) and (2.13), and the coupling equations, Eqn. (2.14) with j=2 and j=3:

$$(\delta^4 u_2) = 4(u_1^A - u_1) - (u_0^A - u_0) = 0, \tag{2.19}$$

and

$$(\delta^4 u_3) = -(u_1^A - u_1) \qquad\qquad = 0, \qquad\qquad (2.20)$$

since $(\delta^4 u_2^A) = (\delta^4 u_3^A) = 0$. Solving these four equations for A_0, A_1, A_2, and A_3, the solutions for the atoms on the free surface of the crack are

$$u_j = u_0 + (u_1 - u_0 - P/6\beta)j + Pj^3/6\beta, \qquad\qquad (2.21)$$

for $j = 0, 1, 2, \cdots, n-1$, where u_0 and u_1 are as yet unspecified.

The general solution of the difference equation (2.16) is obtained by assuming a solution of the form $u_i = [\rho \exp(i\lambda)]^j$. Substitution into Eqn. (2.16) gives the determinantal equation

$$(\rho e^{i\lambda} - 1)^4 = - (2\alpha/\beta)\rho^2 e^{2\lambda i}, \qquad\qquad (2.22)$$

for which both the real and imaginary parts must be satisfied. After some algebraic manipulation, one of the four solutions for ρ and λ is

$$\rho = \left(\frac{\xi-c}{\xi+c}\right)^{1/2} \quad \text{and} \quad \lambda = \sin^{-1}(c/\xi), \qquad\qquad (2.23)$$

where ξ is a length defined by the spring constant ratio, β/α, according to

$$\xi/c = [(1 + \sqrt{1 + 8\beta/\alpha})/2]^{1/2},$$

or

$$(2\beta/\alpha) = \xi^2(\xi^2 - c^2)/c^4. \qquad\qquad (2.24)$$

It is easily seen from Eqn. (2.22) that when one solution for ρ and λ has been obtained, the three remaining solutions for $\rho e^{i\lambda}$ are obtained from this solution by taking both its inverse and its complex conjugate. Thus, the general solution of Eqn. (2.16) can be written as

$$u_j^B = (B_1 e^{i\lambda j} + B_2 e^{-i\lambda j}) \left(\frac{\xi-c}{\xi+c}\right)^{j/2}$$

$$+ (B_3 e^{i\lambda j} + B_4 e^{-i\lambda j}) \left(\frac{\xi+c}{\xi-c}\right)^{j/2}. \qquad\qquad (2.25)$$

Since the displacements must become vanishingly small a large distance from the crack tip in the "uncracked" portion of the chains [i.e., $\lim(u_i) = 0$, as $j \to \infty$], and since $\xi \geq c$, the coefficients B_3 and B_4 must vanish as another boundary condition.

The solutions for the "cracked" and "uncracked" portions of the chains are coupled together by the crack-tip region. The pertinent equations are Eqn. (2.14) with $j=n-2$ and $j=n-1$, and Eqn. (2.16) with $j=n+1$ and $j=n+2$:

$$(\delta^4 u_{n-2}) = (u_n - u_n^A) = 0 \qquad (2.26)$$

$$(\delta^4 u_{n-1}) = (u_{n+1}^B - u_{n+1}^A) - 4(u_n - u_n^A) = 0 \qquad (2.27)$$

$$(\delta^4 + 2\alpha/\beta)u_{n+1} = (u_{n-1}^A - u_{n-1}^B) - 4(u_n - u_n^B) = 0 \qquad (2.28)$$

$$(\delta^4 + 2\alpha/\beta)u_{n+2} = (u_n - u_n^B) = 0. \qquad (2.29)$$

Basically, these equations required that the solutions for the two regions, Eqn. (2.21) and Eqn. (2.25), are both satisfied for $j = n-1$, n, and $n+1$. The four resulting equations can be solved for the constants u_0, u_1, B_1, and B_2 to give the following solutions for the displacements:

(i) In the "cracked" region ($j = 0,1,2,\cdots$, $n-1$)

$$u_j = (a+\xi-x_j)u_n/\xi$$
$$+ P(a-x_j)[2a^2 + 3a\xi + c^2 - x_j(a+x_j)]/6\beta c^3, \qquad (2.30)$$

where $x_j = jc$ and $a = nc$; and

(ii) In the "uncracked" region ($j = n+1$, $n+2, \cdots$)

$$u_{n+\ell} = [u_n \cos(\lambda\ell)$$
$$- (Pa\xi^2/2\beta c^3)\cos(\lambda)\sin(\lambda\ell)] \left(\frac{\xi-c}{\xi+c}\right)^{\ell/2}, \qquad (2.31)$$

where $\ell = j-n = 1,2,3,\cdots$, and $\sin(\lambda) = c/\xi$.

The simplified form of these expressions compared to those of Thomson et $al.$ [6] is due in large part to the simple, but obscure, definition of ξ, Eqn. (2.24).

Finally, the displacement u_n of the crack-tip atoms, that interact through the nonlinear cohesive force $f_I(u_n)$, is determined from the nonlinear coupling equation (2.15). Substitution of the displacements from Eqns. (2.30) and (2.31) into Eqn. (2.15) gives this relation as

$$P(a+\xi)/\xi = K_I(u_n)/\xi \equiv f_I(u_n) + 2\alpha u_n \left(\frac{\xi-c}{2c}\right). \qquad (2.32)$$

The left-hand side of this equation determines the driving force for fracture of the chains, and as will be discussed in Section 2.3, is related to a one-dimensional stress intensity factor.

For a given driving force, solutions of Eqn. (2.32) for u_n give the equilibrium configurations of the crack. These points are discussed in more detail in Section 2.3.

2.2 Mode II Lattice Model

This model of a sliding shear fracture also consists of two semi-infinite chains of atoms; but now the atomic displacements are constrained to be parallel to the chains instead of transverse to them, and are produced by a force couple acting on the free ends of the chains at the zeroth atoms. As illustrated in Fig. 2, one shearing force P tends to compress its chain of atoms, whereas the other force P tends to extend its chain. In addition, transverse constraining forces are implicitly assumed, in order to resist transverse displacements of the atoms and rotation of the chains. However, since these forces do no work, they need not be considered.

Two types of interatomic force interactions resist the deformation from the applied force couple in this mode II crack model. Force interactions, that are modeled as stretchable spring elements, now resist compression or extension of each chain of atoms, and interactions that are modeled as shearable spring elements, provide a sliding shear resistance of the atoms in one chain from passing over their nearest neighbor in the other chain. These shearable elements up to the nth atoms are assumed to be sheared beyond their finite range of interaction, again forming a crack of finite length, a = nc. The equilibrium lattice spacing both parallel and perpendicular to the crack is denoted by c. The displacement of the jth atom in one chain from its equilibrium position is denoted by u_j, and that of its companion atom in the other chain is $-u_j$.

The total potential energy for this lattice model also consist of three contributions:

(i) *Potential energy of external loading system.* As before, this contribution is given by

$$U_{external} = -2Pu_0. \tag{2.33}$$

(ii) *Strain energy of the stretchable spring elements.* These spring elements are similar to the cohesive elements described for the mode I lattice crack, except that here they are taken to be only linear elastic with a spring stiffness α. The strain energy of this interaction for the spring element connecting atoms j and j-1 is given by

$$\frac{1}{2} \alpha(u_{j-1} - u_j)^2. \tag{2.34}$$

The total strain energy is the summation of these contributions for both chains of atoms,

$$U_{stretch} = \alpha \sum_{j=1}^{\infty} (u_{j-1} - u_j)^2. \tag{2.35}$$

(iii) Surface energy of the shear bonds. The interatomic interactions that oppose sliding shear of the crack faces are each modeled as a nonlinear bendable spring element, whose arbitrary restoring force, $f_{II}(u)$, is a function of the relative displacement $2u = u-(-u)$, of the two atoms that it bonds. In the linear elastic region these spring elements have a shear stiffness, μ, given by

$$\mu = [df_{II}(u)/d(2u)]_{u=0}. \tag{2.36}$$

Similar to the cohesive bonds of the mode I lattice model, these shear bonds are also assumed to have a finite range of interaction which is determined by a critical displacement $2u_c$. If a surface energy density and shear bond energy are defined in an analogous manner to Eqns. (2.5) and (2.6), the total shear surface energy is given by an expression like Eqn. (2.7). Making the simplifying assumption that only the crack-tip bond has extended into its nonlinear regime, this contribution to the potential energy becomes

$$U_{shear} = 2\gamma_{IIo}nc + 2\gamma_{II}(u_n)c + 2\mu\sum_{j=n+1}^{\infty} u_j^2. \tag{2.37}$$

Combining these contributions, Eqns. (2.33), (2.35) and (2.37), the total potential energy for this system becomes

$$U_{II} = U_{external} + U_{stretch} + U_{shear}$$

$$= -2Pu_0 + \alpha\sum_{j=1}^{\infty} (u_{j-1} - u_j)^2$$

$$+2\gamma_{IIo}nc + 2\gamma_{II}(u_n)c + 2\mu\sum_{j=n+1}^{\infty} u_j^2. \tag{2.38}$$

The equations of stability, $(\partial U_{II}/\partial u_j) = 0$, for this model are:

$$\alpha(u_0 - u_1) = P; \tag{2.39}$$

$$\alpha(\delta^2 u_j) = 0, \qquad \text{for } j=1,2,\cdots, n-1; \tag{2.40}$$

$$\alpha(\delta^2 u_n) - f_{II}(u_n) = 0; \tag{2.41}$$

$$\alpha(\delta^2 u_j) - 2\mu u_j = 0, \qquad \text{for } j=n+1, n+2,\cdots; \tag{2.42}$$

where the difference operator δ^2 is given by

$$(\delta^2 u_j) = u_{j+1} - 2u_j + u_{j-1}. \tag{2.43}$$

The second-order difference equations, Eqn. (2.40) and Eqn. (2.42), and related boundary conditions are easily solved in a manner analogous to that use for the mode I fracture model. The solutions for the displacements are:

(i) In the "cracked" region $(j = 0,1,2,\cdots, n-1)$

$$u_j = u_n + P(a-x_j)/\alpha c \tag{2.44}$$

where $x_j = jc$ and $a = nc$; and

(ii) In the "uncracked" region $(j = n+1, n+2, \cdots)$

$$u_{n+\ell} = u_n \left(\frac{\zeta-c}{\zeta+c}\right)^\ell , \tag{2.45}$$

where $\ell = j-n = 1,2,3,\cdots$.

The characteristic length ζ is defined by the spring constant ratio, α/μ, according to

$$\zeta/c = \sqrt{1 + 2\alpha/\mu} ,$$

or

$$(2\alpha/\mu) = (\zeta^2-c^2)/c^2 . \tag{2.46}$$

Using the above solutions for the displacements, the nonlinear equation for the crack-tip atoms becomes,

$$P = K_{II}(u_n)/\zeta \equiv f_{II}(u_n) + 2\mu u_n \left(\frac{\zeta-c}{2c}\right). \tag{2.47}$$

The similarity of this equation to the mode I expression, Eqn. (2.32), puts the two models on an equivalent footing so that their solutions can be discussed on a more general basis as in the next section.

2.3 Lattice Trapping

Equilibrium configurations of a lattice crack are specified by a set of atomic displacements that satisfy the equations of stability. Since this set of displacements is determined solely by the crack-tip displacements, we turn our attention to the nonlinear coupling equation, $(\partial U/\partial u_n) = 0$, which determines u_n. This equation for both modes of fracture [Eqns. (2.32) and (2.47)] is of the form

$$K = \xi [f(u) + 2\alpha u \left(\frac{\xi-c}{2c}\right)], \tag{2.48}$$

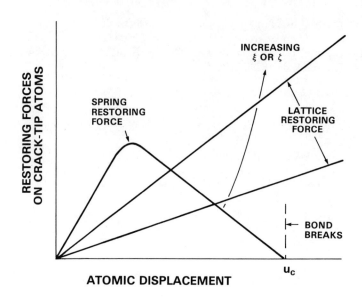

Figure 4.   Restoring forces on the crack-tip atoms. The spring
            force represents the interatomic interaction between
            the crack-tip atoms. The lattice force results from
            the interatomic interactions of the rest of the lattice,
            and is plotted here for two values of ξ, or ζ, which
            determine the linear elastic response of the lattice.

where for convenience the subscript on u_n is omitted and the mode
I expression is used. All conclusions, unless specifically
stated otherwise, will also apply for the mode II model.

     The right-hand side of Eqn. (2.48) is composed of two terms
that act as restoring forces on the crack-tip atoms. The first of
these terms is the nonlinear cohesive force, $f(u)$, which is
supplied by the crack-tip spring element. This spring restoring
force is illustrated schematically in Fig. 4 for an idealized
nonlinear atomic force law. The second of these terms is the
restoring force on the crack-tip atoms that is supplied by the
rest of the lattice. A similar restoring force acts on all the
atoms in the "cracked" portion of the chains. This lattice
restoring force is also illustrated in Fig. 4 for two values of ξ,
or equivalently, for two ratios of flexible to stretchable spring
stiffness, β/α. This restoring force is linear elastic with
atomic displacement because of the assumed linearity of the
lattice interactions, except for the crack-tip bond.

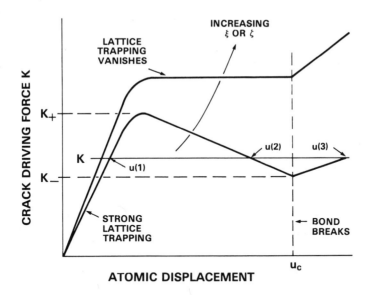

ATOMIC DISPLACEMENT

Figure 5.   Graphical solution of the nonlinear coupling equation
            (2.48) for the crack-tip displacements. The crack
            driving force K is resisted by the spring and lattice
            restoring forces of Fig. 4. The two values of ξ, or ζ,
            vary between strong and vanishing lattice trapping.
            For given elastic properties, the lattice-trapping limits
            are K_+ and K_-.

       The summation of these two restoring forces is balanced by a
driving force for fracture, K. As will be shown in Section 2.4,
this crack driving force is equivalent to a one-dimensional
stress intensity factor. It is proportional to the external
applied load P and for the mode I crack model is proportional to
the crack length as well:

$$K_I = P(a+\xi), \text{ for mode I;} \qquad (2.49a)$$

      and

$$K_{II} = P\zeta, \qquad \text{for mode II.} \qquad (2.49b)$$

       A graphical solution of Eqn. (2.48) is shown in Fig. 5 for
the restoring forces illustrated in Fig. 4. For given elastic
properties, namely, cohesive force f and stiffness ratio β/α,
there exists in general a range of applied stress intensity
factors K for which Eqn. (2.48) has three solutions for the

crack-tip displacement u. Within this range, the first and third solutions, denoted on Fig. 5 by u(1) and u(3), respectively, correspond to stable equilibrium configurations of the lattice crack. In the first stable equilibrium configuration, the crack-tip atoms still feel the influence of the cohesive bond connecting them, and the crack length is a = nc. In contrast, a crack-tip displacement of u(3) corresponds to the nth bond being "broken", so that the crack has advanced by one lattice spacing.

Viewed in multi-dimensional configuration space, the total potential energy of the system as a function of all the atomic displacements, $U = U(u_0, u_1, \cdots)$, has a relative minimum at configurations that correspond to both u(1) and u(3). Topological arguments require that at least one saddle point exists between these two minima. Since the configuration corresponding to the crack-tip displacement u(2) is the only remaining equilibrium configuration, this configuration is the required saddle point.

When the applied stress intensity factor is increased to a critical value K_+ where solutions u(1) and u(2) coalesce, unstable bond rupture occurs and rapid fracture ensues. Similarly, when the applied stress intensity factor is decreased to a critical value K_- where solutions u(2) and u(3) coalesce, spontaneous bond healing occurs. In the intermediate range of applied stress intensity factors, $K_- < K < K_+$, the crack is mechanically stable or "lattice trapped".

This phenomenon of lattice trapping is a general characteristic of crack stability in a lattice. In addition to one-dimensional lattice models, lattice trapping has been observed in calculations of stability for two-dimensional lattice cracks in mode I loading [20, 21] and in mode III loading [22], and for three-dimensional lattice cracks (kink stability) in mode I loading [21]. The nature of Eqn. (2.48) suggest that this phenomenon of crack trapping may not depend as much on long-range crystalline order as on local atomic-bond-rupture processes in the crack-tip region. As such the results might also be applicable to polycrystalline materials and to noncrystalline structures such as glasses. Indeed, lattice trapping has been invoked by Wiederhorn et al. [38] and by Tyson et al. [39] as a possible mechanism to describe subcritical crack growth of certain glasses in vacuum [38].

Lattice trapping also depends to a large extent on the linear elastic properties of the material. As seen from Figs. 4 and 5, increasing the characteristic length ξ, or ζ, results in a stronger lattice restoring force and a corresponding decrease in the lattice-trapping range. This diminishing lattice-trapping effect, as first noted by Thomson et al. [6] and later discussed

by Smith [8], results from "stiffer" lattice spring elements in
conjunction with "softer" bonding across the crack plane. This
combination of elastic properties gives a broad crack-tip core
which is insensitive to the atomicity of rupturing individual
bonds. Furthermore, if the bond rupturing process itself is
softened by replacing the precipitous "bond-snapping" cut-off
force of earlier models [6,8] with a more gradual descent from
maximum force, then lattice trapping can vanish for some finite,
critical value of ξ, or ζ (the upper curve of Fig. 5). In general,
softening the bond rupture process reduces the range of lattice
trapping in comparison to a "bond-snapping" rupture [21, 7, 9].

2.4   Griffith Criterion

In recent theoretical literature [21, 7, 9, 10], increased
attention has been given to the relationships between the limits
of lattice trapping, Griffith's criterion for fracture, and the
fundamental role of surface energy in brittle fracture. This
attention resulted from the lattice-statics calculations of
Esterling [21] for two- and three-dimensional lattice cracks with
nonlinear elastic cohesive forces in the crack-tip region. His
computer results gave the following inequality between the surface
energy density, γ_0, and the strain-energy-release rate determined
by the lower lattice-trapping limit, G_-: $2\gamma_0 < G_-$. Thus, the
classical Griffith criterion of fracture fell in a regime where
the crack was mechanically unstable with regard to spontaneous
healing. Esterling concluded from this result that the surface
energy density was a global property of the material and that the
use of this surface energy of separation, via the Griffith thermodynamic approach, was inadequate as an accurate fracture criterion.
Similar results were demonstrated for one of the cohesive force
laws used in the one-dimensional lattice model of Smith [9, 10].
In contrast, however, calculations for another one-dimensional
lattice using similar cohesive forces [7] have refuted these
findings. This section will reconsider these relationships from
the viewpoint of the one-dimensional lattice models and from
arguments of a more general nature.

In order to relate the lattice trapping limits, K_+ and K_-, to
the surface energy density, γ_0, for the one-dimensional lattice
models, it is necessary to define more precisely the driving force
for fracture, or strain-energy-release rate. The simplest approach
is to define a strain-energy-release rate that has the correct
form in the limit of linear elastic fracture mechanics. In these
nonlinear elastic models, however, the strain energy and surface
energy are not easily separated. Accordingly, the rate of change
of the total potential energy of the system with crack length will
be related to a "strain-energy-release rate" G according to,

$$- (\partial U/\partial a)_P = (G - 2\gamma_0), \qquad\qquad (2.50)$$

for these one-dimensional lattice models.

With this definition in mind, the total potential energy is separated into the following three contributions:

(i) Potential energy of the external loading system, which is the same as previously defined, namely, $-2Pu_0$;

(ii) "Surface energy" of the broken atomic bonds across the crack plane, $2\gamma_o a$;

(iii) Strain energy, and possibly surface energy, in the unbroken spring elements.

In the linear elastic limit, this last contribution becomes the elastic strain energy, and accordingly, this potential energy term is used to define the strain-energy-release rate. This "strain energy" is given by

$$U_{strain} = U_{flexural} + U_{cohesive} - 2\gamma_{Io}a, \qquad (2.51)$$

for the mode I lattice model, and by

$$U_{strain} = U_{stretch} + U_{shear} - 2\gamma_{IIo}a, \qquad (2.52)$$

for the mode II lattice model. Substitution of the appropriate equilibrium displacements [Eqns. (2.30) and (2.31) for mode I and Eqns. (2.44) and (2.45) for mode II] into these expressions gives in both cases,

$$U_{strain} = Pu_0 + [2\gamma(u_n)c - u_n f(u_n)]. \qquad (2.53)$$

For linear elasticity [see Eqns. (2.8), for example], the term in square brackets vanishes, and the resulting expression is the usual linear elastic fracture mechanics relationship that the strain energy is one-half the work done by the external loading system,

$$U_{strain} = W_{ext}/2 = -U_{ext}/2.$$

The strain-energy-release rate, as defined from Eqn. (2.50), is

$$G = -[\partial(U_{external} + U_{strain})/\partial a]_P. \qquad (2.54)$$

Using the appropriate expression for u_0, for the crack-tip equation (2.48), and for the definition of stress intensity factor, Eqn. (2.49), this equation becomes:

$$G_I = K_I^2/\beta c^3, \qquad \text{for mode I;} \qquad (2.55)$$

and

$$G_{II} = K_{II}^2/\alpha\zeta^2 c, \text{ for mode II.} \qquad (2.56)$$

The second expression is obtained directly from Eqn. (2.54) without taking either the linear elastic or the continuum limit. The first expression, however, simplifies to this form only when both the linear elastic and continuum limits are taken. The continuum model is obtained by taking the limit as $n \to \infty$ and $c \to 0$ in such a manner that $nc \to a$, and βc^3 and α/c approach a constant value [7]. The resulting linear-elastic, continuum expression, Eqn. (2.55), is analogous to the force-crack-length relationship for a double cantilever beam, when $P(a+\xi)$ is substituted for K_I.

Using these expressions, four critical values of strain-energy-release rate can be defined for the quasi-one-dimensional lattice models. Two of these critical values, denoted by G_+ and G_-, are defined by the lattice-trapping limits K_+ and K_-, respectively, for the appropriate mode of fracture. The other two critical values of strain-energy-release rate are determined from a macroscopic and a microscopic thermodynamic criterion for fracture. These fracture criteria and the strain-energy-release rates which they define will now be discussed.

The macroscopic fracture criterion is that of the classical thermodynamic treatment of fracture by Griffith [1]; namely, for fracture to occur the strain-energy-release rate must equal or exceed a critical value given by twice the macroscopic surface energy density. In the present lattice models this macroscopic value has been denoted $2\gamma_o$ or U_b/c [see Eqn. (2.6), for example], and as such is related to the thermodynamic work of separation per unit length of crack. Thus, the critical value of strain-energy-release rate defined by the classical Griffith condition is $2\gamma_o$.

The other thermodynamic fracture criterion is the condition where thermal fluctuations cause a crack to advance and to recede at equal rates. As mentioned in Section 2.3, when a crack is lattice trapped, there exists an unstable equilibrium configuration, or saddle point, between two stable configurations of the crack that differ by one lattice spacing. Accordingly, this saddle-point configuration gives a forward activation energy barrier ΔU_+ that retards crack growth and a backward activation energy barrier ΔU_- that retards crack closure. Thermal fluctuations can cause thermally activated subcritical crack propagation, or crack healing, over these activation energy barriers. Thus, the microscopic thermodynamic condition that the thermally activated crack velocity vanishes, or equivalently, that the forward and backward activation energy barriers are equal, defines another critical value of strain-energy-release rate. Since this thermodynamic approach is in the spirit of a microscopic Griffith criterion for fracture, this critical value is denoted G_G.

The relationship of the microscopic Griffith value, G_G, to the lattice trapping limits G_+ and G_- is easily established by recalling the definition of the lattice-trapping limits. Restated from the previous section, catastrophic rapid fracture ensues when the crack driving force reaches a critical value G_+ where the forward activation energy barrier vanishes; and spontaneous crack healing occurs when the backward barriers vanishes at G_-. Since the microscopic Griffith value, G_G, is determined by the position of equal activation barriers, it is always bounded by the lattice-trapping limits G_- and G_+.

On the other hand, the "paradox" of Esterling implies that the macroscopic Griffith value, or surface energy, $2\gamma_o$ is not bounded by the lattice-trapping limits, namely,

$$2\gamma_o < G_-. \tag{2.57}$$

Without recomputing his lattice calculations, it is not possible to examine quantitatively this relationship for his two- and three-dimensional lattice cracks. However, the same result has been obtained [7, 9, 10] for both of the one-dimensional lattice models discussed in this section, and can be examined analytically for them. For the mode I lattice calculation, the model has been shown to be inconsistent for the particular cohesive force law that led to this relationship [7]. This inconsistency resulted because the assumptions required to obtain Eqn. (2.9) from Eqn. (2.7) were not satisfied. In a similar manner, the mode II lattice calculation, that obtained this result [9], can also be shown to violate these assumptions.

The nature of this violation is that the *same* cohesive interactions are assumed to bond all atom pairs across the crack plane, but that only the crack-tip interaction is assumed to have extended into the nonlinear elastic regime. These two assumptions are not necessarily self-consistent. When they are violated, it is reasonable for $2\gamma_o$ to lie below the lower lattice-trapping limit G_-, since this macroscopic surface energy density is now determined by a crack-tip interaction which is weaker than the remaining cohesive interactions that determine G_-. Thus, Esterling's result indicates that a surface energy density determined from weaker bonds in the crack-tip region is less than a lower lattice-trapping strain-energy-release rate which has been determined predominantly from stronger bonds in the lattice.

The question raised by Esterling about the position of the quiescent point in the fracture of a discrete lattice can also be addressed on more general grounds. We divide the internal potential energy of the crystal into three terms,

$$U = U_{tip} + U_{elastic} + U_{surface}. \tag{2.58}$$

In the usual way, the various terms represent the differences in energy between a crystal that contains a crack of length, a, and a crystal at the same stress but without a crack. The total potential energy, U, is composed of: the elastic free energy, $U_{elastic}$, including contributions due to the external forces acting on the body; the surface energy, $U_{surface}$; and the energy of the core region surrounding the crack tip, U_{tip}. The core contains a specified, finite number of atoms, and is sufficiently large that outside it, the crystal can be considered a continuum. The contributions $U_{elastic}$ and $U_{surface}$ refer to this region outside the core.

As a result of lattice-trapping solutions for a discrete lattice, the total potential energy, U, as a function of the length of the crack is actually a discrete function, defined only for increments of crack length measured in terms of the lattice spacing, as illustrated by the closed circles in Fig. 6. This is because the crack is not in stable mechanical equilibrium under the simultaneous action of both an external stress and the internal atomic forces for crack lengths of fractional atomic spacings. Between each of these points of mechanically stable equilibrium, there exists one point of unstable equilibrium, that corresponds to an activated state of the system, as just described for the one-dimensional models.

We can analytically continue the discrete function in a variety of ways to make a continuous curve. If we simply draw the smoothest curve between the points, we find the result illustrated by the dashed line in Fig. 6. This curve we call the coarse grain function, $U_c(a)$. However, U is actually a function of the atomic displacements defined in the configuration space of all the atoms of the lattice, and a "reaction coordinate" always exists between the points of stable equilibrium that passes through the saddle-point states of unstable equilibrium. The function so defined we shall call the fine grain function, and it is sketched in Fig. 7.

We shall now argue that this function always has a broad maximum within the lattice-trapping regime, and that it does indeed have the shape of Fig. 7. Consider the crack at two neighboring discrete lattice points of equilibrium, and consider the difference in the various terms contained in U between these two positions,

$$\Delta U = \Delta U_{tip} + \Delta U_{elastic} + \Delta U_{surface}. \qquad (2.59)$$

We claim that in the limit where the ratio of the size of the crack to the size of the core region becomes increasingly large, that $\Delta U_{tip} \to 0$. This result is due to the fact that the atoms at the crack tip are in a configuration determined entirely by the

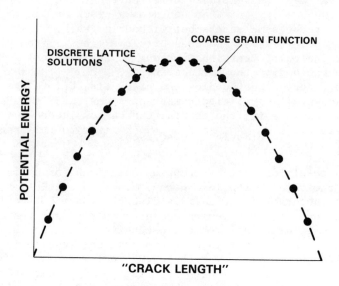

Figure 6.   Potential energy of a lattice as a function of crack
length for a given applied stress. The discrete lattice
solutions are represented by closed circles, and the
coarse grain function is a smooth, analytical continuation
of the discrete solution.

force laws of the critically stretched atoms at the tip, and are
not explicitly dependent upon the size of the crack or the external
stress. In this case, the equilibrium in the coarse grained
function, $U_c(a)$, is determined by

$$\Delta U = (\partial U_c/\partial a)\Delta a = 0 = \Delta U_{elastic} + \Delta U_{surface} , \qquad (2.60)$$

which is exactly the Griffith condition.

On the other hand, there exists a quiescent point defined on
the fine grain function $U_f(a)$ which is the position where the
forward and reverse activation energy barriers between one point
of stable equilibrium to its neighboring equilibrium points are
equal. This condition is also given by

$$\Delta U = \Delta U_f = 0, \qquad (2.61)$$

where again $\Delta U_{tip} \to 0$ in the limit of large cracks. Thus, the
Griffith condition is exactly equilvalent to the quiescent point.

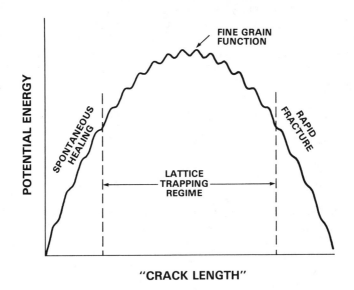

Figure 7. Fine grain potential energy of a lattice as a function
           of "crack length," or "reaction coordinate" in config-
           ation space. This function passes through the discrete
           lattice solutions that correspond to saddle-point
           configurations as well as to the energy minima.

     The proof is now complete when we note that in order for
there to be a quiescent point at all, we must be in a region of
lattice trapping; that is, there must exist local minima in the
fine grain function like those on the "dinosaur" function of Fig. 7.
Thus, quite generally, to satisfy the first and second laws of
thermodynamics, the Griffith condition must lie within the lattice-
trapping regime.

     We note that the surface energy density defined in this
manner is not necessarily equal to half the cohesive bond energy
per unit area of crack. Instead, the surface energy is the true
thermodynamic surface energy in a classical Gibbs sense, because
it is determined in the region outside the core of the crack where
the surfaces are well defined. In this region, surface relaxations,
or rearrangements, may occur which will lower the surface energy
below that determined by the bond cohesive energy. In the limit
of only nearest-neighbor forces, half the cohesive bond energy
coincides with the thermodynamic surface energy.

We believe the previous results are of interest when considering the distinction often made between a "global" fracture criterion
and a "local" criterion. The first is often identified with the
Griffith energy-balance criterion, and the second with an Orowan-
type force condition at the crack tip. We see here that the
Griffith criterion is identified with the quiescent point for the
crack, while a stress necessary to rupture the interatomic bonds
at the crack tip is identified with the upper limit of the lattice-
trapping regime, where the crack becomes dynamic. The first of
these is that identified with "K_{ISCC}", or a stress corrosion
cracking limit, while the second is that of a critical fracture
toughness, K_{IC}.

2.5 Thermally Activated Crack Propagation

When a crack is trapped by a lattice, thermal fluctuation can
cause the crack to advance or to recede subcritically. In previous
treatments [20, 40, 33, 39], the stress dependence of this thermally
activated crack propagation has been developed from modified
continuum models. The nature of this modification was the introduction of a sinusoidally varying surface energy density to
account for the lattice-trapping effects. Although this approach
is conceptually useful, it is not requisite, since the forward and
backward activation energy barrier can be calculated directly from
the equilibrium configurations of the lattice models [7]. However,
this calculation is usually analytically intractable, except for
all but the simplest of cohesive force laws and for the one-
dimensional lattice models. For these instances, however, the
results are instructive, and serve to guide the development of *a
priori* approaches to thermally activated crack growth.

Considering the one-dimensional lattice models, the forward
activation energy barrier is given by the difference in potential
energy between the saddle-point configuration, determined by the
crack-tip displacement u(2), and the stable equilibrium configuration, defined by the crack-tip displacement u(1); namely,

$$\Delta U_+ = U[\cdots u(2) \cdots] - U[\cdots u(1) \cdots]. \tag{2.62}$$

And, similarly, the backward activation energy barrier is given by

$$\Delta U_- = U[\cdots u(2) \cdots] - U[\cdots u(3) \cdots]. \tag{2.63}$$

The microscopic Griffith value of strain-energy-release rate G_G,
that defines the quiescent position of the crack, is determined
from the condition that $\Delta U_+ = \Delta U_-$, or equivalently,

$$U[\cdots u(1) \cdots] = U[\cdots u(3) \cdots]. \tag{2.64}$$

As discussed in the previous section, subcritical crack growth
can occur by thermal fluctuations from this quiescent position, or
stress cracking limit, to the position of catastrophic rapid
fracture.   The range of applied strain-energy-release rates for
which this crack growth occurs is between G_G to G_+.   These values
are related through Eqn. (2.55), or Eqn. (2.56), to a range of
applied stress intensity factors from K_{ISCC} to K_{IC}.

To illustrate these results, consider the simple cohesive
force law shown in Fig. 8.   The cohesive restoring force rises in
a linear manner to a maximum value at an atomic displacement, u_m,
and then decreases linearly to zero.   For illustrative purposes,
the corresponding surface energy density, as defined by Eqn.
(2.5), is shown in Fig. 9.   The cohesive energy necessary to
rupture this interatomic bond is given by

$$U_b = 2\gamma_o c = 2\alpha u_m u_c . \tag{2.65}$$

For rapid fracture of a lattice that is cohesively bonded by
this force law, a critical value of stress intensity factor,

$$K_+ = 2\alpha\xi \left(\frac{\xi+c}{2c}\right) u_m , \tag{2.66}$$

as determined by Eqn. (2.48), must be attained; and for spontaneous
healing of the crack, the applied stress intensity factor must
decrease to a lower critical value,

$$K_- = 2\alpha\xi \left(\frac{\xi-c}{2c}\right) u_c . \tag{2.67}$$

The corresponding strain-energy-release rates, which have been
normalized by the macroscopic surface energy density, are

$$(G_{\underset{-}{+}}/2\gamma_o) = \left[\left(\frac{\xi+c}{\xi-c}\right)\left(\frac{u_m}{u_c}\right)\right]^{\pm 1} . \tag{2.68}$$

Thus, for this cohesive force law, twice the macroscopic surface
energy density is not only bounded by the lattice-trapping limits,
but is equal to their geometric mean,

$$2\gamma_o = \sqrt{G_+ G_-} . \tag{2.69}$$

This result appears to be quite general for this cohesive
force law, and for specific cases of it such as the linear "bond-
snapping" cohesive force.   In addition to being satisfied for both
of the one-dimensional lattice models, this relation is easily
demonstrated to be satisfied for the two-dimensional mode III
lattice model of Smith [22].   Interestingly, however, numerical
results calculated from the two- and three-dimensional mode I
lattice models [20, 21] give $2\gamma_o$ as approximately one-half the
geometric mean.

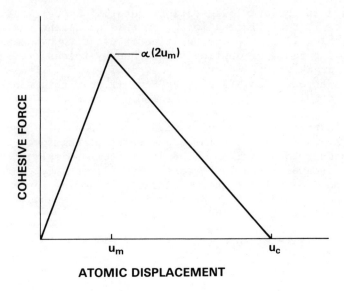

Figure 8.   Simple interatomic cohesive force law of finite range u_c with a maximum at u_m.

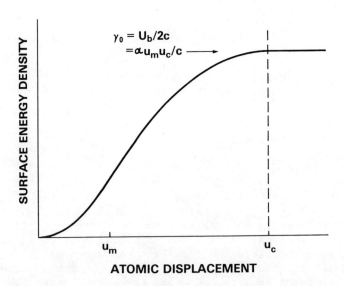

Figure 9.   Surface energy density corresponding to the cohesive force of Fig. 8.

The forward and backward activation energy barriers are easily determined for this cohesive force law. Using the separation of total potential energy given in the previous section, the activation energy between an equilibrium configuration determined by crack-tip displacement u and the saddle-point configuration determined by u(2) is given by

$$\Delta U \equiv U[\cdots u(2) \cdots] - U[\cdots u \cdots]$$

$$= - K(\Delta u)/\xi + 2\int_{u}^{u(2)} f(u)\,du$$

$$- u(2)f(u(2)) + uf(u), \qquad (2.70)$$

where $\Delta u = u(2) - u$, and K is the applied stress intensity factor. The crack-tip displacements, that define the equilibrium configurations and the saddle-point configuration are determined from Eqn. (2.48) as:

$$u(1) = (K/K_+)u_m,$$

$$u(2) = [(K_+-K)u_c + (K-K_-)u_m]/[K_+-K_-],$$

$$u(3) = (K/K_-)u_c. \qquad (2.71)$$

Substitution of these expressions and the equation for the cohesive force law into Eqn. (2.70) gives after some algebraic manipulation:

$$\Delta U_+ = (K_+-K)(\Delta u_+)/\xi$$

$$= U_b \left(\frac{K_+}{K_+-K_-}\right)\left(1 - \frac{K}{K_+}\right)^2, \qquad (2.72)$$

and

$$\Delta U_- = (K-K_-)(\Delta u_-)/\xi$$

$$= U_b \left(\frac{K_-}{K_+-K_-}\right)\left(\frac{K}{K_-} - 1\right)^2, \qquad (2.73)$$

where the activation distances Δu_+ are given by

$$\Delta u_+ = u(2) - u(1) = \left(\frac{U_b\xi}{K_+}\right)\left(\frac{K_+-K}{K_+-K_-}\right), \qquad (2.74)$$

and

$$\Delta u_- = u(3) - u(2) = \left(\frac{U_b\xi}{K_-}\right)\left(\frac{K-K_-}{K_+-K_-}\right). \qquad (2.75)$$

The quiescent position of the crack, as defined by the equality of the forward and backward activation energy barriers, is attained for an applied stress intensity factor K_G that is given by

$$K_G = \sqrt{K_+ K_-} \, , \qquad\qquad (2.76)$$

or equivalently for a strain-energy-release rate given by

$$G_G = \sqrt{G_+ G_-} \quad . \qquad\qquad (2.77)$$

Comparing Eqn. (2.77) with Eqn (2.69), one obtains for this cohesive force law the Griffith criterion that the crack driving force at the quiescent position, as determined by microscopic crack-tip processes, is balanced by a resistance force which is twice the macroscopic surface energy density.

Since a complete dynamic-elasticity solution is not available for these lattice models, reaction-rate theory can be used to describe the bond-rupture processes in the crack-tip region. (See, for example, a review by Laidler [41]). In terms of activated complex theories, thermal flucutations enable the interatomic bonds to pass over an activated complex state (the saddle-point configuration) and to fracture. The rate of this occurrence is proportional to the rate of subcritical crack growth. Thus, using the standard Arrhenius equation, an expression for the velocity of subcritical crack growth is given by

$$v = v_c \left[\exp(-\Delta U_+/kT) - \exp(-\Delta U_-/kT) \right], \qquad\qquad (2.78)$$

where v_c is a proportionality constant that is obtained from a complete dynamic-elasticity solution for the normal modes of the lattice. For the present, this constant is assumed to depend only weakly on applied stress so that the predominant stress dependence of the crack velocity is given by that of the activation energy barriers in Eqns. (2.72) and (2.73). If backward thermal fluctuation are neglected, this expression becomes,

$$v = v_c \exp \left\{ -\frac{U_b}{kT} \left[\frac{1}{1-(K_{ISCC}/K_{IC})^2} \right] \left[1-(K_I/K_{IC}) \right]^2 \right\}, \qquad (2.79)$$

where the equivalences between K_+ and the critical fracture toughness K_{IC}, and between K_G and the stress corrosion cracking limit K_{ISCC} have been assumed.

This expression for the subcritical crack growth rate has several interesting features. The "zero-stress" activation energy is given directly by the cohesive energy to rupture a single interatomic bond (except for the factor $1/[1-(K_{ISCC}/K_{IC})^2]$, which is approximately unity). The activation volume is proportional

to $2U_b/K_{IC}$. And the stress dependence in the exponential has both
a linear and a quadratic term in stress intensity factor. In
contrast, continuum models that are modified to include lattice-
trapping effects [20, 33, 39, 40] have usually assumed only a K_I^2,
or G_I, stress dependence for the lattice-trapping energy barriers.
Similarly, stress corrosion cracking mechanisms, as reviewed in
these proceedings for glass [42], have usually predicted only a K_I
stress dependence.

       To illustrate that the present stress dependence is reasonable,
the subcritical crack propagation data of Wiederhorn *et al.* [38]
for a 61% lead glass in vacuum is compared with Eqn. (2.79).
Their experimental data is plotted in Fig. 10 as $\ln(v)$ versus K_I
for temperatures from 297 K to 625 K. The solid curves are
obtained from a linear least squares regression of $[1-(K_I/K_{IC})]^2$
upon $\ln(v)$ for each temperature. For reference, this regression
is compared in Table 1 to a regression of $[1-(K_I/K_{IC})]$ upon $\ln(v)$,

$$v = v_* \exp\left\{-(U_*/kT)\ [1-(K_I/K_{IC})]\right\}, \qquad (2.80)$$

which is essentially the expression used by Wiederhorn *et al.*
[38]. As recently noted for a variety of functional dependences on
stress intensity factor [32], the crack propagation data is not
able to discriminate between these functional forms. We note in

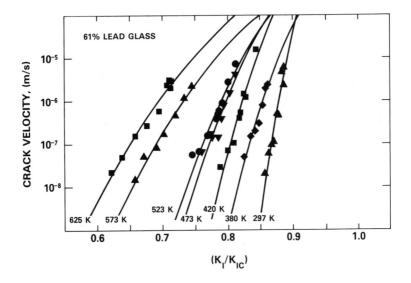

Figure 10.   Subcritical crack growth data for a 61% lead glass in
             vacuum as a function of temperature [38].

Table 1.   Parameters from a least squares regression analysis of
           Eqns. (2.79) and (2.80) for the crack propagation data
           of a 61% lead glass in vacuum [38]. Activation energies
           U_b and U_* are in units of kcal/mole, crack velocities v_c
           and v_* are in m/s, and temperatures are in kelvin.
           "Corr. Coef." stands for the correlation coefficient of
           the fit and the $\pm$ values represent one standard deviation.

Temp.	Corr. Coef.	$\ln(v_c)$	U_b	Corr. Coef.	$\ln(v_*)$	U_*
297	0.996	-2.4 ± 1.3	457 ± 46	0.971	10.0 ± 2.3	116 ± 11
380	0.975	-6.3 ± 0.9	268 ± 27	0.980	2.3 ± 1.6	83 ± 8
420	0.972	-3.9 ± 1.1	265 ± 26	0.981	6.7 ± 1.7	97 ± 8
473	0.942	-5.4 ± 1.4	193 ± 28	0.951	3.7 ± 2.5	82 ± 11
523	0.969	-6.0 ± 0.8	184 ± 18	0.979	2.5 ± 1.3	81 ± 6
573	0.992	-7.0 ± 0.5	107 ± 6	0.994	1.3 ± 0.8	64 ± 3
625	0.988	-6.4 ± 0.5	99 ± 6	0.991	2.1 ± 0.9	65 ± 3

passing, the nonclassical temperature dependence of the subcritical
crack growth, namely, the activation energy as determined from
Eqn. (2.79) varies as approximately $1/kT$. The value obtained at
higher temperatures of approximately 100 kcal/mole is a reasonable
value for the Si-0 cohesive bond energy.

3.   QUANTUM MECHANICAL TUNNELING IN FRACTURE

     Several years ago Gilman and Tong [30] first suggested that
quantum mechanical tunneling might occur as a fracture process.
Their theory was based on the premise that the large stresses in
the vicinity of the crack tip would decrease the potential energy
of the crack-tip atoms, so that the interatomic potential which
bonded these crack-tip atoms (a Morse-type function, for example)
would be biased, or bent down, at large atomic displacements by

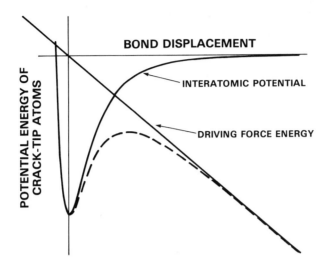

Figure 11.   Potential energy of the crack-tip atoms (dashed curve)
              following the development of Gilman and Tong [30].
              The potential energy consist of two contributions:
              the interatomic potential between the crack-tip atoms;
              and the external driving force energy as transmitted
              to the crack-tip atoms. A contribution from the lattice
              restoring forces was not included.

this driving force potential. This effect is illustrated in Fig.
11 where these components of the potential energy are plotted
separately and then combined into the dashed curve. Thus, at the
crack tip not only is the interatomic bond stretched nearly to its
maximum force of cohesion, but the atoms there are bounded by a
potential barrier through which they can tunnel and rupture the
bond. The height, but most significantly, the width of this
potential barrier is stress dependent. Gilman and Tong applied
WKB tunneling theory to this single most distended bond, and
derived an approximate expression for the rate of tunneling at a
temperature $T = 0$ K. Their expression depended on crack-tip
stress, σ, according to

$$\nu = \nu_o \exp [\sigma_o/\sigma], \qquad\qquad (3.1)$$

where σ_o and ν_o are a combination of the parameters from their
model. The parameter σ_o is proportional to the square root of the
barrier height, and as such has a slowly varying dependence on
stress, but can be approximated by a constant.

Gilman and Tong also derive expressions for the temperature dependence of the tunneling, which shows a natural transition into the normal classical activated process in which the system jumps over the top of the barrier by thermal fluctuations. The expressions are complicated and are not given here, but lead to jump rates which are slowly varying with temperature as compared to the Arrhenius behavior of the classical process.

The theoretical prediction of Gilman and Tong have been applied to experimental results in polymers by the original authors and to glasses both by the original authors and by Doremus [31] and Wiederhorn [32]. In the experimental comparisons, the stress dependence (termed "stress activation" by Gilman and Tong) given by Eqn. (3.1) is the crucial matter, and is claimed to be *prima facie* evidence for quantum tunneling.

Quantum tunneling processes have been invoked for other atomic processes in solids, most notably for interstitial diffusion [43], but also for dislocation motion [44]. Although the general treatments developed by Flynn and others for diffusion are very elegant, the general framework presented there is not appropriate for the fracture problem, because the excitations involved in the fracture process are in the rupture of a stretched bond between two atoms at the crack tip, and not in the transport of atoms across barriers in the crystal. Fluctuations leading to these two types of processes are sufficiently dissimilar to make it necessary to start anew in the fracture problem.

In the light of the previous discussion on the one-dimensional fracture models, the model used by Gilman and Tong is too restricted because more than two atoms are involved when a bond snaps and the crack advances. Strictly, the model of Gilman and Tong only applies to two atoms which are stretched apart by an external massless spring whose force does not depend upon the distance between the atoms. No account is taken of the interaction of the other atoms in the crystal as the bond breaks. Since the force stretching the crack-tip atoms open is transmitted by the other atoms, their behavior should be expected to be important as well. These interactions are the ones that were designated as lattice restoring forces for the one-dimensional models (see Fig. 4, for example).

In an attempt to assess the effect of the lattice, Thomson [45] and Shie [46] estimated the tunneling in a very different way than that of Gilman and Tong. In their work, Thomson and Shie calculated the probability of obtaining an amplitude in the harmonic oscillations of the crystal which was both localized at the crack tip and of sufficient size to break a bond. To distinguish it from Gilman and Tong, we shall term this model the quantum fluctuation model. In the following development, linear

force laws are assumed to be valid up to some critical value, at
which the bond snaps. The wave functions of the normal modes of
the crystal are known in this approximation, and in the ground
state are given by simple gaussian functions,

$$\psi_\ell = A \exp\left(-q_\ell^2/q_{o\ell}^2\right) .$$
(3.2)

In this expression, q_ℓ is the normal-mode amplitude of the ℓth
mode. The vibrational amplitude of a given atom (at the crack
tip) must be obtained by summing over all the normal modes,

$$x_i = \Sigma_\ell \; q_\ell \; e^{iq_\ell x_i} .$$
(3.3)

From these expressions, one can obtain an approximate rate at
which an amplitude of sufficient size is attained at the crack tip
to break the bond when the crystal is in its ground state (T = 0
K). The result is

$$\nu = 4\pi^2\bar{\nu}x\left(\frac{m\bar{\nu}}{h\pi}\right)^{\frac{1}{2}} \exp\left(-4\pi^2 m\bar{\nu}x^2/h\right) ,$$
(3.4)

where x is the size of the critical amplitude to break a bond at
the crack tip, m is the atomic mass, $\bar{\nu}$ is a suitable average of
the frequency over the oscillators, and h is Planck's constant.
Since the Debye spectrum is peaked at high frequencies, a fair
approximation for $\bar{\nu}$ is the Debye frequency.

At the crack tip, the size of the critical fluctuation is
stress dependent, but a detailed atomic model of fracture is
necessary in order to obtain a functional form for this dependence.
However, on a heuristic level, one might make a guess in the
following manner. If tunneling is occurring, it must be taking
place from one lattice-trapped position to the next. But the
lattice-trapping regime occurs only over a limited region in
stress. For a given crack length, this stress regime is given by

$$\sigma_- < \sigma < \sigma_+ ,$$
(3.5)

and presumably there is an intermediate stress σ_G between σ_- and σ_+
where the forward and reverse fluctuations occur at equal rates. We
shall assume in a simple manner that $\sigma_G = (\sigma_+ + \sigma_-)/2$. At σ_+, the
size of the critical amplitude to break the crack-tip bond must go
to zero. Hence, the simplest function for x is a linear one; namely,

$$x_+ = x_o(\sigma_+ - \sigma)$$
(3.6)

for the forward fluctuations, and

$$x_- = x_o(\sigma - \sigma_-),$$
(3.7)

for the backward fluctuations. We thus write heuristically,

$$\nu = \nu_o \left\{ \exp\left[-\left(\frac{\sigma_+ - \sigma}{\sigma_o} \right)^2 \right] - \exp\left[-\left(\frac{\sigma - \sigma_-}{\sigma_o} \right)^2 \right] \right\}, \qquad (3.8)$$

where σ_o and ν_o are constants not yet determined. Near σ_G, we thus have a linear region where

$$\nu \simeq \nu_o' \, (\sigma - \sigma_G), \qquad (3.9)$$

and in the regime where the reverse fluctuations can be neglected, we have

$$\nu = \nu_o \, \exp\left[-\left(\frac{\sigma_+ - \sigma}{\sigma_o} \right)^2 \right] . \qquad (3.10)$$

   This equation for tunneling can be given a more solid basis by a detailed rendition of the one-dimensional models which were presented in the previous sections. In a one-dimensional bond-snapping model, the critical fluctuation amplitude is given by the expression,

$$x = 2(\Delta u_+) = 2\left(\frac{K_+ - K}{K_+} \right) u_c , \qquad (3.11)$$

in the notation of the previous sections (note in particular, $u_m = u_c$ for the bond-snapping cohesive force). In this case, x does in fact depend linearly on $(K_+ - K)$ as assumed in the above derivation. Using the expressions for the bond energy in this model, $U_b = \frac{1}{2}\alpha(2u_c)^2$, and for the frequency, $\bar{\nu} = (1/2\pi) \sqrt{\alpha/m}$, we replace Eqn. (3.10) by

$$\nu = \nu_o \, \exp\left\{ -\left(\frac{2U_b}{h\bar{\nu}} \right) [1 - (K/K_+)]^2 \right\} . \qquad (3.12)$$

This expression for crack growth by quantum fluctuations has a form that is similar to that obtained in the previous section for activated complex theory.

   In a fundamental sense, however, the model for quantum fluctuations outlined above is still not an adequate solution to the problem, because the rate at which the system returns to the initial position from its critical fluctuations is still unknown. Although one suspects that this rate does not change the form of the results, the problem can be considered from yet another approach to verify this point.

   The fracture of a discrete lattice has been shown to lead to lattice trapping when the elastic properties of the material have the proper relationships between one another. This means that two adjacent positions of the crack, separated in crack length by one lattice spacing, are positions of local minima in energy, with at least one saddle point connecting them in configuration space. This physical situation can be modeled by the quantum mechanical problem of a double harmonic oscillator connecting these

two positions of local energy minima. As in the quantum fluctuation
model, all the normal modes of the crystal should be considered in
order to do the problem adequately, but in the following treatment
only the one mode that transports the crack by one lattice spacing
is considered. The reason for this approach is that the summation
over all the normal modes was performed for the quantum fluctuation
model. Here the only concern is whether an important reflection
term has been neglected for the portion of the fluctuation when
the crack-tip bond has broken, but the system has not yet settled
into the next equilibrium valley. Hence we consider the problem
of a one-dimensional double harmonic oscillator. This problem is
relatively well known, and the reader is referred to the book on
quantum mechanics by Merzbacher [47] for details.

The potential function is given in Fig. 12 with the origin at
the mid-point. The maximum barrier potential is V_0 and the
ground state energy level is given as E, $E < V_0$. A particle in
such a potential well will oscillate between the wells with a
frequency given by

$$\nu = 2\nu_e \ \sqrt{2V_0/h\nu_e}\,\pi \ \exp\left[-\left(\frac{2V_0}{h\nu_e}\right)\right] , \qquad (3.13)$$

where ν_e is a frequency given by $E = h\nu_e/2$. The maximum potential

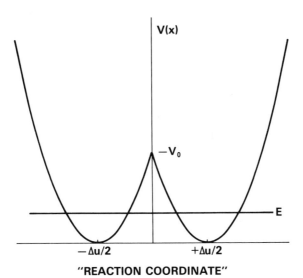

"REACTION COORDINATE"

Figure 12.   Potential energy function V(x) for a one-dimensional
             double harmonic oscillator. The ground-state energy
             level E is less than the barrier height V_0: $E < V_0$.

energy V_0 defines the lattice-trapping barrier, and from the previous section, is given by

$$V_0 = \Delta U_+ = U_b \left(\frac{\xi+c}{2c}\right) [1-(K/K_+)]^2, \qquad (3.14)$$

so that

$$\nu = \nu_d \exp\left\{-\left(\frac{2U_b}{h\nu_e}\right)\left(\frac{\xi+c}{2c}\right) [1-(K/K_+)]^2\right\}, \qquad (3.15)$$

where ν_d is of the order of the Debye frequency for the lattice. When $G \rightarrow G_G = 2\gamma_0$, the characteristic frequency of the process is given by

$$\nu(G=2\gamma_0) = \nu_d \exp\left[-\left(\frac{2U_b}{h\nu_e}\right)\left(\frac{\xi - \sqrt{\xi^2-c^2}}{c}\right)\right]. \qquad (3.16)$$

The exponential form for the stress dependence in Eqn. (3.15) is the same as before for Eqn. (3.12), but with somewhat different numerical constants. Even more striking is the similarity of Eqn. (3.15) to Eqn. (2.79) of the previous section. The exponential expressions are identical except for the driving force energy for subcritical crack propagation. In the previous section, the energy for overcoming the lattice-trapping barriers was supplied by thermal excitations of the order of kT. Whereas, in the present quantum fluctuation treatment at T = 0 K, this energy is supplied from the zero-point quantum vibrations of energy $\frac{1}{2}h\nu_e$.

Upon comparing the quantum fluctuation model and the double-potential-well model with the "stress activation" tunneling model of Gilman and Tong, we believe that the stress dependence predicted by Gilman and Tong was incorrect. The reason is that the fracture path in configuration space of a cracked lattice is not that corresponding to the decohesion of two atoms in a constant force field. Instead, the energy function is more like that which is familiar for solid state diffusion and which is given by our one-dimensional model. The physical reason why the potential function does not have the character of Fig. 11 is that when the atom pair at the crack tip exceeds the maximum cohesive force in the bond, the lattice snaps in a nonlinear way into the adjacent lattice-trapping configuration, with a potential function similar to the double-well model of Fig. 12. In Fig. 11, the main stress dependence comes from the decrease in tunneling distance as the stress increases, while in Fig. 12, the main effect is in the lowered value of V_0.

We are cautious about applying our results in an uncritical numerical way to experimental results. First, the steep exponential stress dependence predicted in Eqns. (3.10), (3.12), and (3.15) necessitate a precise evaluation of the various coefficients of the stress in the exponential, and of course our calculation is

based on a very simple model. Second, the ratio $2U_b/h\nu_e$ which
appears in both Eqn. (3.12) and Eqn. (3.15) is probably an overesti-
mate of the stress coefficient because the bond-snapping model has
been used instead of a more realistic and softer bond energy
function. Such a function would have the effect of rounding off
the cusp in Fig. 12, and of increasing the rate of tunneling. In
spite of these deficiencies, however, our arguments leading up to
Eqn. (3.10) make us believe that the stress dependence we have
derived on the basis of the specific models is correct, and we
suggest it as a characteristic for tunneling.

Gilman and Tong have extended their work to include thermal
effects. We have not yet done so, but have noted the nonclassical
thermal behavior of subcritical crack growth when the results of
Section 2.5 have been used to explain one set of experimental
results.

REFERENCES

1. A. A. Griffith, "The phenomena of rupture and flow in
   solids," Phil. Trans. Roy. Soc. (London) A221, 163 (1921);
   and "The theory of rupture," in Proc. First Intl. Congr.
   Appl. Mech., edited by C. B. Biezeno and J. M. Bergers
   (Delft, Holland, 1924), p. 55.

2. H. A. Elliott, "An analysis of the conditions for rupture due
   to Griffith cracks," Proc. Phys. Soc. (London) 59, 208
   (1947).

3. E. Orowan, "Fracture and strength of solids," Rep. Progr.
   Phys. 12, 185 (1949).

4. G. I. Barenblatt, "The mathematical theory of equilibrium
   cracks in brittle fracture," Advan. Appl. Mech. 7, 55 (1962).

5. J. Frenkel and T. Kontorova, "On the theory of plastic
   deformation and twinning," Physik. Zeit. Sowjetunion 13, 1
   (1938); J. Phys. (Moscow) 1, 137 (1939).

6. R. Thomson, C. Hsieh, and V. Rana, "Lattice trapping of
   fracture cracks," J. Appl. Phys. 42, 3154 (1971).

7. E. R. Fuller, Jr. and R. Thomson, "Nonlinear lattice theory
   of fracture," in Fracture 1977, edited by D. M. R. Taplin
   (Univ. of Waterloo Press, Canada, 1977), Vol. 3, p. 387.

8. E. Smith, "The effect of the discreteness of the atomic
   structure on cleavage crack extension: use of a simple one-
   dimensional model," Matls. Sci and Eng. 17, 125 (1975).

9.   E. Smith, "Survey of recent work on the effect of the atomic
     structure's discreteness on cleavage crack extension in
     brittle materials," in Fracture 1977, edited by D. M. R.
     Taplin (Univ. of Waterloo Press, Canada, 1977), Vol. 4, p.
     65.

10. E. Smith, Matls. Sci. and Eng., to be published.

11. J. N. Goodier and M. F. Kanninen, "Crack propagation in a
     continuum model with nonlinear atomic separation laws." ONR
     Tech. Rep. No. 165 (Stanford Univ., Div. of Eng. Mech.,
     1966); discussed by J. N. Goodier, "Mathematical theory of
     equilibrium cracks," in Fracture, edited by H. Liebowitz
     (Academic Press, New York, 1968), Vol. 2, p. 1.

12. P. C. Gehlen and M. F. Kanninen, "An atomic model for
     cleavage crack propagation in α-iron," in Inelastic Behavior
     of Solids, edited by M. F. Kanninen, W. A. Adler, A. R.
     Rosenfield, and R. I. Jaffee (McGraw-Hill, New York, 1970),
     p. 587.

13. M. F. Kanninen and P. C. Gehlen, "Atomic simulation of crack
     extension in b.c.c. iron," Int. J. Fract. Mech. 7, 471 (1971).

14. M. F. Kanninen and P. C. Gehlen, "A study of crack propagation
     in α-iron," in Interatomic Potentials and Simulation of Lattice
     Defects, edited by P. C. Gehlen, J. R. Beeler, and R. I.
     Jaffee (Plenum Press, New York, 1972), p. 713.

15. P. C. Gehlen, G. T. Hahn, and M. F. Kanninen, "Crack extension
     by bond rupture in a model of BCC iron," Scripta Met. 6, 1087
     (1972).

16. A. J. Markworth, M. F. Kanninen, and P. C. Gehlen, in Proc.
     Intl. Conf. on Stress Corrosion Cracking and Hydrogen Embrittle-
     ment of Iron Base Alloys, edited by R. W. Staehle (Natl.
     Assoc. Corr. Engrs., Houston, Texas, 1973).

17. J. E. Sinclair, "Atomistic computer simulation of brittle-
     fracture extension and closure", J. Phys. C.: Solid State 5,
     L271 (1972).

18. J. E. Sinclair and B. R. Lawn, "An atomistic study of cracks
     in diamond-structure crystals," Proc. Roy. Soc. (London)
     A329, 83 (1972).

19. J. E. Sinclair, "The influence of the interatomic force law
     and of kinks on the propagation of brittle cracks," Phil.
     Mag. 31, 647 (1975).

20. C. Hsieh and R. Thomson, "Lattice theory of fracture and crack creep," J. Appl. Phys. $\underline{44}$, 2051 (1973).

21. D. M. Esterling, "Lattice theory of three-dimensional cracks," J. Appl. Phys. $\underline{47}$, 486 (1976).

22. E. Smith, "Effect of the discreteness of the atomic structure on cleavage crack extension in brittle crystalline materials," J. Appl. Phys. $\underline{45}$, 2039 (1974).

23. W. T. Sanders, "On the possibility of a supersonic crack in a crystal lattice," Eng. Fract. Mech. $\underline{4}$, 145 (1972).

24. J. H. Weiner and M. Pear, "Crack and dislocation propagation in an idealized crystal model," J. Appl. Phys. $\underline{46}$, 2398 (1975); Erratum, J. Appl. Phys. $\underline{47}$, 5494 (1976).

25. W. T. Ashurst and W. G. Hoover, "Microscopic fracture studies in the two-dimensional triangular lattice," Phys. Rev. $\underline{B14}$, 1465 (1976).

26. A. Kelly, W. R. Tyson, and A. H. Cottrell, "Ductile and brittle crystals," Phil. Mag. $\underline{15}$, 567 (1967).

27. J. R. Rice and R. Thomson, "Ductile versus brittle behavior of crystals," Phil. Mag. $\underline{29}$, 73 (1974).

28. W. R. Tyson, "Atomistic simulation of the ductile/brittle transition," in Fracture 1977, edited by D. M. R. Taplin (Univ. of Waterloo Press, Canada, 1977), Vol. 2, p. 159.

29. An idea suggested in private communications with E. Smith and V. Vitek.

30. J. J. Gilman and H. C. Tong, "Quantum Tunneling as an elementary fracture process," J. Appl. Phys. $\underline{42}$, 3479 (1971).

31. R. H. Doremus, "Static fatigue in brittle solids," J. Appl. Phys. $\underline{47}$, 540 (1976).

32. S. M. Wiederhorn, "Dependence of lifetime predictions on the form of the crack propagation equation," in Fracture 1977, edited by D. M. R. Taplin (Univ. of Waterloo Press, Canada, 1977), Vol. 3, p. 893.

33. B. R. Lawn and T. R. Wilshaw, Fracture of Brittle Solids (Cambridge Univ. Press, 1975), Chaps. 7 and 8.

34. R. M. Thomson, "Some microscopic and atomic aspects of fracture," in The Mechanics of Fracture, edited by F. Erdogan (Amer. Soc. Mech. Engrs., New York, 1977), Appl. Mech. Div., Vol. 19, p. 1.

35. R. M. Thomson, "The fracture crack as an imperfection in a nearly perfect solid," Ann. Rev. Matl. Sci. 3, 31 (1973).

36. W. R. Tyson and L. C. R. Alfred, "Crack propagation on an .atomic scale," in Corrosion Fatigue (Natl. Assoc. Corr. Engrs., Storrs, Conn., 1972), p. 281.

37. M. Kh. Blekherman and V. L. Indenbom, "The Griffith criterion in the microscopic theory of cracks," Phys. Stat. Sol. (a) 23, 729 (1974).

38. S. M. Wiederhorn, H. Johnson, A. M. Diness, and A. H. Heuer, "Fracture of glass in vacuum," J. Amer. Ceram. Soc. 57, 366 (1974).

39. W. R. Tyson, H. M. Cekirge, and A. S. Krausz, "Thermally activated fracture of glass," J. Mater. Sci. 11, 780 (1976).

40. B. R. Lawn, "An Atomistic model of kinetic crack growth in brittle solids," J. Mater. Sci. 10, 469 (1975).

41. K. J. Laidler, Theories of Chemical Reaction Rates (McGraw-Hill Book Company, New York, 1969).

42. S. M. Wiederhorn, "Mechanisms of subcritical crack growth in glass," in Fracture Mechanics of Ceramics, edited by R. C. Bradt, D. P. H. Hasselman, and F. F. Lange (Plenum Press, New York, 1978), these proceedings.

43. C. P. Flynn, Point Defects and Diffusion (Oxford University Press, 1972).

44. J. J. Gilman, "Escape of dislocations from bound states by tunneling," J. Appl. Phys. 39, 6086 (1968).

45. R. Thomson, "Bond breaking at low T," Report, ARPA Materials Research Council Proceedings.

46. J. Shie, "Quantum reactions in solids," Thesis, Dept. of Materials Science, State University of New York at Stony Brook, 1972.

47. E. Merzbacher, Quantum Mechanics (John Wiley & Sons, Inc., 1961), pp. 64-77.

MECHANISMS OF SUBCRITICAL CRACK GROWTH IN GLASS

S. M. Wiederhorn

Institute for Materials Research
National Bureau of Standards
Washington, D. C.   20234

ABSTRACT

Mechanisms of subcritical crack growth are divided into two groups: those that occur near the crack tip; and those that occur some distance from the crack tip. In the first group are diffusion mechanisms, plastic flow mechanisms, and chemical mechanisms of crack growth. In the second group are physical processes that control the transport of active agents to the crack tip, and chemical processes that control the composition of the crack tip environment. Other factors that influence subcritical crack growth are the structure of the crack tip and the type of rate limiting reaction that occurs during crack growth. In this paper, the above phenomena are discussed with regard to data on crack growth in glass. It is noted that although water is the main cause of subcritical crack growth in glass, there is still considerable controversy as to the dominant mechanism of fracture in glass.

1.   INTRODUCTION

Subcritical crack growth is one of the most important reasons for failure of structural materials. It is often caused by low concentrations of reactants which are apparently inert in the absence of applied stresses. Generally known as stress-corrosion cracking, the phenomenon of subcritical crack growth occurs in metals, plastics, ceramics and glasses. The phenomenon is characterized by a time-delay-to-failure, so that structural components fail unexpectedly after having supported loads for some time. It is widely recognized that improved reliability in structural design requires a thorough understanding of the physical conditions that control subcritical crack growth so that this type of failure can be avoided.

In glass, subcritical crack growth is due primarily to water in the environment (1-4). Relatively low concentrations of water can cause substantial reductions in the strength of glass. Because of subcritical crack growth, safety factors of greater than three times the service stress are not uncommon for structural components made of glass. In fact, one manufacture recommends 1,000 psi as the design strength of glass, even though strengths of 8,000 psi are commonly measured in the laboratory in short term strength tests. The reason for this conservative approach to design, is that subcritical crack growth causes time dependence of strength so that under sustained load, the strength of glass is very much less than that measured in a short term laboratory test.

This paper discusses some of the mechanisms that have been proposed to explain subcritical crack growth in glass. These mechanisms may be divided roughly into two groups: those that occur at the crack tip; and those that occur some distance away from the crack tip. Both types of processes must be considered in order to obtain a complete picture of fracture in glass. Therefore, physical processes such as ion exchange, and viscous flow, which indirectly effect the fracture process, are considered on a par with processes such as lattice trapping, chemical reactions and plastic flow which occur at the crack tip. These mechanisms of fracture are discussed in the light of experimental evidence that is used to support them. Special attention is paid to chemical mechanisms and to plastic flow mechanisms of fracture since these are most often suggested as the rate controlling mechanisms for subcritical crack growth.

2. SUMMARY OF FRACTURE MECHANISMS

Before going into a detailed discussion of fracture mechanisms, a summary of the various types of processes that can occur during crack growth is presented in order to obtain a total picture of the fracture process. A summary of the types of physical processes that effect subcritical crack growth in glass is presented in figure 1. As is noted, the stress corrosion environment must contain water, since water is the only known stress corrosion agent for glass. If water is present in the gaseous state, then the partial pressure of the water vapor is an important parameter that controls crack growth. In the liquid state, dissolved ions can also affect the fracture process. Since the concentration of hydroxyl ions is known to play a part in crack growth, any chemical agent in the solution that control the pH will also be important to the fracture process.

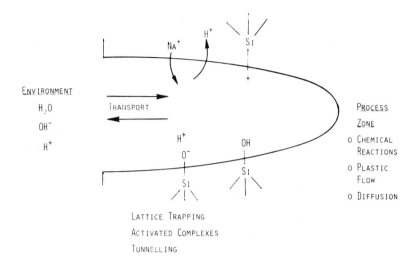

Figure 1.   Schematic diagram of the types of physical and chemical
            processes that effect subcritical crack growth in glass.

     The rate of transport of the bulk environment to the crack tip
can be a dominant factor in the stress corrosion process.   In
a gas, the rate of transport of water vapor to the crack tip may be
the rate limiting step in fracture.   In a liquid, the concentration
of solutes in the solution can be modified by a chemical reaction
with the glass fracture surface as the liquid moves to the crack
tip.   Since freshly fractured glass is highly reactive with water,
these chemical reactions can significantly change the aqueous
environment so that the active components at the crack tip differ
from those in the bulk environment.   In addition to these chemical
reactions, physical processes resulting from viscous drag of the
corrosive environment on the crack surface can affect the rate of
fracture by significantly altering the stress concentration at the
crack tip.

     Once the reactive environment has reached the crack tip,
details of the bond breaking reactions that occur in the process
zone must be considered.   The bond breaking reactions will be
discussed in three broad categories: chemical reactions, plastic
flow, and diffusion.   A number of fracture mechanisms have been
proposed for each of these categories.   In addition to these bond
breaking reactions, other processes that have been factored into
descriptions of fracture are lattice trapping and the type (activated complex or tunneling) of rate limiting reactions that occur at
the crack tip.

In considering fracture processes it should be recognized that
any one of the above processes can control the fracture of glass.
To decide which one of the processes controls fracture in a given
circumstance it is important to know the order in which these
processes occur. Some of them occur independently of one-another,
while others must occur in sequence. When two or more processes
can occur independently, then the fastest of the processes will
control fracture. Thus, if a bond may be broken either by plastic
flow, or by chemical reaction, the fastest one of the two processes
will determine the fracture rate. By contrast, if two or more
processes must occur in sequence, then the slowest of the processes
will control the fracture rate. Chemical reactions, for example,
are sequential reactions since the corrosive environment must first
be transported to the crack tip before a bond can rupture by chemical reaction. In this case, the slowest of these two processes
(transport or chemical reaction) will determine the rate of crack
propagation.

Considerations such as those discussed in the preceeding
paragraph were used to analyze crack propagation data obtained in
nitrogen gas (5). As can be seen in fig. 2 crack motion in soda
lime silicate glass depended on the relative humidity of the gas
and on the applied stress intensity factor. For a given humidity,
trimodal curves that contained three regions of crack propagation
were obtained. Regions I and II were attributed to a stress-
enhanced chemical reaction between the water and the glass. In

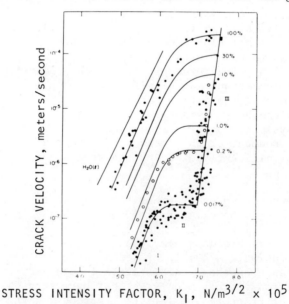

STRESS INTENSITY FACTOR, K_I, $N/m^{3/2}$ x 10^5

Figure 2.   Effect of relative humidity on subcritical crack growth
in soda-lime-silicate glass[5].

region I, crack motion was reaction rate limited, so that the rate
of transport of water to the crack tip was greater than the rate at
which water was consumed at the crack tip. In region II, however,
the water was consumed by the chemical reaction more rapidly than
it could be supplied to the crack tip, causing the crack propagation
rate to be transport limited. Thus, the slowest of the two processes controlled the rate of crack propagation.

Crack propagation in region III was attributed to an environment free fracture process. Although fracture could also have
occurred by a chemical reaction, the stresses at the crack tip were
high enough that spontaneous rupture of the bonds occurred more
rapidly than the chemical reaction. Because of this, fracture
occurred as if there were no environment present.

3. TRANSPORT OF REACTIVE ENVIRONMENTS TO THE CRACK TIP

As the reactants are carried to the crack by the fluid from
the bulk environment, two processes occur that significantly
effect the crack motion. In one, the composition of the fluid is
changed by chemical reaction with the fracture surfaces, so that
the composition of the fluid at the crack tip is not the same as
that in the bulk environment. In the second process, viscous
forces arising from the fluid flow increase the forces required for
crack motion. For crack growth in glass, the chemical processes
are probably more important than the hydrodynamic processes. There
is, however, good evidence for hydrodynamic effects on crack
growth in glass for cracks moving at high velocities, $>10^{-1}$ m/s.

3.1 Effect of Viscous Drag on Crack Propagation. As a crack
advances, it sucks fluid from the bulk environment into the crack,
setting up a viscous flow pattern between the two surfaces formed
by the fracture process. This viscous flow carries reactants from
the bulk environment to the crack tip where they react to cause
crack propagation. Flow of fluid from the bulk environment to the
crack tip requires a pressure drop, so that the pressure at the
crack tip must be lower than that in the bulk environment. If the
bulk environment is at atmospheric pressure, for example, the crack
tip environment must be at a pressure that is less than 1 atomsphere
It can be shown that for viscous liquids, negative pressures occur
near the crack tip. These can effect crack motion both by causing
cavitation near the crack tip and by reducing the effective stress
intensity factor at the crack tip. Again, these effects are important only when cracks move at fairly rapid velocities $>10^{-1}$ m/s.

Two mathematical models of the effect of fluid flow on crack
motion have been published. Perrone and Liebowitz (6) treat the
crack as a two dimensional channel with parallel walls, and use the
Poiseuille model of fluid flow to describe mass transport to

the crack tip. They determined the pressure at the crack tip, but
did not consider cavitation, or evaluate the effect of pressure on
the stress intensity factor. Perrone and Liebowitz suggested that
fluid flow limits crack growth rates in the process of liquid metal
embrittlement.

Newman and Smyrl (7) used a different approach to describe the
flow of fluids to the crack tip. Treating the crack as a wedge
shaped notch, they determined the pressure drop due to fluid flow,
and concluded that large negative pressures exist at the tips of
moving cracks. These pressures are low enough to cause cavitation
of the fluid at the crack tip for relatively low crack velocities.
Newman and Smyrl suggest that cavitation can occur in liquid metal
systems. The effect of the negative pressure on the crack tip
stress intensity factor was not considered by these authors.

The subject of fluid flow to crack tips was considered further
by Wiederhorn and Swain (8) who assumed the crack to be a parabolic
slit. The pressure, P, within the crack is related to the crack
velocity, v, the viscosity of the liquid, μ, the shear modulus of
the solid, G, and the applied stress intensity factor, K_I:

$$P = P_a + 6\pi v\mu (G/K_I)^2 \ln(x/c), \tag{1}$$

where P_a is the ambient pressure, x is the distance from the crack
tip and c is the crack length. Because (x/c) is always less than
1 the pressure within the crack is always less than the ambient
pressure.

For the crack configuration and physical conditions assumed in
table 1, large negative pressures are easily achieved for rapidly
moving cracks. For crack velocities greater than approximately 10^{-2} m/s
m/s, these pressures are low enough in water for cavitation to
occur near the crack tip. These negative pressures can also cause
a decrease in the effective stress intensity at the crack tip for a
given applied load. Therefore, an increase in the applied load is
required to achieve the same crack tip stress as if the fluid were
not present.

The decrease in the stress intensity factor, ΔK, due to the
presence of a liquid can be calculated by the following equation:

$$\Delta K = -13.1 \, v \, \mu \, (G/K_I)^2 (c\pi)^{\frac{1}{2}} \tag{2}$$

At crack velocities greater than $\sim 5 \times 10^{-1}$ m/s this estimate of ΔK is
of the same order of magnitude as K_I required to achieve this
velocity (table 1). Therefore, a hydrodynamic effect on crack
growth in glass is expected at velocities greater than $\sim 10^{-1}$ m/s.

Table 1.   Effect of Viscous forces on Crack Motion Calculated from
           equations 1 and 2 for a crack 1mm long in water. Pres-
           sure is estimated at a distance 1 μm from the crack tip.

Crack Velocity (m/s)	P-P$_a$ (eq. 1) (MPa)	ΔK (eqn. 2) (MN/m$^{3/2}$)
1	-3.7×10^2	2.09×10^0
10^{-1}	-3.9×10^1	2.2×10^{-1}
10^{-2}	-4.5×10^0	2.54×10^{-2}
10^{-3}	-5.78×10^{-1}	3.26×10^{-3}

An effect of this sort may explain crack propagation behavior at
high velocities in water. As noted both by Schönert, Umhauer and
Klemm (9), and by Weidman and Holloway (10) crack propagation
curves obtained in water bend over as the crack velocity exceeds
$\sim 10^{-2}$ m/s, (figure 3).   This decrease in the slope of the crack
propagation curve is probably due to viscous drag on the crack by
the water during crack propagation.

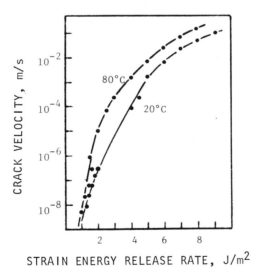

STRAIN ENERGY RELEASE RATE, J/m^2

Figure 3.   Effect of water on subcritical crack growth in
            soda-lime-silicate glass[9].

3.2 Effect of Chemistry on the Crack Tip Environment. Commonly
believed to be chemically inert to water, glass in fact is slowly
attacked by water (11-17). The rate of attack depends on the glass
composition. Glasses that contain high concentrations of alkali
ions are attacked more readily than glasses that contain no alkali
ions (14). The pH of the aqueous solution is also very important
to the rate of attack: basic solutions with a pH greater than ∿9
attack glass very readily, resulting in massive removal of material
from the glass surface (17). Temperature also plays an important
role in promoting the attack of water on glass. At temperatures in
excess of 200°C, glass is rapidly attacked by superheated steam (11).
Considering these effects, it is not surprising to find that the
solution contained in a crack is modified by these chemical processes. Furthermore, because the volume of fluid within a crack is
small compared to the total surface area of the crack (high surface
to volume ratio), the crack tip solution is rapidly modified by the
glass and contains significant concentrations of elements that are
normally present in the glass. As will be shown, these surface
reactions between the glass and the water have a strong effect on
subcritical crack growth in glass.

    The first reaction to occur when a freshly formed glass
surface comes into contact with water is ion exchange between the
alkali ions in the glass and the hydrogen ions in the solution.
(15, 16) This exchange process occurs rapidly even at relatively
low temperatures. Pantano, Dove and Onoda have shown that sodium
ion depletion occurs to a depth of ∿300Å in a soda lime silicate
after only 5 minutes exposure to air (18). The exchange rate
depends on the glass composition, and occurs more quickly as the
temperature is increased. The effect of this exchange process is
to produce a basic solution at the glass surface. If this exchange
process occurs in a restricted volume of liquid, such as exists in
a crack, than the pH of the solution can reach high values. Once
the pH exceeds ∿9, direct attack of the silicate network occurs.

    A secondary effect of the exchange process is the formation of
a stressed layer at the glass surface. Because hydrogen ions are
smaller than the alkali ions they replace, the stressed layer is
tensile. Metcalf et al. have shown that the tensile stresses may
be great enough to result in spontaneous fracture of modified
E-glass fibers (19,20).

    Ions in the glass other than alkali ions can also influence
the pH of the solution in the crack. These behave either as acids
or as bases and modify the crack tip solution accordingly. Alka-
line earth elements, such as calcium and magnesium, behave as bases
when placed in water and produce a basic crack tip solution. By
contrast, network modifiers such as phosphorous and boron, behave
as acids and therefore tend to form acidic crack tip solutions.

These modifiers are fixed to the network structure and cannot move
at low temperatures. Consequently, any effect they may have on the
crack tip pH requires the crack to pass very close to them. At
high pH, however, dissolution of the network permits these modifiers
to enter the solution and to play a more direct role in establishing
the crack tip pH.

For very pure silica glass, the pH at a crack tip is controlled by a chemical reaction between the glass and the solution.
Silica glass is normally covered by silanol groups, which behave as
weak acids when glass is brought in contact with water (21, 22).
The hydrolysis of these groups adds hydrogen ions to the crack tip
solution, which tends to produce a slightly acidic crack tip solution.

In basic environments, an entirely different type of reaction
occurs between the glass and the solution in the crack. Hydroxyl
ions tend to react with the bonds of the network structure (11,
15). As shown by the following equation, cleavage of these bonds
removes hydroxyl ions from the solution and replaces them with
hydrogen ions:

$$OH^- + \equiv Si-O-Si \equiv \rightarrow \equiv Si-O^- + \equiv SiOH \qquad (3)$$

$$\equiv SiOH \rightarrow \equiv SiO^- + H^+ \qquad (4)$$

In basic solutions the reactions indicated by equations go to
completion so that the entire surface of the glass is covered by
silanolate groups. In acidic environments, however, the high
concentration of hydrogen ions force the reaction given by equation
(4) to the left and the glass surface will be covered by silanol
groups. The equilibrium constant of reaction (4) determines the
chemical structure of the glass surface. Other network formers
such as aluminum and boron can exert a similar influence on the
crack tip solution.

Since the crack tip environment is open to the external
environment, ion transport between the two environments modifies
the crack tip environment. The degree of modification depends on
the crack velocity. For a static crack, concentration differences
between the crack tip and the external environment gradually
disappear as the glass bordering the crack tip solution becomes
depleted in reactive ions. For a rapidly moving crack, however,
freshly formed crack surfaces provide new sources of reactive ions,
so that the composition of the crack tip solution is controlled
primarily by the composition of the glass. Thus, one might expect
two regimes of crack growth: at high velocities crack growth is
controlled by the glass composition; at low velocities crack growth
is controlled by the external environment.

Experimental studies have confirmed many of the features so
far described. Measurements of the pH of ground-glass water
slurries indicate the importance of glass composition to the
composition of the liquid at a crack tip (23, 24). The slurry
simulates the high surface to volume ratio that exists at crack
tips. As indicated in table 2, pH measurememts on glass slurries
by different experimental techniques give similar pH values for
each glass. Slurries of glasses containing high concentration of
alkali ions were found to be basic; pH values ranged from 11.5 to
12.5. By contrast, silica glass, which contains only about 1ppm
alkali impurity, was found to be acidic; pH values ranged from 4.3
to 5.3. Other glasses containing lower concentrations of mobile
alkali ion were found to have a pH intermediate between these two
types of glasses. The results of these studies suggest that the pH
of the crack tip solution depends on the glass composition. High
alkali glasses result in basic solutions at the crack tip, whereas
low alkali glasses have crack tip solutions that are mildly acidic.

Additional support for the effect of glass composition on the
crack tip environment has been obtained from fracture mechanics
studies in which crack velocity measurements were made on three
glasses (soda-lime-silicate, chemical borosilicate, and silica) as
a function of pH (25). Studies were done in buffered solutions
ranging in pH from −0.8 to 14.8. For all three glasses, the slope
of the crack propagation curve was observed to decrease as the pH

Table 2.   Room Temperature pH Values of Glass Slurries[25]

Glass	Slurry pH (by four methods of measurement)
Silica	4.3 - 5.3
Aluminosilicate (no alkali)	9.2 - 10.2
Aluminosilicate (high alkali)	11.2 - 11.7
Chemical Boro-silicate	8.2 - 8.6
Lead Alkali	11.4 - 12.0
Soda-Lime-Silicate	11.5 - 12.3

of the test solution was increased, figure 4. Furthermore, the
slopes of the crack propagation curves were found to vary monotonically with the pH of the test solution, figure 5. By plotting
the slope of the crack propagation curve that was obtained in
water, against the pH of a ground glass slurry, data points were
obtained that fell on the general curve for each glass (see triangle in figure 5). This result suggests that the pH of the crack
tip solution is determined by the pH of the slurries. Because
similar results were obtained for each of the three glasses studied
it is felt that the slurry technique is generally applicable for
determining the pH of the crack tip solution.

Ion transport between the crack tip environment and the bulk
environment can also influence crack motion in glass. This effect
is illustrated in figure 6 which shows crack propagation data on a
high silica glass (7.5% TiO_2, 92.5% SiO_2) obtained in 1 N NaOH and
1 N HCl (26). For 1N HCl, the data can be represented over the
entire range of variables by a single curve. The position and
slope of this curve is identical to that obtained in water, suggesting that the crack tip environment of this glass is normally
acidic. In support of this finding, measurement of the pH by the
slurry technique also gave a slightly acidic reading. In contrast
to the experimental results obtained in HCl, two curves were
required to represent the data obtained in 1 N NaOH. At high crack
velocities, $>10^{-6}$ m/s, the data was represented by a curve that was
identical to the one obtained in 1 N HCl. At low crack velocities,

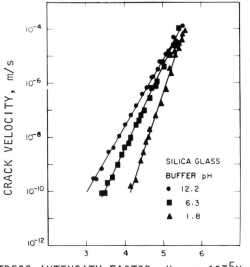

Figure 4.   Effect of pH on subcritical crack growth in silica
            glass[25].

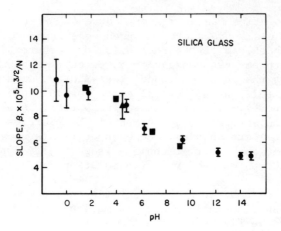

Figure 5.   Dependence of the slope of crack propagation data for
silica glass on environment pH. Crack propagation
data such as those shown in figure 4 were represented
by the following equation: $v = v_o \exp(-\beta K_I)$. ● Concentrated buffer solutions, ■ dilute buffer solutions,
▲ water. Brackets give standard deviation where more
than one measurement was made[25].

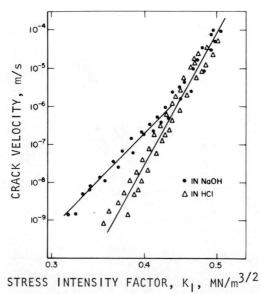

Figure 6.   Crack velocity data in 1 N HCl and 1 N NaOH. Glass
composition 92.5 wt % SiO_2; 7.5 wt % TiO_2 [26].

however, the slope of the curve in 1 N NaOH was about one-half that
obtained in 1 N HCl. This result suggests that at high crack
velocities the constituents of the glass control the pH of the
crack tip solution. New fracture surfaces are formed at a rate
that is fast enough to create an acid environment at the crack tip.
At slow crack propagation rates, however, the rate of production of
hydrogen ions at the crack tip is not sufficient to reverse the
effect of hydroxyl ion diffusion from the bulk solution to the
crack tip. As a result of this diffusion, the crack tip solution
becomes basic.

4. REACTION ZONE KINETICS

Most theories of subcritical crack growth in glass are concerned with reaction processes that occur at the crack tip. These
theories can be divided into three general types: those that
involve chemical reactions; those that involve plastic flow; and
those that involve diffusion. The nature of the crack tip structure, and the type of the rate limiting step that controls fracture
have also been considered in developing crack growth theories.

4.1 The Structure of the Crack Tip. In glass, the structure of
the crack tip is not precisely known because of its submicroscopic
size. Estimates of the reaction zone size at the crack tip place
it at ~1 to 10 nm depending on the type of glass and the theory
used to estimate the zone size (27-29). This size lies well below
the limits of resolution of most instruments that can be used to
examine the crack tip structure. Transmission electron microscopy,
which has a high resolution and has been used successfully to
examine the crack tip structure of crystals (30, 31), cannot be
used on glass because it is amorphous. The crack tip structure in
glass is usually inferred from measurements made at relatively
large distances from the actual crack tip (32, 33). As a result,
it is doubtful that the structure is known to a resolution of
better than 20 nm. Because of the lack of information on the real
structure of the crack tip, various models of the crack tip have
been incorporated into theories of subcritical crack growth. The
model selected depends on the theory used to explain crack growth.

Many theories of crack growth assume that glass is a continuum,
therefore the detailed atomic structure of the glass is ignored in
considering crack growth. The first crack tip model was proposed
by Griffith (34), who suggested that the crack could be approximated by an elliptical slot in the glass. For a given load, the
stress at the crack tip depends on the dimensions of the ellipse,
increasing as the crack length increases and as the radius of
curvature of the crack tip decreases. Griffith recognized that
atomic forces are present near the crack tip where the crack
surfaces approach to within atomic distances of one another.
Griffith incorporated these forces into a term that accounted for

the general resistance of the material to crack motion. In
Griffith's derivation of the conditions for crack stability, this
term was equated to the surface energy of the material.

Perhaps the best known application of the Griffith model for
the crack tip structure is the Charles-Hillig (35, 36) theory of
stress corrosion cracking. This theory assumes that crack growth
results from a stress-enhanced chemical reaction. Since stresses
along the crack surface depend on the local radius of curvature of
the surface, the rate of reaction also depends on the curvature and
as a result the curvature changes as the reaction proceeds, figure
7. When the applied load is high, the stresses along the crack
surface cause the curvature at the crack tip to decrease as the
crack grows, resulting in an acceleration of the rate of reaction.
By contrast, at low applied loads the curvature at the crack tip
increases as the reaction proceeds, resulting in a decrease in the
rate of reaction at the crack tip. The load for which the curvature remains constant is defined as the static fatigue limit.
Regardless of how long this load is applied to the glass, failure
will not occur. The predicted increase of the radius of curvature
at applied loads below the fatigue limit, is used to explain the
strength increase observed when glass is permitted to age in a
moist environment (37).

A continuum model for the crack tip structure is also assumed
in fracture mechanics theory (38). Cracks are envisioned as two
dimensional slits in an elastic continuum. When subjected to an
applied stress, these slits open to form parabolic shaped crack
tips. In the fracture mechanics model of a crack the stresses near
the crack tip have a relatively simple mathematical form, which

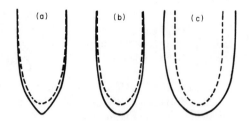

Figure 7.   Hypothetical changes in flaw geometry due to corrosion
            or dissolution [35].   (a) flaw sharpening as a result
            of stress corrosion.   (b) flaw growth such that the
            rounding of the tip by stress corrosion balances the
            lengthening of the flaw. (c) rounding by corrosion or
            dissolution.

depends on the type of loading, but not on the crack size or
shape. Furthermore, the crack tip stresses may be easily related
to the applied load through a parameter called the stress intensity
factor. For opening mode loading (all loads perpendicular to the
crack plane) the stresses, σ_{ij}, near the crack tip are given by the
following equation

$$\sigma_{ij} = K_I \, f_{ij} \, (\theta) \, r^{-\frac{1}{2}} \tag{5}$$

where r and θ are the cylindrical polar coordinates of the point of
interest relative to the crack tip. The function $f_{ij}(\theta)$ depends on
the polar angle θ, but not on the crack geometry. As can be seen
from this equation, a stress singularity occurs at the crack tip.
The slit model of the crack tip is used in any crack propagation
theory that employs fracture mechanics to describe crack growth.
To avoid the stress singularity, most crack propagation theories
assume (usually implicitly) that there is a process zone at the
crack tip, in which the stresses are nonlinear. Outside of the
process zone, the stresses are assumed to be elastic and are
described by the fracture mechanics formulation.

A mathematical approach that eliminates the stress singularity
from the crack tip was proposed separately by Dugdale and by
Barenblatt (39, 40). These authors recognized that the stresses at
the crack tip cannot be infinite and suggested that a more realistic
solution of the stress distribution could be obtained by adding
stresses to the crack surface in the vicinity of the crack tip.
These stresses, $\sigma(x)$, are summed over a small region near the
crack tip to obtain a stress intensity factor, $K(z)$:

$$K(z) = -(2/\pi)^{\frac{1}{2}} \int_{C'-D}^{C'} (C'-x)^{-\frac{1}{2}} \, \sigma(x) \, dx \tag{6}$$

D is the length of crack surface over which $\sigma(x)$ is summed, and C'
is the length of the crack tip prior to application of the surface
stresses. The stress singularity at the crack tip is now eliminated
by setting K_z equal to $-K_I$, where K_I is the stress intensity factor
in the absence of surface tractions. In these models, the surface
stresses given by $\sigma(x)$ are equated to the nonlinear stresses normally present in the vicinity of a real crack tip. After application of the surface stresses, the crack tip is viewed as lying a
distance $-D$ from the original crack position C', so that the position of the crack tip, C, is now given by $C = C'-D$.

The Dugdale-Barenblatt model of the crack tip can be used to
obtain a new level of insight into the process of subcritical crack
growth. With this model it is possible to estimate the size of a

Table 3.   Characteristic dimensions of nonlinear zone at the tips
            of cracks in brittle solids theoretical strength
            $\underset{\sim}{} E/10$ (41).

Material	K_{IC} $(MN/m^{3/2})$	E (GPa)	D (nm)
Si	0.62	130	0.97
Silica Glass	0.80	72	4.9
Mica	4.5	200	2.0
LiF	2.7	91	0.35

plastic zone in the vicinity of the crack tip. For $\sigma(x) = \sigma_y$, the
yield stress of the material, Dugdale has shown that the length of
the plastic zone, D, has the following dependence on the critical
stress intensity factor, K_{IC}:

$$D = (\pi/8) \ (K_{IC}/\sigma_y)^2 \tag{7}$$

The crack opening displacement, $2 U_c$, at the tip of the crack is
also given by the Dugdale approximation:

$$U_c = 4\sigma_y \ D/\pi E \tag{8}$$

Using the above equation and theoretical estimates for σ_y,
Lawn and Wilshaw (41) have shown that these plastic zones are small
(D=1-5 nm) for brittle materials, Table 3. The Dugdale-Barenblatt
model of the crack tip structure has been used by Williams and
Marshall and by Weidman and Holloway in their theories of subcritical
crack growth by plastic flow (42, 43). While the Dugdale-Barenblatt
model of the crack tip adds an element of atomicity to the crack
tip structure, it is still basically a continuum model of fracture
and a different approach is required to achieve a truly atomistic
model of fracture.

Thomson and his colleagues have recently formulated two
atomic models of fracture (44, 45). To simplify the mathematics
required to solve the crack problem, linear atomic force laws were
used. Once the displacement between two atoms exceeded a critical
displacement, bonds were assumed to be broken. In one model, a
infinite, two-dimensional, simple-cubic lattice of atoms was
considered (45). A crack, n atoms long, was placed at the center
of the plane of atoms, and the conditions for fracture were determined. As with the Griffith theory, Thomson found that a critical

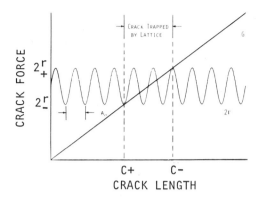

Figure 8.   Generalized crack-force diagram, illustrating that
            within the limits $2\Gamma_- \leq G \leq 2\Gamma_+$, $C- \leq C \leq C+$, where
            $G = K_I^2/E$. Outside this equilibrium range crack either
            extends or retreats dynamically [41].

stress was required for crack extension. However, a second lower
critical stress was required for crack healing. Between these two
stresses, the crack is "lattice-trapped" (figure 8). Crack growth
occurs when the stress intensity factor, K_+, for growth is exceeded.
Crack healing occurs when the stress intensity is less than a
critical stress intensity, K_-, for crack healing. Therefore, there
is a range of stress intensity factors for which the crack is
stable. This finding contrasts with that of the Griffith theory
which predicts only one value of the stress intensity factor for
crack stability. Subcritical crack growth can occur for a crack
that is "lattice-trapped" if the crack has sufficient thermal
energy to move from one position of stability to the next (41).
Other atomic models of fracture also confirm the possibility of
lattice trapping when cracks grow in brittle materials.*

4.2 <u>Rate Limiting Reactions</u>. The development of quantitative
models of subcritical crack growth requires an assumption to be
made as to the mechanism of the rate limiting step in the fracture
process. The rate limiting step is the slowest step in the
fracture process and its rate determines the rate of fracture.
Theoretical models of reaction rates usually assume the existence
of a potential barrier that separates the initial from the final
states of the reaction. For a reaction to go to completion, the
reactants must overcome this barrier. In the field of fracture,
two general reaction rate theories have been used to explain
subcritical crack growth: activated complex theory; and quantum
tunneling theory. In the activated complex theory the reactants

---

*See the article by E. R. Fuller and R. Thomson in this volume for
a review of the atomic model theories of fracture (48).

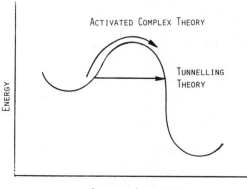

Figure 9.   Schematic diagram of two mechanisms of overcoming a
            potential barrier.

are assumed to go over the potential barrier; whereas in the
tunneling theory, the reactants are assumed to penetrate through
the barrier, figure 9.   Both of these theories provide a basis for
explaining the kinetics of crack motion.

    Activated complex theory, sometimes called "absolute rate
theory" or "transition-state theory," assumes that the reactants
pass over a potential barrier through the formation of an activated
complex (46).   The internal energy of the complex determines the
energy at the top of the barrier, whereas the energy of the
reactants and products of reaction determine the energy of the
potential wells on either side of the barrier.   The reactants are
assumed to be in a state of equilibrium with activated complexes
that were reactants in the immediate past.   Similarly, the products of reaction are assumed to be in equilibrium with those
complexes that were products of reaction.   The reaction rate, r,
for the reaction, $A + B \rightarrow (A-B)^{\ddagger} \rightarrow C$, is given by

$$r = (kT/h) \; [A] \; [B] \; (Q_{\ddagger}/Q_A Q_B) \; \exp \; (-E_o/RT) \qquad (9)$$

where [A] and [B] are the concentratious of the reactants; k is
Boltzman's constant; h is Plank's constant; T is the absolute
temperature; and Q_A, Q_B and $Q_{\ddagger}$ are the partition functions of A,
B and the activated complex.   The activation energy, E_O, is
calculated from the height of the potential barrier.   A similar
reaction rate formula controls the reaction in the reverse
direction.

In principle, the reaction rate can be determined from the molecular structure of both the reactants and the activated complex by using quantum and statistical mechanics to calculate $Q_{\ddagger}$, Q_A, Q_B and E_o. In practice, however, these parameters are difficult to determine, even for the simplest reaction (46). Because of this difficulty activated complex theory cannot be used from first principles, and is normally used as a semiempirical guide to relate the reaction rate to physical parameters such as pressure, temperature and composition. For this purpose, a thermo-dynamic formulation of the reaction rate equation is useful:

$$r = (kT/h) \ [A] \ [B] \ \exp \ (-\Delta G^{\ddagger}/RT) \qquad (10)$$

where $\Delta G^{\ddagger}$ represents the change in free energy of reaction as the reaction goes to activated state. The free energy of activation can be expressed in terms of the activation energy, $\Delta E^{\ddagger}$, the activation entropy, $\Delta S^{\ddagger}$, and the activation volume, $\Delta V^{\ddagger}$:

$$\Delta G^{\ddagger} = \Delta E^{\ddagger} - T \ \Delta S^{\ddagger} - P \Delta V^{\ddagger} \qquad (11)$$

This form of the reaction rate equation has been used by Charles and Hillig (35, 36), by Wiederhorn (24), and by Pollet and Burns (47) in discussions of the effect of environment on crack growth kinetics.

The theory of tunneling is entirely different from activated complex theory. Based on quantum mechanical concepts, the theory admits the possibility that reacting systems with insufficient energy to go over the barrier can penetrate through the potential barrier (46). Quantum tunneling as a rate process is most prob-able for systems that involve small particle-mass and low, narrow barriers. Hence, it finds application in reactions that involve electrons, protons and deuterons. In contrast to the activated complex theory, there is no simple equation that describes tunneling. There are however techniques for estimating the probability of a transition between the initial state and the final state of a reaction so that the fraction of reactants able to pass through the potential barrier can be calculated. To perform a tunneling calculation it is necessary to determine the quantum mechanical wave function of the reacting molecules, and the shape and height of the potential barrier. The complexity of this calculation is about as great as that encountered in using activated complex theory. Because of this difficulty and the lack of a simple equation to describe tunneling, there have been only two attempts to apply this theory to fracture problems. The one by Fuller and Thomson is included in this volume (48); the other was presented by Gilman and Tong (49). As with activated complex theory, it is possible to obtain the dependence of motion on applied stress and temperature.

4.3 <u>Mechanisms of Fracture</u>. Theories that explain crack motion
in glass can be classified into three broad categories: diffusional
theories; plastic flow theories; and chemical reaction theories.
Many of these theories have been developed to the point that
quantitative predictions of crack growth behavior can be made as
a function of environment, stress, and temperature. As a consequence, these theories can be compared directly with experimental
data, and their validity can be ascertained from this comparison.
In this section, a number of the proposed theories will be
discussed with reference to the available data on crack propagation in glass.

<u>Diffusion Mechanisms</u>. Mechanisms of subcritical crack growth
involving diffusion have been proposed by Hasselman (50) and by
Stevens and Dutton (51). These authors discussed the diffusion
process as it related to the motion of vacancies and atoms near
the crack tip. They propose that crack growth occurs by diffusion
through the elimination of the plane of atoms that constitute the
projected crack plane. Stevens and Dutton (51) assume that the
driving force for crack motion is the difference in chemical
potential between the highly stressed atoms at the tip of the
crack and those in the bulk of the solid. As a result of this
difference in chemical potential, a gradient in the concentration
of vacancies develops at the crack tip. The diffusion of vacancies
to the crack tip then controls the rate of crack growth. Detailed
calculations for crack growth by bulk diffusion, surface diffusion,
and vapor phase transport were compared with high temperature
(950 to 1900°C) data on alumina. The authors felt that their
theory gave a reasonable representation of their data and attributed subcritical crack growth to surface diffusion at low levels
of applied load, and to vapor phase transport at high levels of
applied load.

     Using somewhat different assumptions, Hasselman (50) obtained
an equation for crack motion that was similar to the one derived
by Stevens and Dutton. His equation provided a description of
the static fatigue and crack propagation data on a variety of
materials. A suggestion made by Hasselman to explain crack
growth by diffusion at room temperature is that water can lower
the diffusion rate of atoms and vacancies within the glass by
absorbing the crack surface. This process results in crack
propagation at very much lower stresses than are observed in
vacuum or in an inert environment. While this effect of water on
the diffusion rate of silicon and oxygen within the glass might
account for the relatively low activation energies observed for
crack propagation at low temperatures, there is no available data
on glass that supports this suggestion.

     The applicability of these theories for subcritical crack
growth in glass can be determined by comparing activation energies

for diffusion with those for crack growth in glass. Table 4
presents a summary of data for crack propagation of several
glasses in water in the temperature range of $\sim0°C$ to 90°C (52).
The activation energies, E*, were determined from a least squares
fit of the following equation: $v = v_o \exp(-E*+bK_I)/RT$. The
constants v_o and b were also determined from the least squares
fit. The activation energies ranged from approximately 25 to 33
Kcal/mole. These values can be compared with values of the
activation energy for diffusion in glass, Table 5.

By comparing table 4 with table 5 we see that with the
exception of sodium (and other akali ions) the activation energy
for crack motion in the presence of water is too low to be explained by a diffusion mechanism. Although alkali ions do have
activation energies for diffusion that are of the correct magnitude, it is doubtful that these ions by themselves can explain
fracture in the way envisioned by Hasselman, or by Stevens and
Dutton. Fracture by the mechanisms proposed by these authors
require migration of the network formers (Si, B, Al, 0) for crack
motion to occur. The activation energies for the diffusion of
these elements are too high to be consistent with the results of
fracture studies. Furthermore, even if a mechanism of bond
rupture could be envisioned that involved only akali ion migration,
such a mechanism could not be used to explain crack motion in
glasses that are normally free of akali ions (silica, for example,
has an alkali concentration of the order of 1 PPM). Thus, a
second mechanism would be required to explain crack motion in
these glasses. Since the crack growth behavior is essentially
the same for these two types of glasses, it is more reasonable to
conclude that crack growth in the presence of water is controlled
by some common mechanism.

Crack motion has also been observed in glass in water-free
environments (59). A summary of crack growth data collected in
vacuum is presented in table 6 for four glass compositions. The
activation energies for crack growth are somewhat higher than the
activation energies for the diffusion of the network formers, but
the difference is not as great as that obtained for crack propagation in water. Therefore, crack growth by a diffusional process
cannot be excluded solely on the basis of a comparison of activation
energies. In all probability, the bond breaking processes required
for the diffusion of network formers is similar to that required
for subcritical crack growth, so that the activation energy of
the two processes should be of similar magnitude.

Plastic Flow Mechanisms. The possibility of plastic flow as a
crack propagation mechanism might seem inconceivable for a
material that is as brittle as glass. However, there is good
experimental evidence to suggest that under special conditions
glass can be plastically deformed. For example, permanent densification has been observed by Bridgman and others on glasses that

Table 4.   Summary of subcritical crack growth data for glass
           tested in water. Temperature range 2 to 90°C.

Glass	E^* Kcal/mol ($\times 10^5$ J/mol)	b (MKS units)	$\ln v_o$
Silica	33.1 (1.39)	0.126	-1.32
Aluminosilicate (no alkali)	29.0 (1.21)	0.138	5.5
Aluminosilicate (high alkali)	30.1 (1.26)	0.164	7.9
Chemical Borosilicate	30.8 (1.29)	0.200	3.5
Lead-alkali	25.2 (1.06)	0.144	6.7
Soda-lime silicate	26.0 (1.09)	0.110	10.3

Table 5.   Diffusion of Elements in glass

Element	Activation Energy $\times 10^5$ J/mol (Kcal/mol)	Reference
Na	$\sim$0.6-1.3 ($\sim$15 - 30)	53
Ca	$\sim$2.3 ($\sim$55)	54
O	3.0 (71)	55
B	3.7 (88)	56
Si	$\sim$3.3 ($\sim$80)	57
Al	$\sim$2.5 ($\sim$60)	58

Table 6.   Data for crack propagation in vacuum. Least squares fit
           to the equation: $v = v_o \exp(-E^* + bK_I)/RT$

Glass	E^* 10^5 J/mol (Kcal/mol)	b MKS Units	$\ln v_o$
61% lead	3.48 (83.1)	0.51	5.1
aluminosilicate	7.05 (176.0)	0.77	14.0
Borosilicate Crown	2.75 (65.5)	0.26	6.6
Soda-lime silicate	6.05 (144)	0.88	-5.7

have been subjected to pressures of the order of 7 GPa (60, 62).
Furthermore, hardness indentation tests on glass leave permanent
marks in the glass surface (18, 63-66). These have been attributed
to plastic flow and densification near the indentor. Permanent
deformation has also been observed at stresses of approximately 7
GPa in compressive tests conducted by Ernsberger on glass rods
that contained oblate bubbles (67). In all of these examples,
high compressive stresses were necessary for the deformation.
There have been, however, no unequivocal reports of permanent
deformation of glass that resulted from tensile stresses.

Plastic flow as a mechanism for fracture was suggested by
Marsh (27,63) and later put into quantitative form by Weidmann
and Holloway (43), and Williams and Marshall (42). Marsh's
assumption that plastic flow was the main cause of fracture in
glass was based on a number of observations. These included the
observation that: (1) permanent deformation occurs at hardness
impressions and scratches; (2) crack healing does not occur when
an applied load is released from cracked glass; (3) glass does
not attain theoretically predicted fracture strengths; (4) the
fracture energy of glass is considerably higher than the predicted
surface energy; (5) during the high speed fracture of glass,
crack velocities fail to reach the theoretically predicted fracture
velocities; and finally, (6) plastic flow occurs in the glass as a
result of high pressures.

In a discussion of the plastic behavior and fracture of
glass, Hillig (68) provided an alternative interpretation of much
of the evidence used to support a plastic flow model of fracture.
He notes that: (1) frictional heating during scratching or
indenting could account for the softening of glass into the
viscous flow range; (2) the wedging of debris between crack
surface could prevent healing of cracks; (3) the theoretical
strength or limit velocities attained during high speed fracture
were not known with enough precision to support the arguments
presented by Marsh; (4) diamond pyramid hardness impressions
result from complex processes that involve densification as well
as plastic flow; (5) even if plastic flow does occur under high
compressive stresses, this cannot be taken as evidence of plastic
flow under similar tensile stresses. Hillig also questions the
fundamental significance of the correlation between fracture and
hardness in view of the complex nature of the indentation process.

Since Marsh's publications, the observation of crack healing
in glass by Wiederhorn and Towsend (69) casts further doubt on
plastic flow as a fracture mechanism in glass. Wiederhorn and
Towsend argued that, since crack healing occurred, plastic flow at
crack tips could not be large enough to wedge cracks open and thus
prevent molecular forces from drawing the surfaces back together.
The possibility of lattice trapping as a fracture mechanism, also
offers a reasonable explanation for the high fracture energy of

glass. Lattice trapping requires loads in excess of those estimated from the usual theoretical treatments of strength, so that high fracture energies would be expected in glass. Despite these arguments, plastic flow as a crack propagation mechanism remains an intriguing possibility, and the fact that the calculated plastic zone size at a crack tip is greater than one might expect from purely elastic considerations suggests that some sort of inelastic process may be playing a role in the fracture of glass.

In his qualitative theory of fracture, Marsh assumed that failure was due entirely to plastic yielding of the glass (27, 63). The yield stress was less than that required for tensile fracture. The fact that a macroscopic yield stress has never been observed in glass was explained by assuming that plastic flow is not of a strain-hardening type. Therefore, no mechanism exists for the stabilization of the plastic flow; once a crack starts growing by the plastic flow mechanism, fracture is catastrophic. Environments are thought to play a role in the fracture of the glass by penetrating the glass at the crack tip and softening it. Marsh presented data to support this theory. By revising the yield stress hardness relationship for hardness indentation measurements, he was able to demonstrate that the calculated yield stress for the glass was equal to the fracture strength. Both the fracture strength and the yield strength depended on the moisture in the environment. Glass was found to be softer at higher relative humidities and the hardness was found to be time dependent.

Recent studies by Gunasekera and Holloway (70) confirm these observations of the dependence of hardness on environment and provide quantitative relationships between diamond pyramidal hardness measurements and the time of load in a variety of environments. In these experiments, loading times from 10^{-3} to 10^{5} s were used for surfaces that were immersed in air, water, and dried liquid paraffin. Gunasekera and Holloway showed that for loading times less than approximately a second, the hardness of the glass depended on the time of loading, but did not depend on the test environment. By contrast, for loading times greater than a second, the hardness depended on both the time of loading and on the test environment. For long term loads, the glass was considerably harder when tested in liquid paraffin than when tested in air or water, and tests in air were slightly higher than those conducted in water. Tests conducted by Marsh (27) in air were consistent with those obtained by Gunasekera and Holloway. These measurements were used by Weidmann and Holloway and by Williams and Marshall to support a quantitative theory of crack propagation for a plastic mechanism.

A quantitative theory of crack propagation was developed from hardness data by considering the plastic zone at the crack

tip. As the crack moved, glass in front of this zone became
strained and stretched until rupture occurred. The strain rate
of the plastic zone was assumed to depend on the crack velocity,
and for a given temperature and environment, a constant zone size
was assumed. As the crack velocity increased, these requirements
necessitated an increase in the stresses in the vicinity of the
crack tip. The assumptions used in this theory are consistent
with hardness observations on glass, which indicate an increase
in strength as the rate of straining is increased.

     Stated in more quantitative terms, Weidmann and Holloway
(43) assume that the rate of crack motion at a constant stress
intensity factor is given by the virtual rate of increase of the
zone length when the plastic zone was of the critical size:

$$v = (dR/dt)_{\text{crit. size}} \qquad (12)$$

The plastic zone size, R, is related to the applied stress
intensity factor, K_I, and the yield strength of the glass, σ_f,
through a Dugdale-Barenblatt type model of the crack tip:

$$R = (1/2\pi) \ (K_I/\sigma_f)^2 \qquad (13)$$

Finally, an experimental relationship between the yield strength,
σ_f, and the time of load application, t, was obtained from hardness
measurements:

$$\sigma_f = (A \ \ln \ (Bt))^{-1} \qquad (14)$$

By combining these three equations, Weidmann and Holloway obtained
a quantitative expression that related crack velocity to hardness
measurements:

$$v = (K_I AB) \ (2R_c/\pi)^{\frac{1}{2}} \ \exp(-(2\pi R_c)^{\frac{1}{2}}/(K_I A)), \qquad (15)$$

where R_c is the critical zone size. As shown in figure 10 the
relation obtained by Weidmann and Holloway gives a good representation of the crack velocity data.

     Equation 15 was used to evaluate the critical zone size, which
was found to be approximately 5 nm for crack propagation in moist
environments, and 25 nm for crack propagation in moisture-free
environments. The critical zone size, determined independently
from estimates of the stress intensity factor and the flow stress,
was found be to approximately 3 nm in moist environments. Considering the assumptions of the theory and the sensitivity of zone
size to selected values of stress intensity factor and flow
stress, this value agrees with those determined from the slopes
of the crack propagation data. The zone size could also be
determined from the intercepts of the crack propagation data

shown in Figure 10. Following this procedure, excessively large
values of the zone size were obtained by Weidmann and Holloway.
However, by adjusting one of the constants from the hardness
data, a consistent set of data was obtained which gave a reasonable
value for the zone size. The adjustment was felt to be necessary
because of differences in the nature of the stress field and the
size of the region undergoing plastic flow at cracks and indenta-
tions.

The theories developed by Weidmann and Holloway, and Williams
and Marshall provide a reasonable description of crack propagation
in glass, and therefore have the potential of explaining static
fatigue. However a considerable amount of work remains to show
that the plastic flow model agrees in detail with the large body
of experimental data that has been obtained on glass. To verify
the theory it will be necessary to show that it predicts the
correct dependence of crack velocity on partial pressure of
water, on temperature in both moist and dry environments, and on
environmental pH. By conducting hardness measurements under
these various experimental conditions, a verification of the
theory should be possible. Once this is done, the plastic flow
model will have to be considered a viable alternative to other
theories of static fatigue.

Chemical Theories of Fracture. All of the chemical theories of
fracture have in common the assumption that the crack growth rate
is determined by the rate of reaction of corrosive agents with
the strained glass at the crack tip. Water, the prime stress
corrosion agent for glass, attacks these strained bonds causing
them to rupture, resulting in crack motion. The observation that

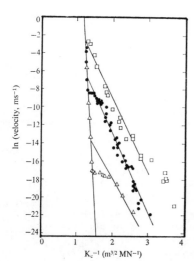

Figure 10.   Comparison of crack growth
data with plastic flow model, eq. 15 [43].
● air; ☐ water; Δ paraffin.

static fatigue is an activated process agrees with the fact that
chemical reactions themselves are activated processes. Thus, the
crack motion predicted by these theories is expected to have an
Arrhenius type of dependence on temperature.

The first quantitative chemical theory to explain the
static fatigue of glass was suggested by Charles, who proposed
that the rate of chemical attack at the crack tip was stress
sensitive (71). The crack velocity was assumed to be a power
function of the applied stress as expressed by the following
equation:

$$v = k \ (\sigma)^n \ \exp \ (-A/RT), \qquad (16)$$

where k, n and A are constants. From experimental measurements,
Charles found that n was approximately 16 for soda-lime-silicate
glass. This value agrees with the results of subsequent investigations by Ritter and Sherburne (72). Since the coefficient of the
stress is proportional to an Arrhenius factor, the equation
predicts the correct temperature dependence for static fatigue.
While this theory provides an adequate fit for most experimental
data on static fatigue, its functional dependence on stress
differs from that usually assumed for reaction rates. Therefore
the theory has the drawback that it is not easily compared with
reaction rate theory, which is used to explain other chemical
kinetic data.

Using reaction rate theory as the fundamental basis for a
description of static fatigue, Charles and Hillig (35, 36) developed
a theory of fracture by considering the rate at which a chemical
reaction occurs on the surface of a Griffith type crack. This
approach is consistent with that usually taken for chemical
reactions. At any point on the crack surface, the rate of reaction
was given by the following equation,

$$v = v_o \ \exp(-\Delta G^{\ddagger}/RT) \qquad (17)$$

where $\Delta G^{\ddagger}$ is the free energy of activation and T the temperature.
Charles and Hillig assumed that $\Delta G^{\ddagger}$ was a function of the applied
stress, and expanded it in the form of Taylors series. The first
two terms of the expansion were retained and a term was added to
account for the surface energy of the glass:

$$\Delta G^{\ddagger} = \Delta E^{\ddagger} - \sigma \Delta V^{\ddagger} + \gamma V_m/\rho, \qquad (18)$$

$\Delta E^{\ddagger}$ is the zero stress free energy of activation, $\Delta V^{\ddagger}$ is the
activation volume, σ is the surface stress, γ is the surface
energy of the crack, V_m is the molar volume of the glass and ρ
is the radius of curvature of the crack tip. Since $\Delta G^{\ddagger}$ is stress
dependent, the Charles-Hillig approach can be used to represent

the dependence of crack velocity on applied stress.

By analyzing the change of crack shape resulting from the chemical reaction along the crack surface, Charles and Hillig predicted a static fatigue limit for glass. At stresses below the fatigue limit, they showed that the radius of curvature of the crack tip increases with time. As a consequence, the stresses at the crack tip decrease with time. This behavior leads to the conclusion that below the fatigue limit, failure will not occur regardless of how long the load is applied to the crack. At stresses above the fatigue limit, however, crack sharpening occurs, so that stresses at the crack tip increase with time, leading to accelerated crack growth.

Since the Charles-Hillig theory is based on continuum mechanics, the theory predicts that crack sharpening will occur indefinitely at stresses greater than the static fatigue limit. As noted by Wiederhorn, however, the crack tip radius is limited by the structure of the glass, and cannot be smaller then a size which is determined by the molecular structure of the glass (52). Therefore, above the static fatigue limit, the crack sharpening proceeds until the radius of curvature of the crack tip reaches a lower limit, at which point the crack propagates without further sharpening.

The Charles-Hillig theory can be used to explain most of the available static and dynamic fatigue data on glass. Since it is based on reaction rate theory, it is felt that the model also predicts the correct dependence of strength on temperature. In addition, the model can account for the aging of glass in which strength is observed to increase as a result of crack tip blunting by stress corrosion. The dependence of crack motion on the partial pressure of water in the environment and on other chemical variables is also explained by the Charles-Hillig model, since the free energy of activation, $\Delta G^{\ddagger}$, will depend on these variables.

In addition to satisfying the experimental data, the Charles-Hillig model is theoretically acceptable because many of the empirical constants used in this model can be explained in physical terms. Thus, the zero stress activation energy can be explained in terms of the zero stress corrosion rate of glass surfaces, while the stress coefficient of the activation energy can be explained in terms of an activation volume for these reactions. The fact that reaction rates depend on surface curvature is also accounted for by the surface energy portion of the activation energy.

Although the Charles-Hillig model for fracture is the most widely accepted chemical theory, others have been suggested. In general, these also provide reasonable descriptions of experimental crack growth data. Doremus has recently questioned the necessity of a surface energy term in the Charles-Hillig model (73). He notes that a detailed consideration of the crack tip geometry is not essential for the derivation of the static fatigue limit. By considering the ratio of the crack width to the crack length, Doremus obtained an explicit function for the static fatigue limit that did not depend on surface energy. This limit is predicted to occur at approximately 15% of the liquid nitrogen strength, which is consistent with the lowest limit measured experimentally on glass.

More recently, Lawn has proposed an atomistic model to explain crack growth in brittle solids (74). Lawn presented the picture of an ideal brittle crack in which sequential bond rupture occurs via the lateral motion of atomic kinks along the crack front. The concept of lattice trapping is introduced by using the theory presented by Thomson and coworkers (44, 45), and the statistical mechanics of kink formation and motion along a crack front are developed. Crack motion is explained in terms of the nucleation and motion of these kinks. Fracture mechanics parameters, such as the strain energy release rate, enter the model in a natural manner. Phenomenologically, the equation developed by Lawn for crack motion differs from the Charles-Hillig equation since the stress dependence of the activation energy is proportional to the strain energy release rate rather than the stress intensity factor. The theory has been used by Lawn and by Tyson et al. (75) to provide an explanation of the dependence of crack propagation on environment, stress and temperature. Furthermore, a static fatigue limit is predicted in a very natural way by this theory.

4.4 General Comments on Theories of Crack Growth and Strength. This review of theories of crack propagation has been relatively brief, covering those theories which are most relevant to currently available data on the fracture of glass. Other theories have been presented to explain this phenomenon: however many are not consistent with the available data, and others are further examples of the theories that have already been discussed. Generally most authors use static fatigue, dynamic fatigue, or crack propagation data to justify their theoretical treatment. Unfortunately, the range of the experimental variables that can be achieved from these types of data is not great enough to distinguish unequivocally between different theories. As a consequence, the same set of data is often used to justify fracture theories that have very different starting assumptions and physical bases. For example, for soda-lime-silicate glass approximately 16 orders of magnitude in time are required to cover one order of magnitude in stress in a static fatigue test. Because of this sensitivity of failure time to stress, a variety of theories will adequately fit the same set of data. The same conclusion holds with regard to dynamic fatigue and crack propagation data (76). Therefore, it follows that these types of data alone cannot be used to distinguish between different failure mechanisms. Conclusions are similar with regard to activation energy measurements for static fatigue or crack growth. The activation energy obtained from an empirical fit of experimental data depends on the form of the equation chosen for the fit, and therefore on the particular model chosen to fit the data. This state of affairs presents a considerable obstacle to the determination of mechanisms of subcritical crack growth in glass.*

REFERENCES

1.   L. H. Milligan, J. Soc. Glass Technol., 13, 351T (1929).
2.   C. Gurney and S. Pearson, Proc. Phys. Soc., 62, 469 (1949).
3.   T. C. Baker and F. W. Preston, J. Appl. Phys., 17, 170 (1946).
4.   T. C. Baker and F. W. Preston, J. Appl. Phys., 17, 179 (1946).
5.   S. M. Wiederhorn, J. Amer. Ceram. Soc., 59, 407 (1967).
6.   N. Perrone and H. Liebowitz, pp. 2065-72 in Proc. 1st Int. Conf. on Fracture, Sendai, Japan, 1965, T. Yokobori, T. Nawasaki and J. L. Swedlow, eds.
7.   J. Newman and W. H. Smyrl, Met. Trans., 5, 469 (1974).
8.   S. M. Wiederhorn and M. V. Swain, to be published.
9.   K. Schönert, H. Umhauer and W. Klemm, Proc. 2nd Int. Conf. on Fracture, Brighton, 1969, p. 474, Chapman and Hall, London (1970).
10.   G. W. Weidmann and D. G. Holloway, Phys. Chem. Glasses, 15, 116 (1974).
11.   R. J. Charles, J. Appl. Phys., 29, 1549 (1958).
12.   M. A. Rana and R. W. Douglas, Phys. Chem. Glasses, 2, 6 (1961).
13.   T. M. M. El-Shamy, J. Lewins, and R. W. Douglas, Glass Technol., 13, 81 (1972).
14.   R. W. Douglas and T. M. M. El-Shamy, J. Am. Ceram. Soc., 50, 1 (1967).
15.   L. Holland, The Properties of Glass Surfaces, Wiley, New York (1964).
16.   L. L. Hench, J. Non-Cryst. Solids, 19, 27 (1975).
17.   D. Hubbard and E. H. Hamilton, J. Research NBS, 27 143 (1941).
18.   C. G. Pantano, Jr., D. B. Dove and G. Y. Onoda, Jr., J. Non-Cryst. Solids, 19, 41 (1975).
19.   A. G. Metcalfe, M. E. Gulden and G. K. Schmitz, Glass Technol., 12, 15 (1971).
20.   A. G. Metclafe and G. K. Schmitz, ibid., 13, 5 (1972).
21.   M. L. Hair, J. Non-Cryst. Solids, 19, 299 (1975).

---

*Ernsberger has recently reached a similar conclusion with regard to experimental justification of static fatigue theories (77).

22. M. L. Hair and W. Hertl, J. Phys. Chem., 74, 91 (1970).
23. S. M. Budd and J. Frackiewicz, Phys. Chem. Glasses, 2, 115 (1961); 3, 116 (1962); 5, 32 (1964).
24. S. M. Wiederhorn, J. Am. Ceram. Soc., 55, 81 (1972).
25. S. M. Wiederhorn, J. Am. Ceram. Soc., 56, 192 (1973).
26. S. M. Wiederhorn, A. G. Evans, E. R. Fuller and H. Johnson, J. Amer. Ceram. Soc., 57, 319 (1974).
27. D. M. Marsh, Proc. Phys. Soc., London, 282A, 33 (1964).
28. S. M. Wiederhorn, J. Am. Ceram. Soc., 52, 99 (1969).
29. K. R. Linger and D. G. Holloway, Phil. Mag., 18, 1269 (1968).
30. S. M. Wiederhorn, B. J. Hockey and D. E. Roberts, Phil. Mag., 28, 783 (1973).
31. B. J. Hockey and B. R. Lawn, J. Matls. Sci., 10, 1275 (1975).
32. S. A. Dozier, D. L. Kinser, and P. F. Packman, pp. 754-759 in Corrosion Fatigue, NACE-2, A. J. McEvily and R. W. Staehle, eds., National Association of Corrosion Engineers, Houston, Texas (1972).
33. S. M. Wiederhorn, pp. 503-28 in Materials Science Research, vol. 3, Edited by W. W. Kriegel and H. Palmour III, Plenum Press, New York, (1966).
34. A. A. Griffith, Phil. Trans. Roy. Soc. (London), 221A, 163-98 (1921).
35. R. J. Charles and W. B. Hillig; pp. 511-27 in Symposium on Mechanical Strength of Glass and Ways of Improving It. Florence Italy, Sept. 25-29, 1961, Union Scientifique Continental du Verre, Charlerai, Belgium (1962).
36. W. B. Hillig and R. J. Charles, pp. 682-705 in High-Strength Materials. Edited by V. F. Zackey John Wiley and Sons Inc., New York (1965).
37. R. E. Mould, J. Am. Ceram. Soc., 43, 160 (1960).
38. P. C. Paris and G. C. Sih, pp. 30-81 in ASTM Special Tech., Publ. No. 381 (1965).
39. D. S. Dugdale, J. Mech. Phys. Solids, 8, 100 (1960).
40. G. I. Barenblatt, Adv. Appl. Mech., 7, 55 (1962).
41. B. R. Lawn and T. R. Wilshaw, Fracture of Brittle Solids, Cambridge University Press, Cambridge (1975).
42. J. G. Williams and G. P. Marshall, Proc. Roy. Soc., London, A342, 55 (1975).
43. G. W. Weidmann and D. G. Holloway, Phys. Chem. Glasses, 15, 68 (1974).
44. R. Thomson, C. Hsieh, and V. Rana, J. Appl. Phys., 42, 3154 (1971).
45. C. Hsieh and R. Thomson, J. Appl. Phys., 44, 2051 (1973).
46. K. J. Laidler, Theories of Chemical Reaction Rates, McGraw-Hill Book Co., New York (1969).
47. J. C. Pollet and S. J. Burns, to be published, Int. J. Fract.
48. E. R. Fuller, Jr., and R. Thomson, in Fracture Mechanics of Ceramics, edited by R. C. Bradt, D. P. H. Hasselman, and F. F. Lange, Plenum Press, New York (1978).
49. J. J. Gilman and H. C. Tong, J. Appl. Phys., 42, 3479 (1971).

50. D. P. H. Hasselman, pp. 297-315 in Ultrafine-Grain Ceramics, J. J. Burke, N. L. Reed and V. Weiss editors, Syracuse University Press, Syracuse, New York (1970).
51. R. N. Stevens and R. Dutton, Mater. Sci. Eng., 8, 220 (1971).
52. S. M. Wiederhorn and L. H. Bolz, J. Amer. Ceram. Soc., 53, 543 (1970).
53. R. H. Doremus, pp 1-71 in Modern Aspects of the Vitreous State Vol. 2, edited by J. D. Mackenzie, Butterworths and Co., Ltd., Washington, D. C. (1962).
54. G. H. Frischat, J. Am. Ceram. Soc.. 53, 625 (1969).
55. G. H. Frischat, Glastech. Ber., 44, 93 (1971).
56. R. H. Doremus, Glass Science, John Wiley and Sons, New York (1973).
57. H. Towers and J. Chipman, J. Trans. Amer. Inst. Min. (Metall.) Engrs., 209, 1 (1957).
58. J. Henderson, L. Yang and G. Derge, Trans. Amer. Inst. Min. (Metall.) Engrs., 221, 56 (1961).
59. S. M. Wiederhorn, H. Johnson, A. M. Diness, and A. H. Heuer, J. Am. Ceram. Soc., 57, 336 (1974).
60. P. W. Bridgman and I. Simon, J. Appl. Phys., 24, 405 (1953).
61. H. M. Cohen and R. Roy, Phys. Chem. Glasses, 6, 149 (1965).
62. J. D. Mackenzie, J. Am. Ceram. Soc., 46, 461 (1963).
63. D. M. Marsh, Proc. Roy. Soc., 279A, 420 (1964).
64. E. Dick and K. Peter, Jr., J. Am. Ceram. Soc., 52, 338 (1969).
65. F. M. Ernsberger, J. Am. Ceram. Soc., 51, 545 (1968).
66. J. E. Neely and J. D. Mackenzie, J. Mat. Sci., 3, 603 (1968).
67. F. M. Ernsberger, Phys. Chem. Glasses, 10, 240-45 (1969).
68. W. B. Hillig, pp. 383 to 411 in Microplasticity, Charles J. McMahon, Jr., Interscience Publishers, New York, (1968).
69. S. M. Wiederhorn and P. R. Townsend, J. Am. Ceram. Soc., 53, 486 (1970).
70. S. P. Gunasekera and D. G. Holloway, Phys. Chem. Glasses, 14, 45 (1973).
71. R. J. Charles, J. Appl. Phys., 29, 1657 (1958).
72. J. E. Ritter and C. L. Sherburne, J. Am. Ceram. Soc., 54, 601 (1971).
73. R. H. Doremus, pp. 743 to 748 in Corrosion Fatigue, A. J. McEvily and R. W. Staehle editors, National Association of Corrosion Engineers, Houston, Texas (1972).
74. B. R. Lawn, J. Mat. Sci., 10, 469 (1975).
75. W. R. Tyson, H. M. Cekirge and A. S. Krausz, J. Mat. Sci., 11, 780 (1976).
76. S. M. Wiederhorn, pp. 893 to 901 in Fracture 1972 ICF4, Vol. 3, D.M.R. Taplin Ed., University of Waterloo Press, Waterloo, Ontario (1977).
77. F. M. Ernsberger, pp. 293-321 in the Proceedings of the 11th International Congress on Glass, Vol. 1 of the Survey Papers, Prague, Czechoslovakia, July 4-8, 1977.

STRESS CORROSION MECHANISMS IN E-GLASS FIBER

Charles L. McKinnis

Owens-Corning Fiberglas Corporation
Technical Center
Granville, Ohio 43023   USA

INTRODUCTION

In a previous paper, the two dominant factors which control the tensile strength of E-glass fiber were described /1/. These factors were, (1) submicroscopic disperse phase* in the glass as determined by laser ultramicroscopy, and (2) the level of ambient humidity during fiber forming and testing. Fiber tensile strength decreased in proportion to an increase in the concentration of either the disperse phase, the ambient humidity, or both. The disperse phase was proposed to function as stress risers in the glass, while the ambient humidity was thought to contribute to stress corrosion of the fiber surface.

However, it was concluded that these were not independent mechanisms. An interaction between the two was proposed. This concept is now developed, resulting in a humidity activated stress corrosion reaction. Though consisting of several interdependent mechanisms, all contribute to the initiation of the microflaw which ultimately produces fracture in E-glass fiber. The technical background for the development of this concept is contained in the section which follows.

TECHNICAL BACKGROUND

The experimental details which resulted in the conclusions contained in the Introduction have been given in the previous publication /1/. They will be summarized here only in sufficient

*A term commonly used to denote light scattering sites

detail to serve as a background for the development of the stress corrosion reaction concept.

A.   Experimental Techniques

E-glass composition was used throughout in these investigations. For glasses of variable disperse phase concentration, this composition was specially melted from batch into cullet; for glasses of constant, low disperse phase concentration, commercial E-glass marbles were used.

All glasses were remelted in an electrically heated, platinum alloy bushing containing a single nozzle, fibers formed for strength testing and beads sampled for disperse phase concentration by laser ultramicroscopy; beads are glass droplets which issue from the bushing nozzle when fiberization is not in process.

For any set condition, the consistent $10\mu m$ diameter E-glass fibers were tested, seven at time, on a multihead tensile tester, at a rate of load elongation of 0.148 inches/minute, a gage length of 3 inches, yielding a strain rate of 0.049 inches/inch/minute. These conditions of constant fiber diameter and test procedures were maintained throughout these investigations. The disperse phase concentration of the beads was subjectively assayed using the light scattering techniques of laser ultramicroscopy; this concentration was then correlated with the measured strength. An inverse relation was always operative between disperse phase concentration and measured strength.

The disperse phase concentration has been estimated at less than 0.5% by volume, with size at least as small as $0.06\mu m$. Tentatively identified as cristobalite from its morphology using dark field microscopy /2/, the disperse phase is thought to be siliceous micro-remnants of the initial batch ingredients or their reaction products. Their presence and concentration is controlled by the glass melting history, which includes the time, temperature and redox conditions of melting and the specific glass batch materials used.

Early in these investigations, it was found that measured strength was inversely proportional to both disperse phase concentration and the ambient humidity of fiber forming and testing. The disperse phase concentration - measured strength correlation was so definite, at constant ambient humidity, that a fairly good estimate of the latter could be made, based on the former. To quantify this correlation, the fibrous strength of glasses having various levels of disperse phase was determined at a constant low humidity of 25% RH @ 70°F.

B.   Variable Disperse Phase - Constant Low Humidity

The resulting correlation is given in Table I, where the first column shows the strength interval, and the second, the subjective disperse phase value or concentration as assayed by laser ultra-microscopy. From the expertise gained, the strength could now be estimated to within 5% of measured values based solely on the disperse phase concentration. This technique was then used to study the effects of variation in the ambient humidity on tensile strength.

C.   Variable Ambient Humidity - Constant Low Disperse Phase

Thirteen separate tensile strength determinations were made as the ambient humidity of fiber forming and testing was varied from 25% to 75% RH @ 70°F. The disperse phase concentration was assessed, and using Table I, the strength estimated. Any measured strength less than the estimate was now the result of humidity degradation. All of these data were mathematically analyzed, transformed using portions of Table I and then converted to yield Figure 1; these details are contained in the previous reference /1/.

Figure 1 shows the linear decrease in the strength (596-448 ksi) of quality, lowest disperse phase containing E-glass fiber, as the ambient humidity is increased over the interval, 25%-75% RH @ 70°F. Extrapolation to 0% and 100% RH yield values of 666 and 376 ksi respectively.

TENSILE STRENGTH INTERVAL −ksi	VALUE (QUANTITY)
>600	0
575-599	1 −
550-574	1
525-549	1 +
500-524	2 −
475-499	2
450-474	2 +
425-449	3 −
400-424	3
375-399	3 +
350-374	4 −
325-349	4
300-324	4 +
275-299	5 −
250-274	5
225-249	5 +

Tab. 1 - Fiber tensile strength interval vs. disperse phase concentration, E-glass @ 25% RH, 70°F.

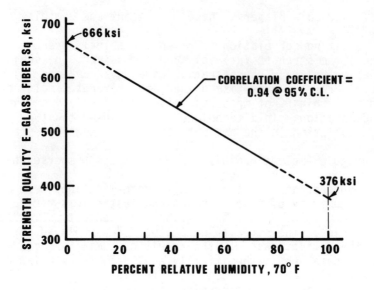

Fig. 1 - Tensile strength, quality E-glass fiber vs. percent relative humidity, 70°F.

The equation describing the linear plot of Figure 1 is,

$$S_q = S_o (1-KH_r)$$

Equation 1

S_q = tensile strength of quality E-glass fiber at some relative humidity, 70°F.

S_o = tensile strength of quality E-glass fiber at 0% relative humidity, 70°F. = 666 ksi.

K = constant, characteristic of E-glass and test procedures = 4.35 X 10^{-3}.

H_r = percent relative humidity, 70°F.

Equation 1 and the information from Table 1 were then empirically combined, resulting in a strength equation incorporating the degrading effects of both disperse phase and the ambient humidity.

  D. Variable Disperse Phase and Ambient Humidity

  The resulting empirical equation is,

Equation 2

$$S = S_0 (1-KH_r) \left(1-\frac{I_{dp}}{24}\right)$$

S   = tensile strength of E-glass fiber of unknown quality at some relative humidity, 70°F.

S_0 = tensile strength of quality E-glass fiber at 0% relative humidity, 70°F. = 666 ksi.

K   = constant, characteristic of E-glass and procedures = 4.35 X 10^{-3}.

H_r = percent relative humidity, 70°F.

I_{dp} = disperse-phase interval, E-glass, (from laser ultramicroscopic analysis and Table 1).

The disperse phase concentration and the ambient humidity are the only data needed to calculate the strength of E-glass fiber using Equation 2. Its validity is shown in Figure 2, where predicted strength is plotted against measured values. This equation incorporates the two factors known to control the strength of E-glass fiber, disperse phase in the glass, a volume effect described by the $\left(1-\frac{I_{dp}}{24}\right)$ term, and the ambient humidity of forming and testing, a fiber surface effect by the $(1-KH_r)$ term.

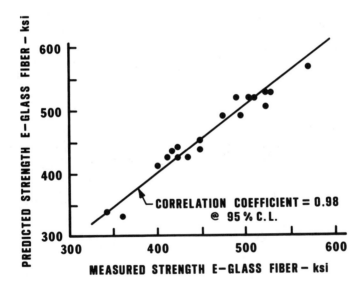

Fig. 2 - Measured tensile strength E-glass fiber vs. strength predicted from Equation 2.

Based on Equation 2 and the results of Figure 2, it was apparent that some type of interaction existed between the disperse phase and the effects produced by the ambient humidity. The concept of a stress field associated with disperse phase, and of its interaction with the fiber surfacesorbed water species to produce a stress corrosion reaction follows.

STRESS MICELLE AND MICROSTRESS CONCEPTS

In this investigation, finite element analysis was utilized /3/. Cristobalite was used as the model since it had been tentatively identified as a source of disperse phase in beads. Considered as a cylinder of aspect ratio 1 X 1, its coefficient of expansion and Young's modulus were compared with similar properties of E-glass.

With no load on the fiber, a tensile stress of about 450 ksi may exist at the disperse phase-glass interface, diminishing to zero intensity at a distance of about three disperse phase units into the glass. Whatever the disperse phase concentration in E-glass, a portion will be located in close proximity to the surface, the remainder in the volume of the fiber; it is highly improbable that any will ever penetrate the fiber surface.

The disperse phase and its stress field have been termed a stress micelle, while those in close proximity to the fiber surface specifically a microstress. These are as shown in Figure 3. The stress field of the microstress should penetrate into the fiber surface. Reaction of the fiber surfacesorbed* water species will be greater with this stressed micro-area, than with adjacent glass /4/.

The result is the initiation of a microflaw, in situ, analogous to the classical Griffith flaw. A typical reaction, microstress-surfacesorbed water species, a stress corrosion mechanism, is as depicted in Figure 4 for a fiber at rest and under test. The length of the tensor under the microstress notation is indicative of the tensile stress intensity within the microstress. Variation in glass structural bond lengths was a matter of convenience in constructing these drawings. Because of the complexity of both the reaction and the E-glass structure, further definition of this reaction is not possible at this time. However, once initiated, the severity of the produced microflaw will quickly increase, this the result of the tensile nature of the stress field.

----

*Term used instead of chemisorbed, since water species-glass surface bond unknown.

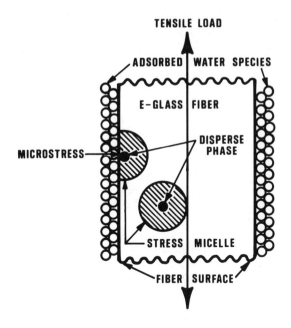

Fig. 3 - Disperse phase-stress micelle-microstress relationship in
E-glass fiber.

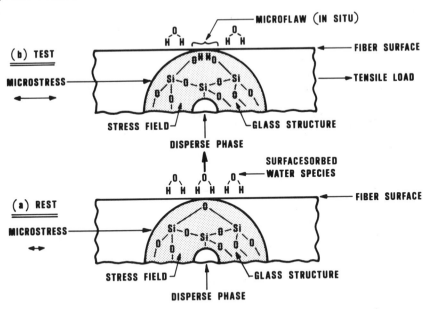

Fig. 4 - Reaction, microstress-surfacesorbed water species, E-glass
fiber.

Application of a tensile load to the fiber will proportionately increase the stress intensity of the microstress and concomitantly increase the probability of the reaction, microstress-surfacesorbed water species, resulting in microflaw initiation.

The commonly accepted theories of flaw initiation, such as surface damage, or reaction of water species with alkali of the glass, are not normally found with E-glass fiber. Fiberizing from a hot melt produces a pristine surface which, if appropriate precautions are taken, is maintained at test. Little, if any, alkali reaction will occur with E-glass, since it contains less than 1.0% of this constituent. This is confirmed by the high correlation coefficient of Figure 2, using Equation 2. Therefore, it is concluded that the dominant flaw producing mechanism is the reaction, microstress-surfacesorbed water species.

Increase in the disperse phase concentration results in a proportionate increase in the number of microstress. This further increases the probability of this reaction, leading to microflaw initiation and finally to fiber fracture. Furthermore, this reaction is enhanced as the level of the ambient humidity is increased. This then leads to the concept of the fiber sorbed water hull, and its role in the stress corrosion reaction.

FIBER SORBED WATER HULL

Water sorption studies on 10μm diameter E-glass fiber have shown a linear increase in the weight of water species sorbed, as the partial pressure water vapor was increased incrementally over the range, $p/p_o = 0.20 - 0.80$ @ 20°C. /5/. Even at the lowest partial pressure, the water sorption was more than sufficient to cover the entire fiber surface. These conditions are comparable to those of Figure 1, where the strength of quality 10μm diameter E-glass fiber decreases with increase in the humidity level, 25%-75% @ 70°F.

From this, it is concluded that the decrease in fiber strength with increase in humidity is due, in part, to a corresponding increase in water species sorbed onto the fiber surface. These sorbed water species have been termed, the water hull, the thickness of which is directly proportional to the ambient humidity. It consists of a surfacesorbed monolayer, followed by subsequent adsorbed layers, the number directly proportional to the water hull thickness. Figure 5, portions of which are not drawn to scale, shows these conditions for a fiber at rest.

The double arrow notation indicates that the outer portions of the water hull are in equilibrium with the water species of the ambient. Very little interaction between the surfacesorbed water species and the microstress will occur; the stress corrosion reaction is minimal. However, application of a tensile load changes these conditions, as shown in Figure 6, for a fiber under test.

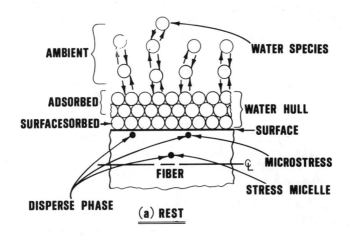

Fig. 5 - Water hull-microstress relationships, E-glass fiber at rest.

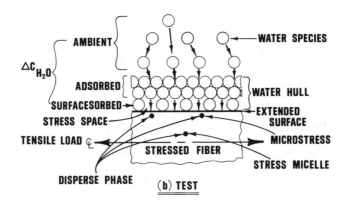

Fig. 6 - Concentration gradient, Δc_{H_2O}, and water species diffusion through water hull for E-glass fiber under test.

     The resulting fiber stress produces a concentration gradient, Δc_{H_2O}, from the ambient, through the water hull, to the fiber surface. This concentration gradient is the chemical potential or driving force for the diffusion of water species to the fiber surface. Its formation is proposed to result from the following interactions.

Incremental tensile loading of a 10μm diameter E-glass fiber will produce a corresponding linear increase in fiber surface area; this is about 4% at 500 ksi, if both Young's modulus and Poisson's ratio are taken into account /6/. Furthermore, in the water sorption studies on E-glass fiber, it was found that the last vestige of surfacesorbed water species could not be removed, except by heat treatment at about 300°C. in a vacuum /5/. Little lateral movement of these surfacesorbed water species would then be expected, as the fiber is tensile loaded, yielding an extended surface.

The net result is the formation of "stress space" among the surfacesorbed water species, producing the concentration gradient, ΔC_{H_2O}. The equilibrium between the water species of the ambient and those of the water hull is now shifted; all contribute to the concentration gradient, ΔC_{H_2O}, as indicated by the single arrow notation of Figure 6.

Diffusion of water species to the extended fiber surface is the result. The diffusion rate will be directly proportional to the concentration gradient, ΔC_{H_2O}, the chemical potential. The gradient is dependent upon the water hull thickness, which is directly proportional to the ambient humidity.

The diffusion rate determines the rate of increase of surface-sorbed water species. As the concentration of the latter increases, so does the probability of the reaction, microstress-surfacesorbed water species, proposed in Figure 4. The reaction time will be inversely proportional to this probability.

The fiber stress, in addition to producing conditions conducive to water species diffusion to the fiber surface, increases the stress intensity within the microstress, (see tensor notation of Figure 4). This further increases the reaction probability with resulting decrease in reaction time. In addition, increase in the disperse phase concentration with the corresponding increase in microstress will further increase this probability of reaction, with concomitant decrease in reaction time. Under constant tensile test procedures, the reduced reaction time yields a lower measured tensile strength.

Thus, the components of the stress corrosion reaction are indicated. The glass melting history controlled disperse phase produce the microstress, and the level of the ambient humidity determines the thickness of the water hull. At tensile test, the stress intensity of the microstress increases, and diffusion of water species to the fiber surface is initiated, the rate dependent upon the water hull thickness. The resultant stress corrosion reaction, microstress-surfacesorbed water species, produces the microflaw, the precursor to fiber fracture.

Various aspects of these developed concepts would appear to be applicable in areas of fracture mechanics. These might include, static fatigue testing, the effects produced in tensile testing by variations in the rate of load elongation, gage length and strain rate, and the effects of the humidity level on all of these. Examination of these areas, however, is beyond the scope of this paper.

There are analogies between these developed concepts and other chemical processes. Furthermore, an estimate may be made using available information as to which mechanism is rate controlling in the overall stress corrosion reaction, diffusion of water species through the water hull, or the reaction, microstress-surfacesorbed water species.

DEVELOPED CONCEPT ANALOGIES AND ANALYSIS

The water species diffusion described in the previous section is not a process where one water specie slips past another. It is, most probably, similar to the process in water solvent osmosis, where water species diffuse through highly hydrated pores of the semipermeable membrane.

This diffusion is complex, and is thought to involve a proton-hydroxyl bond rupture, transfer and re-coupling process within the hydrated pores of the membrane, in the direction of the chemical potential. The chemical potential, the driving force for osmosis, is produced when a one molar solution of solute in water, for example, is placed on one side of the membrane, and water on the other side. The resulting osmotic pressure is the measure of this chemical potential.

An increase in the molar concentration of the solute produces a corresponding increase in the osmotic pressure, Π. Only the molar concentration of the solute and its solubility in the solvent are of significance in osmosis; the identity of the solute is immaterial. It is proposed that the conditions for water solvent osmosis may be re-stated in terms of a water concentration gradient.

A one molar solute-water solution may be considered as <u>deficient</u> in one mole of water, compared to the all water solution on the other side of the osmotic membrane. The chemical potential is now determined by the water concentration gradient, ΔC_{H_2O}, water-water minus one mole, as measured by the osmotic pressure, Π. This new technique has been successfully used in the solution of several problems in osmosis; it has not been subjected to rigorous thermodynamic analysis, however.

From these arguments, an analogy between the water species diffusion in water solvent osmosis, and that through the water hull in fiber tensile testing may be made. In water solvent osmosis, the driving force for diffusion is the chemical potential produced by the concentration gradient, ΔC_{H_2O}, as measured by the osmotic pressure, Π. In water species diffusion through the water hull, the driving force is the chemical potential produced by the concentration gradient, ΔC_{H_2O}, as measured by the water hull thickness.

An activation energy difference should exist between these two processes, however, equivalent to the heat of vaporization of water. In water solvent osmosis, it is proposed that this energy barrier must be overcome if liquid water is to achieve the appropriate activity, and thus diffuse via proton-hydroxyl transfer through the membrane. This proposed activation energy and water species activity difference suggest that the effects of liquid water on E-glass fiber tensile strength might be different than that of 100% RH water vapor.

In an earlier investigation, the strength of 10μm diameter E-glass was determined first under ambient conditions and then immersed in liquid water at ambient temperature /7/. The fiber test conditions were equivalent to those of this investigation. The ambient strength was 500 ksi and water immersed strength 444 ksi, yielding a strength retention of about 89%.

From Figure 1, the ambient strength at 45% RH is 536 ksi and that extrapolated to 100% RH is 376 ksi, or a 70% strength retention. The higher reactivity of water vapor (species) with the E-glass fiber surface, and thus its more detrimental effects on strength, compared to liquid water, are apparent.

Up to this point, no determination has been made as to which of the water species related mechanisms, (1) diffusion through the water hull, or (2) the reaction, microstress-surfacesorbed water species, is rate controlling in the overall stress corrosion reaction. Interpretation of information available in the literature allows this estimate to be made.

The strength of 10μm diameter E-glass fiber has been determined in intervals, from room temperature down to that of liquid nitrogen (-196°C); the test strain rate was 0.39 inches/inch/minute /8/. These data have been recalculated to yield an Arrhenius type plot as shown in Figure 7. The ordinate of this figure is the logarithm of the normalized strength, $\ln S_0/_s$, where S_0 is the strength at -196°C. and S the strength at more elevated temperature. This ratio was used in order that the resulting linear plots would have the proper sign. The abscissa is the reciprocal absolute temperature of each test.

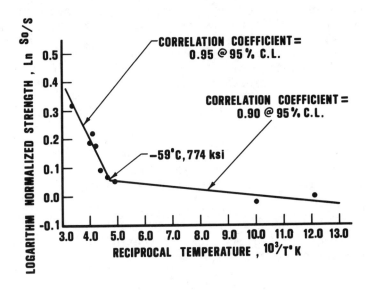

Fig. 7 Ln S_0/s vs. $10^3/T^\circ K$, E-glass fiber, ambient temperature to -196°C.

Two intersecting straight lines are the result, each of good correlation coefficient. The intersection point occurs at -59°C, where the estimated strength is 774 ksi. This linearity with inter-section is characteristic of certain types of Arrhenius plots, the inflection point indicative of a change in controlling reaction mechanism. The activation energies for the reactions are determined by the slope of each linear portion.

These slopes have been determined, yielding -0.5 kcal/mole for the steep slope (ambient to -59°C.) and -0.2 kcal/mole for the shallow slope (-59°C. to -196°C.). Both of these activation energies are low. These low values may be explained in the following manner.

In most Arrhenius plots, the logarithm of the equilibrium constants of a reaction are plotted against the reciprocal absolute temperature. In the present case, however, the logarithm of strength, a secondary response to a primary reaction was used. Furthermore, a ratio of strengths was used in the logarithmic term. From this, it is evident that the absolute magnitude of these calculated activation energies cannot be correct. However, they may be used, in a realtive sense, to indicate rate controlling mechanisms, and the transition from one mechanism to the other. The following interpretation of these plotted data is offered.

From ambient down to -59°C., both the water species diffusion through the water hull, and the reaction, microstress-surfacesorbed water species, are operative in the stress corrosion reaction. The diffusion mechanism is rate controlling, however. From -59°C. down to -196°C., the activity of water species is so diminished that diffusion is highly improbable. Now, only the reaction, microstress-existing surfacesorbed water species, is operative in the stress corrosion reaction. The inflection at about -50°C. has been observed by others /8, 9/.

Static fatigue has recently been observed in glass at -196°C. /10/. The stress corrosion reaction, though limited to one mechanism of diminished rate, is still functional at this low temperature.

CONCLUSIONS

The strength of E-glass fiber is controlled by a stress corrosion reaction consisting of two mechanisms, water species diffusion through the sorbed water hull to the fiber surface, and the reaction, microstress-surfacesorbed water species. This reaction produces the microflaw which is the precursor to fiber fracture.

The diffusion mechanism is controlled by the ambient humidity and the microstress concentration by the melting history of the glass. At ambient temperature, the diffusion mechanism is rate controlling, but at more reduced temperatures, only the reaction, microstress-surfacesorbed water species is operative.

It is proposed that the diffusion mechanism is analogous to that in water solvent osmosis, differing by an activation energy equivalent to water heat of vaporization. This energy barrier concept is substantiated by the lower reactivity of liquid water compared to water species in E-glass fiber strength measurements.

REFERENCES

1. McKinnis, C. L. "Disperse Phase in and Stress Corrosion of E-glass Fiber and Their Combined Effects on Tensile Strength". XI International Congress on Glass, Prague, Czechoslovakia, July 4-8, 1977. To be published.

2. Cohen, C. I. "Observation of Phase Growth in Fibers". Rolla Fiber Conference, University of Missouri, Rolla. March 16-18, 1971.

3. Crume, S. V. "Finite Element Modelling of Stress Fields Around Disperse Phase in Glass Fibers". Glass Division, American Ceramic Society, Bedford Springs, Penna. October 9-11, 1974.

4. Charles, R. J. "Static Fatigue of Glass, I and II,"J. Appl. Phys. 29 11 1549-60 (1959).

5. Huang, R. F. Demirel, T and McGee, T. D., "Adsorption of Water Vapor on E-Glass". J. Amer. Ceram. Soc. 55 (8) 399-405 (1972).

6. Crume, S. V., Owens-Corning Fiberglas, Technical Center, Granville, Ohio, Private Communication.

7. Metcalf, A. G. and Schmitz, G. K. "Mechanism of Stress Corrosion in E-glass Filaments", Glass Tech. 13 (1) 5-16 (1972).

8. Cameron, N. M. "The Effect of Environment and Temperature on the Strength of E-Glass Fiber, Part I, High Vacuum and Low Temperature". Glass Tech 9 (1) 14-21 (1968).

9. Charles, R. J. "Fracture" Ed. Averback, Felbeck, Hahn and Thomas, John Wiley (1959).

10. Ceriani, J. M. "Dynamic Fatigue Studies on Tempered and Untempered Soda-Lime-Silica Glasses", M. S. Thesis, Pennsylvania State University (1975).

A MULTIBARRIER RATE PROCESS APPROACH TO SUBCRITICAL CRACK GROWTH

S. D. Brown

University of Illinois at Urbana-Champaign

Department of Ceramic Engineering, Urbana, Il   61801

NOMENCLATURE

$\Delta A_j^{\bigstar}$ change in the surface area of the system that occurs when the crack advances through N bonds in the solid.

a proportionality constant in Eq. (18) ($\equiv \#\lambda_k^2$).

$a_{g,j}$ dimensionless parameter in Eq. (1) pertaining to the $g\underline{th}$ subset of the $j\underline{th}$ set (i.e., sequence), defined explicitly in Ref. 28.

b_z coefficient of the applied stress in Zhurkov's equation.

$c_{1,j}$ lumped parameter in Eq. (1), defined explicitly in Ref. 28.

$\Delta G_{g,j}^{\ddagger}$ Gibbs free energy of activation for the individual forward steps in the $g\underline{th}$ subset of the $j\underline{th}$ set (i.e., sequence).

$G_{i,j}$ Gibbs free energy of the $i\underline{th}$ stable position in the $j\underline{th}$ sequence. See Fig. 9.

$(-\Delta G_j)$ Gibbs free energy driving force for the entire $j\underline{th}$ sequence. See Fig. 9.

g an integer ($1 \leq g \leq m_j$), used as a subscript, that denotes a subset of the $j\underline{th}$ set (i.e., sequence).

$\Delta H_{g,j}^{\ddagger}$ activation enthalpy for the individual forward steps in the $g\underline{th}$ subset of the $j\underline{th}$ set (i.e., sequence).

h Planck constant

i an integer ($1 \leq i \leq n_j$), used as a subscript, that denotes a step and/or a stable position in the $j\underline{th}$ sequence. Fig. 9.

j an integer ($1 \leq j \leq \ell$), used as a subscript, that denotes an entire set (i.e., sequence) of consecutive reaction steps. See Fig. 9.

K stress intensity factor

K_j^* threshold stress intensity factor, defined by Eq. (8).

K_I stress intensity factor for the crack opening mode

k Boltzmann constant.

k_s system conductivity for a resistive network analog.

k_I conductivity of a resistive element, corresponding to Region I of crack growth, in a network analog.

k_{II} conductivity of a resistive element, corresponding to Region II of crack growth, in a network analog.

k_{III} conductivity of a resistive element, corresponding to Region III of crack growth, in a network analog.

k_{IV} conductivity of a resistive element, corresponding to Region IV of crack growth, in a network analog.

L parameter in Eq. (5), defined by Eq. (7).

ℓ total number of sequences j acting to propagate a crack.

m_j total number of subsets g in the $j\underline{th}$ set.

N Avogadro's number.

n_j total number of single, forward steps in the $j\underline{th}$ sequence.

p absolute pressure in the solid at the crack tip.

q_0 experimental activation energy associated with Eq. (3) and the environment-independent term in Eq. (4).

q_1 experimental activation energy associated with the environment-dependent term of Eq. (4).

q_x experimental activation energy associated with the environment-dependent term of Eq. (4).

q_z activation energy in Zhurkov's equation.

q an integer $(1 \le q \le \zeta)$, used as a subscript, that denotes a chemical species involved in the bond-rupturing process.

R gas constant.

$\mathcal{R}_{g,j}$ $\#\lambda_k^2 r_{g,j}/\theta_{g,j}$. See Eq. (9).

$r_{g,j}$ unidirectional forward rate corresponding to a single step in the gth subset of the jth set (i.e., sequence).

$r_{(i,i+1),j}$ unidirectional forward rate corresponding to the ith step in the jth set (i.e., sequence). See Fig. 9.

$r_{(i+1,i),j}$ unidirectional backward rate corresponding to the ith step in the jth set (i.e., sequence). See Fig. 9.

$\mathcal{r}_{i,j}$ net forward rate corresponding to the ith step in the jth set (i.e., sequence), defined by Eq. (22).

$\mathcal{r}_j$ net forward rate corresponding to the jth set (i.e., sequence).

ΔS_j change in the entropy of the system that occurs when the crack advances through N bonds in the solid.

$\Delta S_{g,j}^{\ddagger}$ activation entropy for the individual forward steps in the gth subset of the jth set (i.e., sequence).

$(\Delta S_{g,j}^{\ddagger})_0$ $\Delta S_{g,j}^{\ddagger}$ at zero applied stress.

T absolute temperature.

t_f time to failure under a static load.

t_0 pre-exponential parameter in Zhurkov's equation.

$U_i^{\ddagger}$ activation energy in the Cekirge-Tyson-Krausz equation.

ΔU_j change in the internal energy of the system that occurs when the crack advances through N bonds in the solid.

$(\Delta U_{EXP})_{g,j}$ experimental activation energy connected with the $g\underline{th}$ subset of the $j\underline{th}$ set (i.e., sequence).

ΔV_j change in the volume of the system that occurs when the crack advances through N bonds in the solid.

V_M molar volume of the solid at the crack tip.

$(^{\delta}\tilde{V}_{q,k})$ partial molar volume of species q in the solid at the crack tip.

$\Delta V^{\ddagger}_{g,j}$ activation volume associated with the individual forward steps in the $g\underline{th}$ subset of the $j\underline{th}$ set (i.e., sequence).

v crack velocity.

$v_{MAX.}$ maximum dynamic crack velocity.

$v_{0,i}$ pre-exponential parameter in the Cekirge-Tyson-Krausz equation.

ΔW_i mechanical energy parameter associated with the activation process in the Cekirge-Tyson-Krausz.

α_0 pre-exponential parameter associated with Eq. (3) and the environment-independent term in Eq. (4).

α_1 pre-exponential parameter associated with the environment-dependent term in Eq. (4), and defined apropos of Eq. (12).

α_1' pre-exponential parameter defined apropos of Eq. (16).

α_x pre-exponential parameter associated with the environment-dependent term in Eq. (4).

β_0 coefficient of the stress intensity factor in Eq. (3) and the environment-independent term in Eq. (4).

β_1 coefficient of the stress intensity factor in the environment-dependent term of Eq. (4), and defined apropos of Eq. (11).

β_x coefficient of the stress intensity factor in the environment-dependent term of Eq. (4).

Γ surface free energy.

$\gamma_{g,j}$ parameter defined by Eq. (14).

ζ total number of chemical species in the system.

$\theta_{g,j}$ number of single, sequential reaction steps in the $g\underline{th}$ subset of the $j\underline{th}$ set (i.e., sequence).

Λ_j parameter defined by Eq. (7).

λ_k average step size of a double kink.

$(\Delta\mu_{q,0})_j$ chemical potential difference between the bulk medium and the crack tip for the species q at K = 0.

ρ crack tip radius

σ applied stress.

Φ_j thermodynamic driving force function, defined by Eq. (2).

Φ_s system driving potential for a resistive network analog.

ϕ_q number of moles of species q involved in the rupture of N bonds in the solid at the crack tip.

Ω_0 $\alpha_0'T\, \exp.(-q_0/RT)$. Defined in more detail in Ref. 28.

Ω_1 $[\gamma_{2,2} + (\beta_0/T)]/R$. Defined in more detail in Ref. 28.

Ω_2 $\alpha_1'T\, \exp.(-q_1/RT)$. Defined in more detail in Ref. 28.

Ω_3 $[\gamma_{2,1} + (\beta_1/T]/R$. Defined in more detail in Ref. 28.

Ω_4 $\alpha_x\, \exp.(-q_x/RT)$. Defined in more detail in Ref. 28.

\#   number of kinks per unit length of crack band.

BACKGROUND

A substantial literature deals with theories of subcritical fracture (e.g., Refs. 1-24). The Tobolsky-Eyring mechanism (Ref. 1) was developed apropos of the fracture of polymeric materials. It makes no provision for environmental effects. Nevertheless, it introduced concepts and mathematic forms remarkably like those associated with later theories of delayed failure and slow crack growth in glasses and ceramics.

The incisive Charles-Hillig theory (Refs. 7,8) has had an enormous impact upon our understanding of environment-dependent fracturing. Its central contention that a stress corrosion reaction at the roots of cracks underlies phenomena such as delayed failure, slow crack growth, and the loading-rate dependence of strength has been substantiated repeatedly. Yet, there is an error in the basic equation for the crack velocity (Ref. 9).

The expression $t_f = t_0 \, exp.[(q_z - b_z\sigma)/RT]$ proposed by Zhurkov (Ref. 10) on the basis of time-to-failure experiments on fifty different materials is compatible with that by Tobolsky and Eyring. Three inferences pertinent to this note were drawn from Zhurkov's work; viz., (1) that q_z is the binding energy on the atomic scale, (2) that b_z is dependent upon the stress magnification and the disorientation of the molecular structure at the crack tip, and (3) that t_0 is the period of the natural atomic oscillations in the solid.

Hsiao (Ref. 11) wrote an expression similar to that of Tobolsky and Eyring. In Hsiao's equation, however, the bond-breaking and bond-mending rates are proportional to the areal concentrations of bonds and broken bonds, respectively, at the crack line. This improvement, incorporated into the Tobolsky-Eyring formulation, led to the Tobolsky-Eyring-Hsiao mechanism (Ref. 12, p. 360).

Wiederhorn (Refs. 13,14) performed crack growth experiments in a soda-lime silicate glass exposed to moist air, and identified three distinct regions in the subcritical stress range of the crack velocity; viz., I, II, and III. This triplex pattern occurs in other systems (e.g., glasses, ceramics, plastics, and metals), although there are exceptions. Region I lies at comparatively low stresses, above an initial threshold. The crack velocity therein is strongly stress- and temperature-dependent. There is an important exception to this Region I behavior wherein a strong stress dependence is exhibited in the absence of any sensible temperature dependence (Ref. 25). At intermediate stresses (Region II), crack velocity is a marked function of the reactive species concentration in the environment (relative humidity in Wiederhorn's experiments), but relatively independent of the stress. Wiederhorn perceived that stress corrosion in which water is a key reactant underlies Region I behavior in glass, and that the transport of water from the environment to the crack tip across an intervening stagnant gas film dominates in Region II. He also recognized that the two processes bear a sequential relationship. Adopting the Charles-Hillig theory, Wiederhorn derived a relationship that conformed to his Region I-Region II crack velocity data (Ref. 14). Region III lies at higher subcritical stresses. Crack propagation in this region is by a highly stress-dependent process that is virtually insensitive to the chemical make-up of the environment (Ref. 15).

Crack growth by the thermally-activated, stress-enhanced movement of vacancies into the crack tip was treated by Hasselman (Ref. 16), Stevens and Dutton (Ref. 17), and Dutton (Ref. 18). Hasselman

considered the near-surface vacancy transport to be speeded by water
adsorption, thus providing an environmental effect. Like Charles
and Hillig, he chose to ignore the reverse rate. Dutton and Stevens
accounted for it through the chemical potential change associated
with the process.

The concept of "lattice trapping" of cracks was developed by
Thomson, et al. (Refs. 19-21). In this theory, a crack is seen as
being mechanically trapped within a stable stress region between
two unstable regions; viz., one in which spontaneous crack healing
occurs, and one in which catastrophic fracture happens. Trapping
stems from the discreteness of the lattice. Propagation of a crack
from one stable position to the next occurs by a thermally-activated
process that entails double kink generation and movement along the
crack line (see also Ref. 13). Lawn (Ref. 22) proposed a similar
model, and derived an equation for the Region I--Region II crack
velocity. In common with Wiederhorn's expression, Lawn's combined
reaction control in Region I with reactant transport control in
Region II.

Krausz, et al. (Refs. 23,24) treated crack growth in terms of
activation over a system of consecutive energy barriers. This
approach followed logically the Tobolsky-Eyring theory (Ref. 1),
and Krausz' earlier work on deformation kinetics (e.g., Ref. 12).
Krausz recognized an analogy between crack growth, on the one hand,
and heat transfer or diffusion processes, on the other (Ref. 23).
The crack velocity was written

$$v = 1/\sum_i \{v_{0,i} \exp.[(-U_i^{\ddagger} + \Delta W_i)/RT]\}^{-1},$$

implying a negligible reverse reaction.

None of the aforementioned theories (Refs. 1-8,10-24) provided
an expression capable of producing the triplex curve needed to fit
the entire range of subcritical crack rate data. Neither did any
deal with the fact that certain data (e.g., Ref. 25) manifest a
strong stress dependence of the crack velocity, but no sensible
temperature dependence in Region I. This note is addressed to
these problems. We also discuss the threshold effect, and a dc
network analog to crack growth.

THE MULTIBARRIER KINETICS THEORY OF SUBCRITICAL CRACK GROWTH

The general equation for the isothermal subcritical crack
velocity,

$$v = \#\lambda_k^2 (kT/h) \sum_{j=1}^{\ell} \{\Phi_j[(c_{1,j}k/h) + \sum_{g=2}^{m_j} (\theta_{g,j}a_{g,j}e^{\Delta G_{g,j}^{\ddagger}/RT})]^{-1}\}, \quad (1)$$

$$\Phi_j \equiv 1 - \exp.(\Delta G_j/RT), \tag{2}$$

was derived (Refs. 26-28) from a model that includes the generation and movement of double kinks within a crack band (or reaction zone) vicinal to the crack line. Steady-state multibarrier kinetics (Refs. 29,30) was adapted for the purpose. An outline of the derivation is given in the APPENDIX. Attention also is called to an independent development by Krausz (Refs. 31,32) which, though somewhat similar, bears certain fundamental differences.

In principle, Eq. (1) can be abridged so as to apply to specific materials, stress states, and environmental conditions. Four of the variations that emerge for $\ell = 2$ are of interest; viz., Eq. (3), and Eq. (4) with $(q_x = q_1, \beta_x = \beta_1)$, $(q_x \neq q_1, \beta_x = \beta_1)$, and $(q_x \neq q_1, \beta_x \neq \beta_1)*$. Eq. (3) pertains to

$$v = \alpha_0 T e^{-(q_0 - \beta_0 K)/RT} \tag{3}$$

$$v = \alpha_0 T e^{-(q_0 - \beta_0 K)/RT} + \frac{\alpha_1 T e^{-(q_1 - \beta_1 K)/RT} [1 - e^{-\Lambda(K-K^*)/RT}]}{1 + \alpha_x e^{-(q_x - \beta_x K)/RT}} \tag{4}$$

cracking which occurs in the presence of an inert gas or liquid, or vacuum. By "inert" is meant that the chemical character of the environment has no sensible effect upon the crack velocity. The variation of Eq. (4) for which $(q_x = q_1, \beta_x = \beta_1)$ corresponds to those instances where the environment is a dilute reactive gas; e.g., cracking of glass in moist air. The mathematic form obtained for $(q_x \neq q_1, \beta_x = \beta_1)$ has wide applicability. It conforms to data taken in either dilute or concentrated reactive liquid environments, or in concentrated reactive gases. Figs. 1-3 are respective examples. Moreover, some instances that apparently involve surface film effects fall within the pale of this variation. The underlying mechanisms, and consequently the definition of the α's, β's, and q's in terms of

---

*In specializing Eq. (1) to obtain Eqs. (3) and (4), $\#$ was treated as a constant. Actually, we expect it to be some function of K, T, and the character of the system. Derivation of the function is outside the scope of this note.

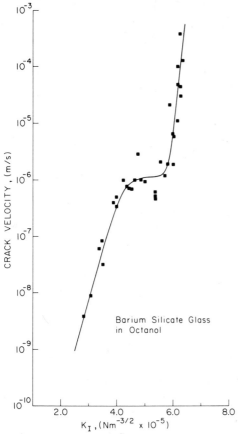

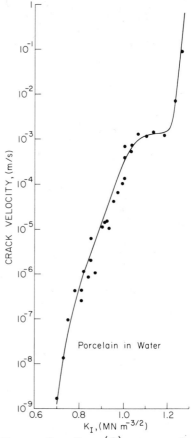

Figure 1. Eq. (5) compared with data of Freiman (Ref. 50) on barium silicate glass in octanol. Points represent the data; the line, the equation. Taken from Ref. 28.

Figure 2. Eq. (5) compared with data of Evans and Linzer (Ref. 51) on porcelain in water. Points represent data; the line, the equation. Taken from Ref. 28

fundamental quantities, vary according to the specifics of each case. The variation for which ($q_x \neq q_1$, $\beta_x \neq \beta_1$) arises whenever two distinguishable, sequential processes that are stress-enhanced and thermally-activated contribute to crack growth. While this represents an interesting set of cases, its treatment is outside the scope of the present note.

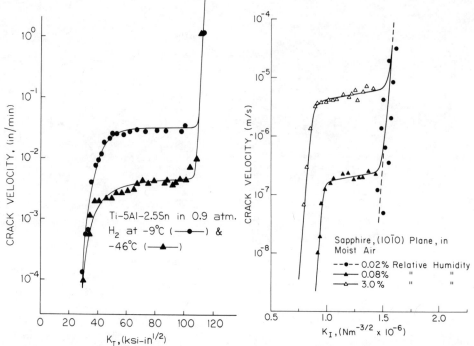

Figure 3. Eq. (5) compared with data of Williams and Nelson (Ref. 25) on Ti-5Aℓ-2.5Sn in hydrogen gas. Points represent data; lines, the equation. Taken from Ref. 28.

Figure 4. Eq. (5) compared with data of Wiederhorn (Ref. 52) on sapphire in moist air. Points represent data; lines, the equation. Taken from Ref. 28

DISCUSSION

Dependence of the Subcritical Crack Velocity upon K and T

The stress dependence of v obtained from Eq. (4) with $\beta_x = \beta_1$ takes the form of Eq. (5):

$$v = \Omega_0 e^{\Omega_1 K_I} + \frac{\Omega_2 e^{\Omega_3 K_I}[1 - e^{-L(K_I - K_I^*)}]}{1 + \Omega_4 e^{\Omega_3 K_I}} \tag{5}$$

This expression conforms well to Region I--Region II--Region III data for a wide variety of systems (Refs. 26-28). Figs. 1-5 are examples.

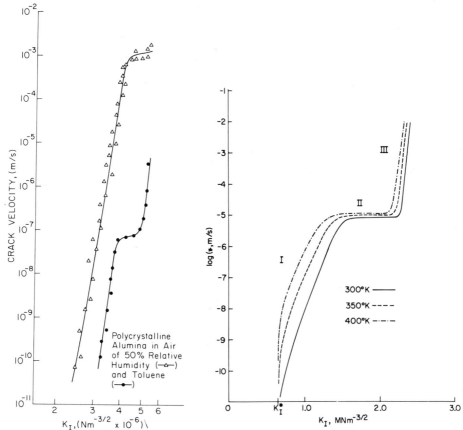

Figure 5. Eq. (5) compared
with data of Evans (Ref. 53)
on polycrystalline alumina
in moist air and toluene.
Points represent data; lines,
the equation. Taken from Ref.
28.

Figure 6. Hypothetical $v = v(K,T)$
plot made using Eq. (4) and parametral values as follows: $\alpha_0 =$
98.0 m s^{-1} °K^{-1}, $\alpha_1 = 5.80 \times 10^{-4}$
m s^{-1} °K^{-1}, $\alpha_x = 2.09 \times 10^4$, $\beta_0 =$
0.150 m$^{5/2}$ mol^{-1}, $\beta_1 = 0.0374$ m$^{5/2}$
mol^{-1}, $q_0 = 95$ Kcal mol^{-1}, $q_1 = 19$
Kcal mol^{-1}, $\Lambda = 0.0187$ m$^{5/2}$ mol^{-1},
and K* = 6.50×10^5 N m$^{-3/2}$.

Values assumed by the parameters K*, L, and Ω_i (i = 1,2,3,4),
Table 1, have been scrutinized vis-à-vis their respective theoretic
definitions, and appear reasonable (Ref. 28).

     The temperature dependency of v corresponding to the conditions
and assumptions that pertain to Eq. (5) can best be estimated using
Eq. (4). The approach will be to introduce a typical but hypothetical set of parametral values into Eq. (4), then examine the behavior
of the resulting expression as the temperature is varied. It would

Table I. Values of Parameters in Equation (5) for Various Systems (Ref. 28)

System	Fig.	Ω_0; m/s	Ω_1, $\frac{m^{3/2}}{N}$	Ω_2; m/s	Ω_3, $\frac{m^{3/2}}{N}$	Ω_4	L, $\frac{m^{3/2}}{N}$	K^*, $\frac{N}{m^{3/2}}$
Barium Silicate Glass in Octanol (Ref. 50)	1	1.41×10^{-33}	1.06×10^{-4}	2.96×10^{-14}	4.17×10^{-5}	2.51×10^{-8}	----	----
Porcelain in Water (Ref. 51)	2	1.99×10^{-64}	1.15×10^{-4}	1.19×10^{-18}	3.32×10^{-5}	7.41×10^{-16}	1.45×10^{-5}	6.90×10^{5}
Ti-5Al-2.5Sn in 0.9 Atm. Hydrogen at −46°C (Ref. 25)	3	3.83×10^{-74}	1.31×10^{-6}	1.36×10^{-9}	2.12×10^{-7}*	6.53×10^{-4}	3.64×10^{-8}	3.19×10^{7}
Ti-5Al-2.5Sn in 0.9 Atm. Hydrogen at −9°C (Ref. 25)	3	3.83×10^{-74}	1.31×10^{-6}	1.36×10^{-9}	2.12×10^{-7}*	1.01×10^{-4}	6.45×10^{-8}	3.19×10^{7}
Sapphire, (10$\bar{1}$0) Plane, in Air of 0.08% Relative Humidity (Ref. 52)	4	2.87×10^{-45}	5.85×10^{-5}	2.13×10^{-34}	6.55×10^{-5}	3.72×10^{-28}	5.27×10^{-7}	3.60×10^{5}
Sapphire, (10$\bar{1}$0) Plane, in Air of 3.0% Relative Humidity (Ref. 52)	4	2.87×10^{-45}	5.85×10^{-5}	6.55×10^{-27}	5.64×10^{-5}	4.85×10^{-22}	5.27×10^{-7}	2.98×10^{5}
Polycrystalline Alumina in Air of 50% Relative Humidity (Ref. 53)	5	----	----	1.66×10^{-21}	1.02×10^{-5}	1.40×10^{-18}	----	----
Polycrystalline Alumina in Toluene (Ref. 53)	5	1.35×10^{-30}	1.05×10^{-5}	1.40×10^{-26}	1.14×10^{-5}	1.93×10^{-19}	----	----

*In these cases, $\gamma_1 \gg \beta_1/T$ so that $\Omega_3 \approx \gamma_1/R$. See Eq. (16).

be preferable to test Eq. (4) against experimental $v = v(K,T)$ data taken on a single system across the entire subcritical range. However, available data pertain to either $v = v(K)$ throughout the said range, or to $v = v(K,T)$ for loosely related or unrelated systems within limited portions of the range. We limit consideration to the case in which the environment is a dilute, reactive gas (e.g., moist air). That is, we take $q_x = q_1$ and $\beta_x = \beta_1$. The specific values assigned the parameters are $\alpha_0 = 98.0$ m s^{-1} °K^{-1}, $\alpha_1 = 5.80$ x 10^{-4} m s^{-1} °K^{-1}, $\alpha_x = 2.09$ x 10^4, $\beta_0 = 0.150$ m$^{5/2}$ mol^{-1}, $\beta_1 = 0.0374$ m$^{5/2}$ mol^{-1}, $q_0 = 95$ Kcal mol^{-1}, $q_1 = 19$ Kcal mol^{-1}, $\Lambda = 0.0187$ m$^{5/2}$ mol^{-1}, and K* $= 6.50$ x 10^5 N m$^{-3/2}$. Results for 300°, 350°, and 400°K are plotted as Fig. 6. The curves exhibit the general features expected. In addition to the familiar trimodal behavior, these features include (1) a threshold effect (Refs. 33, 34), (2) convergence of the isotherms for $v = v(K)$ as K approaches K* from K > K* (Refs. 35,36), (3) appropriate temperature effects in Regions I and III (Refs. 15,35), and (4) virtual independence of v with regard to K in Region II. The small dependence of v on T in Region II stems from the fact that transport of the reactive species to the crack tip is via the gas phase. In cases where the environment is a liquid, or the dominant diffusion process occurs in the solid and is thermally-activated, $q_x \neq q_1$, and v of Eq. (4) will bear an exponential dependency upon T in Region II. This agrees with empirical findings.

The temperature range within which the foregoing treatment is valid is limited. Glass rods exposed to moist air, for example, exhibit a minimum strength at about 200°C (Ref. 37). For the said glass, Eq. (4) would be expected to hold at temperatures lower, but not higher than this. The scope of the treatment could be extended to the higher temperatures, however, by inserting a suitable term for the activated desorption of the reactive species into the denominator of the expression in Eq. (4) that pertains to environment-dependent processes. It would be necessary that the activation energy associated with the said term be larger than that connected with the stress corrosion reaction.

The Threshold Effect

Treatments of the threshold effect have been made by others (e.g., Refs. 7,8,17,18,31,38). Here, we consider the effect from the viewpoint of multibarrier kinetics.

The threshold stress is that for which $v = 0$. We assume that this corresponds to a zero thermodynamic potential; i.e., to $\Delta G_j = 0$, and consequently to $\Phi_j = 0$ in Eqs. (1) and (2). From Ref. 28,

$$\Delta G_j / RT = -L_j (K - K_j^*) = -\Lambda_j (K - K_j^*)/RT, \tag{6}$$

$$L_j \equiv \Lambda_j / RT \equiv 2\{\Delta V_j + \sum_{q=1}^{\zeta} [\phi_q(^{\delta}\tilde{V}_{q,k})]\}/3\sqrt{\pi\rho} \; RT, \tag{7}$$

$$K_j^* \equiv \frac{3\sqrt{\pi\rho}\,\{\Delta U_j - T\Delta S_j + (\frac{\Gamma V_M}{\rho}) + \Gamma\Delta A_j^{\bigstar} + \sum_{q=1}^{\zeta} [\phi_q(\Delta\mu_{q,0})_j]\}}{2\{\Delta V_j + \sum_{q=1}^{\zeta} [\phi_q(^{\delta}\tilde{V}_{q,k})]\}} \tag{8}$$

Eq. (8) indicates that K_j^* ordinarily will exhibit a substantial dependence upon the environment, but only a weak, perhaps inconsequential dependence upon T. This is in practical agreement with experimental findings (Refs. 36,39), and Krausz (Ref. 31) who claims from his theory that K_j^* is independent of T.

Environment-dependent processes swamp others for values of K vicinal to the threshold. As K is increased above K_j^*, the ϕ_j's in Eq. (1) rapidly approach unity. Therefore, in specializing Eq. (1) to Eqs. (3) and (4), the ϕ_j associated with the environment-dependent term was retained in an explicit form. That connected with the environment-independent term was set equal to unity.

Stress Dependence of the Pre-exponential Term in $v = v(K,T)$

Region I $v = v(K,T)$ data sometimes exhibit an exponential stress dependency that does not originate in the same term of the rate equation as the major temperature dependency (e.g., Refs. 25,36,40). The hydrogen-induced, Region I cracking of Ti-5Al-2.5Sn, for example, is strongly stress-dependent, but sensibly unaffected by temperature (Ref. 25). That is, the dominant K dependence in these cases is associated not with the activation enthalpy, but with the pre-exponential term; viz., α_1 in equation (4).

That the exponential stress dependence of α_1 stems from the activation entropy is perceived through Eqs. (9) - (12).

$$\mathcal{R}_{2,1} \equiv \#\lambda_k^2 r_{2,1}/\theta_{2,1} = \#\lambda_k^2 (kT/h)(\theta_{2,1}a_{2,1})^{-1}\exp.(\Delta S_{2,1}^{\ddagger}/R)$$
$$\exp.(-\Delta H_{2,1}^{\ddagger}/RT) \tag{9}$$

From the absolute rate theory (Ref. 41), we write

$$\Delta H_{2,1}^{\ddagger} = (\Delta U_{EXP})_{2,1} - RT + p\Delta V_{2,1}^{\ddagger} \tag{10}$$

We identify $q_1 \equiv (\Delta U_{EXP})_{2,1}$, and assume plane stress conditions at the crack tip, an elliptical through crack, and the corresponding Inglis estimate of the crack tip stress magnification (Ref. 42). Then, using the first stress invariant, and the definition of the stress intensity factor for the crack opening mode (Refs. 43,44), we write $p = -2K_I/3\sqrt{\pi\rho}$, and identify $\beta_1 \equiv 2\Delta V^{\ddagger}_{2,1}/3\sqrt{\pi\rho}$. Eq. (10) becomes Eq. (11),

$$\Delta H^{\ddagger}_{2,1} = q_1 - RT - \beta_1 K_I , \tag{11}$$

and Eq. (9), Eq. (12),

$$\mathcal{R}_{2,1} = \alpha_1 T \exp.[-(q_1 - \beta_1 K_I)/RT] , \tag{12}$$

with $\alpha_1 \equiv (\#\lambda_k^2 ke/h\theta_{2,1}a_{2,1})\exp.(\Delta S^{\ddagger}_{2,1}/R)$. This result indicates that $\Delta S^{\ddagger}_{2,1}$ is the most plausible source of the exponential stress dependence of α_1.

We expand $\Delta S^{\ddagger}_{2,1}$ as a Maclaurin series in K_I, and neglect all terms beyond the second as a first approximation to obtain

$$\Delta S^{\ddagger}_{2,1} = (\Delta S^{\ddagger}_{2,1})_0 + K_I (\partial \Delta S^{\ddagger}_{2,1}/\partial K_I)_T \Big|_{K_I=0} . \tag{13}$$

It is easily shown that

$$(\partial \Delta S^{\ddagger}_{2,1}/\partial K_I)_T = \left[2(\partial \Delta V^{\ddagger}_{2,1}/\partial T)_{K_I} \right] / \left[3\sqrt{\pi\rho} \right] \equiv \gamma_{2,1} \equiv \gamma_1 \tag{14}$$

We combine Eqs. (13) and (14) to obtain

$$\Delta S^{\ddagger}_{2,1} = (\Delta S^{\ddagger}_{2,1})_0 + \gamma_1 K_I . \tag{15}$$

The value of Ω_3 that corresponds to the Region I data in Fig. 3 is 2.12×10^{-7} m$^{3/2}$N^{-1}. This requires that $\gamma_1 = 1.76 \times 10^{-6}m^{5/2}$mol^{-1} °K^{-1}. For $\rho = 5 \times 10^{-10}$m, it is necessary that $(\partial \Delta V^{\ddagger}_{2,1}/\partial T)_K \approx 10^{-10}$m^{3}mol^{-1}°K^{-1} which appears reasonable.

Eq. (15) is introduced into Eq. (12) and Eq. (16) is the result:

$$\mathcal{R}_{2,1} = \alpha_1' T \exp.(\gamma_1 K_I/R) \exp.[-(q_1 - \beta_1 K_I)/RT] , \tag{16}$$

with $\alpha_1' \equiv (\#\lambda_k^2 ke/h\theta_{2,1}a_{2,1})\exp.[(\Delta S^{\ddagger}_{2,1})_0/R]$. It is important to realize that even when $\Delta V^{\ddagger}_{2,1}$ in β_1 is so small that the stress dependence associated with $\Delta H^{\ddagger}_{2,1}$ is negligible, $r_{2,1}\theta^{-1}_{2,1}$ can exhibit a substantial stress dependence that stems from $\Delta S^{\ddagger}_{2,1}$ if $(\partial \Delta V^{\ddagger}_{2,1}/\partial T)_{K_I}$ in γ_1 is sufficiently large.

A DC Network Analog

The forms of Eqs. (1) and (4) suggest that slow crack growth may be analogous to dc flow through a resistive network (cf., Refs. 23,28,31,45). For instance, the network

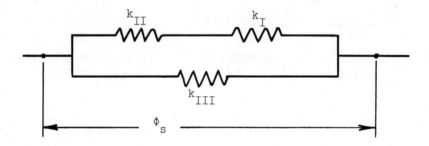

gives Eq. (17) and by analogy, Eq. (18) (Ref. 45):

$$k_s = k_{III} + k_I [1 + (k_I/k_{II})]^{-1} \tag{17}$$

$$v = a \, k_s \Phi_s \tag{18}$$

Eqs. (4) and (18) are virtually equivalent if the following identifications are made: $a \equiv \#\lambda_k^2$, $\Phi_s \equiv 1 - \exp.(\Delta G_s/RT)$, $ak_I \equiv \alpha_1 T \exp.[-(q_1 - \beta_1 K)/RT]$, $ak_{II} \equiv (\alpha_1/\alpha_x)T \exp.\{-[(q_1 - q_x) - (\beta_1 - \beta_x)K]/RT\}$, and $ak_{III} \equiv \alpha_0 T \exp.[-(q_0 - \beta_0 K)/RT]$.

The network analogy can be used as an heuristic device in analyzing other crack growth situations. For instance, the transition from slow to fast crack propagation in glassy materials has been reported to follow a pattern like that depicted in Fig. 7 (e.g., Refs. 46,47). The network

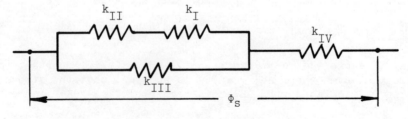

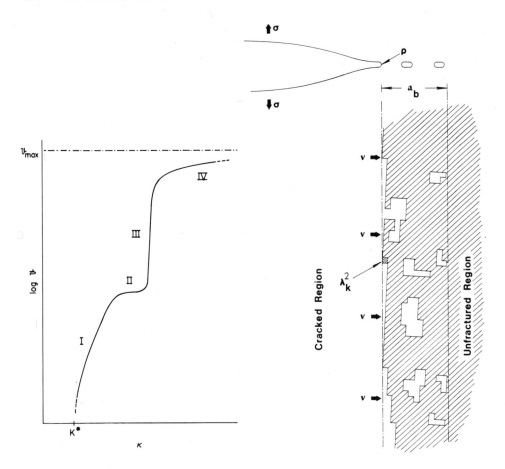

Figure 7. Schema of the transition from slow to fast crack growth in a glass, showing Regions I, II, III, and IV.

Figure 8. Schematic representation of a crack band. The crack moves to the right as a result of the movement of the double kinks. Fracture surface area created by the movement of one double kink through a distance λ_k is $2\lambda_k^2$. Kinks generated at flaws in the bulk of the material lie anterior to the crack line; those generated at the intersection of the crack band with a solid-environment interface move away from the said intersection along the crack band (lower left corner of the band). Taken from Ref. 28.

conforms to this sort of behavior. That is,

$$k_s = \frac{k_{IV}(k_I k_{II} + k_{II} k_{III} + k_I k_{III})}{k_I k_{II} + k_I k_{III} + k_I k_{IV} + k_{II} k_{III} + k_{II} k_{IV}}$$
(19)

and

$$v \sim k_s^{\Phi_s} \; .$$
(20)

The sequence of change in the relationships among the k's in Eq. (19) as K is increased from K* is about as follows:

Region I: $k_{IV} \gg k_{II} \gg k_I \gg k_{III}$

Region II: $k_{IV} \gg k_I \gg k_{II} \gg k_{III}$

Region III: $k_{IV} \gg k_1 \gg k_{III} \gg k_{II}$

              $k_{IV} \gg k_{III} \gg k_I \gg k_{II}$

Region IV: $k_{III} \gg k_{IV} \gg k_I \gg k_{II}$

The analogy suggests that crack propagation in Region IV is controlled by some transport process. We surmise that it is the transport of strain energy into the reaction zone of the dynamic crack.

APPENDIX

Derivation of Eq. (1)

We merely sketch the derivation of Eq. (1). Details are given elsewhere (Ref. 28). We consider a crack to grow by the generation and movement of double kinks within a crack band (or reaction zone) located at the crack front (cf., Refs. 13, 19-22). Moreover, crack growth is perceived as comprising ℓ competitive (i.e., parallel) reaction sequences, each made up of n_j consecutive (i.e., series) steps. Schemata of the crack band and reaction coordinate for the $j\underline{th}$ sequence are given in Figs. 8 and 9. The system is the reaction zone at and vicinal to the crack line. We write the crack velocity as Eq. (21):

$$v = \#\lambda_k^2 \sum_{j=1}^{\ell} r_j$$
(21)

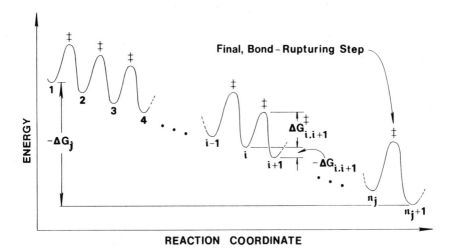

<u>Figure 9.</u> Reaction coordinate vs. energy curve for an hypothetical $j\underline{th}$ sequence showing the $i\underline{th}$ step. Taken from Ref. 28.

We focus on the $i\underline{th}$ step in the $j\underline{th}$ sequence, $i = 1,2,\ldots,n_j$. The <u>net</u> forward rate at the $i\underline{th}$ step, $r_{i,j}$, is given by Eq. (22):

$$r_{i,j} = r_{(i,i+1),j} - r_{(i+1,i),j} = r_{(i,i+1),j} \left[1 - e^{(G_{i+1,j} - G_{i,j})/RT} \right]$$

(22)

It follows from the steady-state assumption, and from a mass inventory of the $i\underline{th}$ position that

$$r_{1,j} = r_{2,j} = r_{3,j} = \ldots = r_{i,j} = r_{n_j,j}$$

(23)

We recognize that $r_j \equiv r_{i,j}$. Then, since i ranges from 1 to n_j, we write Eq. (24) from Eq. (22):

$$\Delta G_j \equiv \sum_{i=1}^{n_j} (G_{i+1,j} - G_{i,j}) = RT\ell n \left\{ \prod_{i=1}^{n_j} [1 - r_j r_{(i,i+1),j}^{-1}] \right\},$$

(24)

Usually, $r_j < r_{(i,i+1),j}$; hence, we obtain Eq. (25) as a first approximation:

$$r_j \simeq \frac{1 - e^{\Delta G_j/RT}}{\displaystyle\frac{1}{n_j}\sum_{i=1}^{n_j} r^{-1}_{(i,i+1),j}} \tag{25}$$

We consider the $j\underline{th}$ sequence to consist of a series of m_j single-barrier and/or multibarrier rate processes. That is, m_j subsets, each of which corresponds uniquely to a different rate process in the series, comprise the set of n_j consecutive steps that make up the $j\underline{th}$ sequence. The $r_{(i,i+1),j}$'s in a subset are assumed equal, and the denominator of Eq. (25) is written

$$\sum_{i=1}^{n_j} r^{-1}_{(i,i+1),j} = \sum_{g=1}^{m_j} [\theta_{g,j} r^{-1}_{g,j}] , \tag{26}$$

where

$$\sum_{g=1}^{m_j} \theta_{g,j} = n_j \geq m_j \tag{27}$$

It is expected (Ref. 41) that the $r_{g,j}$'s will take the form

$$r_{g,j} = a^{-1}_{g,j} k_{g,j} = a^{-1}_{g,j} (kT/h) e^{-\Delta G^{\ddagger}_{g,j}/RT} \tag{28}$$

for the thermally-activated processes in any series. In principle, the $a_{g,j}$'s can be calculated, and the $\Delta G^{\ddagger}_{g,j}$'s expanded as functions of flaw structure, surface energy, strain energy, activation volume, etc., as appropriate and necessary. Eq. (28) generally will apply to processes occurring on surfaces or in condensed phases. The kinetic theory of gases (e.g., Refs. 48,49) was used to write the rate expression apropos of pertinent gas phase transport processes. Ordinarily, only one of the m_j subsets will pertain to these processes. We arbitrarily specify this as the subset for which $g = 1$, and write

$$\theta_{1,j} r^{-1}_{1,j} = c_{1,j} T^{-1} \tag{29}$$

Finally, Eqs. (21), (25), (26), (28) and (29) are combined to yield Eq. (1).

ACKNOWLEDGMENT

    Support of the National Science Foundation through Grant No.
DMR 76-03893 is gratefully acknowledged. Special thanks are due
A. L. Friedberg and G. P. Wirtz for helpful advice.

REFERENCES

1. A. Tobolsky and H. Eyring, "Mechanical Properties of Polymeric
    Materials", J. Chem. Phys. 11 [3] 125 (1943).

2. P. Gibbs and I. B. Cutler, "Fracture of Glass Which Is Subjected
    to Slowly Increasing Stress", J. Amer. Ceram. Soc. 34 [7] 200
    (1951).

3. D. A. Stuart and O. L. Anderson, "Dependence of Ultimate
    Strength of Glass Under Constant Load on Temperature, Ambient
    Atmosphere, and Time", J. Amer. Ceram. Soc. 36 [12] 416 (1953).

4. G. M. Bartenev, Izv. Akad. Nauk. S.S.S.R. Otdel. Tekh. Nauk.
    1955, No. 9, 53 (as cited in Ref. 40).

5. G. M. Bartenev, I. V. Yudina, and P. A. Rehbinder, "Theory of
    Spontaneous Dispersion of Solids", Kolloid Zh. 20, 655 (1958)
    (as cited in Ref. 40).

6. G. M. Bartenev and I. V. Razumovskaya, Proc. First Int'l Conf.
    on Fracture, Sendai, Vol. 3, 1871 (1966); Japanese Society for
    Strength and Fracture of Materials (as cited in Ref. 40).

7. R. J. Charles and W. B. Hillig, "The Kinetics of Glass Failure
    by Stress Corrosion", p. 511 in Symposium on the Mechanical
    Strength of Glass and Ways of Improving It, Florence, Italy,
    25-29 September 1961. Union Scientifique Continentale du
    Verre, Charleroi, Belgium (1962).

8. W. B. Hillig and R. J. Charles, "Surfaces, Stress-Dependent
    Surface Reactions, and Strength", p. 682 in High-Strength
    Materials; edited by V. F. Zackay; John Wiley & Sons, New
    York (1965).

9. S. D. Brown, "The Charles-Hillig Subcritical Crack Velocity
    Reconsidered", submitted to the J. Amer. Ceram. Soc.

10. S. N. Zhurkov, "Kinetic Concept of the Strength of Solids",
    Int'l J. Fract. Mech. 1 [4] 311 (1965).

11. C. C. Hsiao, "Fracture", Phys. Today 19 [3] 49 (1966).

12. A. S. Krausz and H. Eyring, Deformation Kinetics, Wiley-Interscience, New York (1975).

13. S. M. Wiederhorn, "Fracture Surface Energy of Soda-Lime Glass", p. 503 in Materials Science Research, Vol. 3; edited by W. W. Kriegel and H. Palmour III, Plenum Press, New York (1966).

14. S. M. Wiederhorn, "Influence of Water Vapor on Crack Propagation in Soda-Lime Glass", J. Amer. Ceram. Soc. 50 [8] 407 (1967).

15. S. M. Wiederhorn, H. Johnson, A. M. Diness, and A. H. Heuer, "Fracture of Glass in Vacuum", J. Amer. Ceram. Soc. 57 [8] 336 (1974).

16. D. P. H. Hasselman, "Proposed Theory for the Static Fatigue Behavior of Brittle Ceramics", p. 297 in Ultrafine-Grain Ceramics; edited by J. J. Burke, N. L. Reed, and V. Weiss, Syracuse University Press, Syracuse, New York (1970).

17. R. N. Stevens and R. Dutton, "Propagation of Griffith Cracks by Mass Transport Processes", Mater. Sci. Eng. 8 [4] 220 (1971).

18. R. Dutton, "The Propagation of Cracks by Diffusion", p. 647 in Fracture Mechanics of Ceramics, Vol. 2; edited by R. C. Bradt, D. P. H. Hasselman, and F. F. Lange, Plenum Press, New York (1974).

19. R. Thomson, C. Hsieh, and V. Rana, "Lattice Trapping of Fracture Cracks", J. Appl. Phys. 42 [8] 3154 (1971).

20. C. Hsieh and R. Thomson, "Lattice Theory of Fracture and Crack Creep", J. Appl. Phys. 44 [5] 2051 (1973).

21. R. Thomson and E. Fuller, "Crack Morphology in Relatively Brittle Crystals", p. 283 in Fracture Mechanics of Ceramics, Vol. 1; edited by R. C. Bradt, D. P. H. Hasselman, and F. F. Lange, Plenum Press, New York (1974).

22. B. R. Lawn, "An Atomistic Model of Kinetic Crack Growth in Brittle Solids", J. Mater. Sci. 10 [3] 469 (1975).

23. A. S. Krausz, "The Theory of Non-steady State Fracture Propagation Rate", Int'l J. Fract. 12 [2] 239 (1976).

24. H. M. Cekirge, W. R. Tyson, and A. S. Krausz, "Static Corrosion and Static Fatigue of Glass", J. Amer. Ceram. Soc. 59 [5-6] 265 (1976).

25. D. P. Williams and H. G. Nelson, "Gaseous Hydrogen-Induced Cracking of Ti-5Aℓ-2.5Sn", Met. Trans. A 3 [8] 2107 (1972).

26. S. D. Brown, "A Multibarrier Kinetics Approach to Subcritical Crack Growth in Glasses and Ceramics", Amer. Ceram. Soc. Bull. 55 [4] 395 (1976); Paper No. 27-B-76, presented at the 78th Annual Meeting of the American Ceramic Society, 1-6 May 1976, Cincinnati, Ohio.

27. S. D. Brown, "The Multibarrier Kinetics of Subcritical Brittle Crack Growth", p. 257 in Proc. Sixth Canadian Cong. Appl. Mech., Vol. 1, Vancouver, B.C. (1977).

28. S. D. Brown, "Multibarrier Kinetics of Brittle Fracture. I. Stress Dependence of the Subcritical Crack Velocity", submitted to the J. Amer. Ceram. Soc.

29. F. H. Ree, T. S. Ree, T. Ree, and H. Eyring, "Random Walk and Related Physical Problems", p. 1 in Advances in Chemical Physics, Vol. 4; edited by I. Prigogine, Wiley, New York (1962).

30. H. Eyring, T. S. Ree, T. Ree, and F. M. Wanlass, "Departures from Equilibrium Kinetics", p. 1 in Symposium on the Transient State, Special Publication No. 16, The Chemical Society, London (1962).

31. A. S. Krausz, "The Deformation and Fracture Kinetics of Stress Corrosion Cracking", Int'l J. Fract., in press.

32. A. S. Krausz, "The Theory of the Thermally Activated Processes of Stress Corrosion Cracking", submitted to Eng'g Fract. Mech.

33. R. E. Mould and R. D. Southwick, "Strength and Static Fatigue of Abraded Glass Under Controlled Ambient Conditions: I., General Concepts and Apparatus", J. Amer. Ceram. Soc. 42 [11] 542 (1959); "Strength and Static Fatigue of Abraded Glass Under Controlled Ambient Conditions: II., Effect of Various Abrasions and the Universal Fatigue Curve", ibid., [12] 582.

34. R. Sedlacek and F. A. Halden, "Static and Cyclic Fatigue of Alumina", p. 211 in Structural Ceramics and Testing of Brittle Materials, edited by S. J. Acquaviva and S. A. Bortz, Gordon and Breach, New York (1968).

35. S. M. Wiederhorn and L. H. Bolz, "Stress Corrosion and Static Fatigue of Glass", J. Amer. Ceram. Soc. 53 [10] 543 (1970).

36. S. J. Hudak and R. P. Wei, "Hydrogen Enhanced Crack Growth in 18Ni Maraging Steels", Met. Trans. A. 7 [2] 235 (1976).

37. B. Vonnegut and J. L. Glathart, "The Effect of Temperature on the Strength and Fatigue of Glass Rods", J. Appl. Phys. 17 [12] 1082 (1946).

38. B. J. S. Wilkins and R. Dutton, "Static Fatigue Limit with Particular Reference to Glass", J. Amer. Ceram. Soc. 59 [3-4] 108 (1976).

39. R. A. Oriani and P. H. Josephic, "Equilibrium Aspects of Hydrogen-Induced Cracking of Steels", Acta Met. 22 [9] 1065 (1974).

40. W. W. Gerberich and M. Stout, "Discussion of Thermally Activated Approaches to Glass Fracture", J. Amer. Ceram. Soc. 58 [5-6] 222 (1976).

41. S. Glasstone, K. J. Laidler, and H. Eyring, The Theory of Rate Processes, McGraw-Hill, New York (1941).

42. C. E. Inglis, "Stresses in a Plate Due to Presence of Cracks and Sharp Corners", Trans. Inst. Naval Architecture, London 55 [Pt. 1] 219 (1913).

43. G. R. Irwin, "Analysis of Stresses and Strains Near the End of a Crack Traversing a Plate", J. Appl. Mech. 24 [3] 361 (1957).

44. P. C. Paris and G. C. Sih, "Stress Analysis of Cracks", p. 30 in Fracture Toughness Testing and Its Applications, ASTM Special Technical Publication No. 381 (1965).

45. S. D. Brown, "Subcritical Crack Growth: A Treatment Based upon Multibarrier Kinetics, and an Electrical Network Analog", in Environmental Effects on Engineering Materials, edited by M. R. Louthan, Jr. and R. P. McNitt, Virginia Polytechnic Institute, Blacksburg, Va., 10-12 Oct. 1977.

46. F. Kerkhof, "Wave Fractographic Investigations of Brittle Fracture Dynamics", p. 3 in Dynamic Crack Propagation; edited by G. C. Sih, Noordhof International Publishing, Leyden, The Netherlands (1973).

47. W. Döll and G. Weidmann, "Transition from Slow to Fast Crack Propagation in PMMA", J. Mater. Sci. 11 [12] 2348 (1976).

48.   T. K. Sherwood and R. L. Pigford, Absorption and Extraction,
      2nd Edition, McGraw-Hill, New York (1952)

49.   D. G. Samaras, Theory of Ion Flow Dynamics, Prentice-Hall,
      Englewood Cliffs, N.J. (1962), p. 318.

50.   S. W. Freiman, "Effect of Alcohols on Crack Propagation in
      Glass", J. Amer. Ceram. Soc. 57 [8] 350 (1974).

51.   A. G. Evans and M. Linzer, "Failure Prediction in Structural
      Ceramics Using Acoustic Emission", J. Amer. Ceram. Soc. 56
      [11] 575 (1973).

52.   S. M. Wiederhorn, "Moisture Assisted Crack Growth in Ceramics",
      Int'l J. Fract. Mech. 4 [2] 171 (1968).

53.   A. G. Evans, "A Method for Evaluating the Time-Dependent Failure
      Characteristics of Brittle Materials--and Its Application to
      Polycrystalline Alumina", J. Mater. Sci. 7 [10] 1137 (1972).

STRESS RUPTURE EVALUATIONS OF HIGH

TEMPERATURE STRUCTURAL MATERIALS

R. J. Charles

General Electric Company

Corporate Research and Development

ABSTRACT

Data is presented on the stress rupture characteristics of a number of metals and ceramics in terms of a model based on diffusion controlled, intergranular crack growth. Due to the limited experimental data available on the stress rupture behavior of ceramics and in order to substantiate the analytical procedures indicated by the model, a number of well documented metal and alloy systems are first analyzed. Estimates of the long term performance of alumina and various forms of silicon nitride and silicon carbide are made from available data and compared with the metal and alloy results.

INTRODUCTION

In the application and development of high temperature, stress rupture resistant materials, two problems concerning lifetime under load are often encountered. For a mature alloy, estimations of lifetime under loading conditions outside of those utilized in setting the alloy specifications are often desired for design purposes and for developmental alloys and other materials, such as ceramics, there are severe economic constraints to minimize the tedious intermediate stress rupture testing necessary for determining optimal changes in the material formulation and processing. Several procedures involving "time-temperature" parameters[1-6] for estimating alloy performance are of long standing, are in wide use and are applicable to both metals and non-metals. These procedures are highly useful, but unfortunately, rely totally on detailed experimentation to relate the time-temperature behavior of a material at one stress loading to the corresponding behavior at another stress loading. A relatively large number of theoretical models have been proposed[7-12] to account for

623

the dependence of the service time to failure on the stress state
for structural materials at high temperature. Much of the experimentation for application of these models tends, however, to concentrate
on long term, high temperature failure behavior generated under relatively low stresses. The present work is concerned with procedures
for the application of a quasi-theoretical model of failure of polycrystalline materials[13] in which the full range of applicable
stresses is addressed and representative illustrations are included
of its applicability to both mature and developmental materials.

BACKGROUND

As pointed out by Clauss[4], virtually all the parametric procedures currently used for extrapolation and correlation of stress
rupture data have a common basis. The well-known Larson-Miller, Orr-
Sherby-Dorn and Manson-Haferd procedures[1,2,3,] entail a fundamental
relation corresponding to the following

$$f(\sigma) = t_f \exp(-\Delta H/kT) \tag{1}$$

where $f(\sigma)$ is a function depending solely on applied stress, ΔH is
an activation energy value, kT has its usual significance and t_f is
the failure time. It is evident that a given stress fixes $f(\sigma)$ as
a constant and thus rupture time and temperature are simply related
for that stress. If the function $f(\sigma)$ were available then failure
times for all stresses and temperatures could be determined. Clearly, $f(\sigma)$ is dependent on the material but may also be affected by the
type of loading, the environment and other factors.

A great deal of experiment has shown that stress-rupture behavior of materials is closely related to their creep behavior and that
the temperature dependencies of these two processes are generally the
same. For the purer metals, the temperature dependence of high temperature creep has been shown to be the same as that for self-diffusion within the metal[14]. As recognized by Orr, Sherby and Dorn[2],
the above relationships allow the identification of the ΔH in Eq. 1
for stress rupture as ΔH_c for diffusional creep and, indeed, the product $t_f \exp(-\Delta H_c/RT)$ is known as the Orr-Sherby-Dorn stress rupture
parameter, Θ_r.

If it is assumed that the diffusional process responsible for
creep involves vacancy defects, models may be developed for rupture
which involve the condensation of vacancies at the highly stressed
regions of a crack in a solid such that the crack extends. If the
extension of the crack is of more importance than relaxations in the
crack tip region then continuously increasing stress concentrations
may arise which ultimately lead to unrestricted crack propagation and
failure and the temperature dependence of this process will be that
of self diffusion.

Following analyses of diffusional creep [e.g., Ref. 15], a model

has been developed for the process of stress rupture in polycrystalline solids[13]. This latter analysis leads to an expression for the
velocity, v, of a crack in a grain boundary in terms of the instantaneous concentrated stress at the crack tip, σ_m, flaw size, x, and
the ratio of surface energy, γ, and the flaw tip radius, ρ. Eq. 2
gives the result,

$$v \simeq \frac{D_s}{x} \exp(\beta(\sigma_m - \gamma/\rho)) \tag{2}$$

where D_s is the self diffusion coefficient given in the usual form
$D_s = D_o \exp(-\Delta H_c/RT)$ and β is a material constant which accounts for
the rate of change of diffusional flux with normal boundary stress.
To describe a crack growth process in terms of diffusional mass transport, a proper rate expression (e.g., velocity) must involve at least
a diffusion coefficient and a gradient in some potential. Eq. 2 satisfies this criterion wherein the potential gradient is approximated
by the exponential of a term involving a stress or pressure coefficient, β, divided by a diffusion distance. The diffusion distance is
further approximated by the instantaneous size of the propagating
crack, x, on the basis that the range of the perturbation in an otherwise uniform stress field, resulting from a stress concentrating flaw
small relative to the sample, would be about the major dimension of
that flaw. The tip stress, σ_m in Eq. 2 may be related, as an upper
bound, to the applied stress by the Inglis relation[16] for a narrow
elliptical crack such that $\sigma_m \simeq 2\sigma_a (x/\rho)^{\frac{1}{2}}$. If one now assumes that the
crack propagates at a constant root radius, ρ, then with simplifying
approximations[13] Eq. 2 may be integrated from an initial tip stress
to the cohesive strength of the boundary yielding an expression for
the time to failure under a given applied stress σ_a. The result has
been shown to be of the following form, [13],

$$e^{-\sigma_a V}(\frac{V^3}{\sigma_a} + \frac{3V^2}{\sigma_a^2} + \frac{6V}{\sigma_a^3} + \frac{6}{\sigma_a^4}) = Z' = \Theta_r/\Theta_o' \tag{3}$$

where Θ_r is the OSD stress rupture parameter, Θ_o' is the normalizing
constant and V is a further material constant. While V can be related
to fundamental quantities of the material such as atomic volume, pressure coefficients of diffusion and the instantaneous strength of the
material, for present purposes it is not necessary to consider values
of V other than those which would be derived from experimental stress
rupture data for systems of interest. It is important to note, however, that Eq. 3 corresponds to Eq. 1 and that the left hand side of
Eq. 3 proposes an expression for a rupture stress function, $f(\sigma)$,
which we will term Z'. It is also of interest to note that as the
applied stress, σ_a, becomes small, Eq. 3 indicates that Θ_r (i.e., t_f
at constant temperature) varies as approximately the inverse fourth
power of the stress. As recently summarized by Burton and Heald[10],
more detailed diffusional models for stress rupture in this regime
yield exponents of 3 or 5 depending on the various diffusion mechan-

isms that may contribute to the failure process. In addition, experiments have recently shown that this inverse fourth power relationship is obeyed for copper and magnesium[17].

Figure 1 is a plot of σ_a versus the logarithm of Z' for various
values of V which appear to encompass stress rupture behaviors of interest. Figure 2 reproduces Dorn's[14] linear plot of the activation energies for creep versus self-diffusion. This plot is of value
for estimating and comparing ΔH_c values which will occur in the OSD
parameter, Θ_r, for specific systems. This work is concerned with
application of such figures to various experimental data.

Applications of the Z' Function

Figure 3 illustrates the application of the curves in Figure 1
to data[2] obtained for a few relatively pure metals. Figure 4 shows
a similar application of Figure 1 to a well documented commercial
alloy which performs at significantly higher stress values than the
materials in Figure 3. In these figures the captions include values
for the constants V, and Θ_o' and ΔH_c which appear in Eq. 2 and which
are appropriate to the various materials considered. In all cases
the rupture stress function, Z', appears to fit the experimental data
well.

The Z Rupture Stress Function

In the formal theory from which the Z" stress function is developed [13], the parameter V, discussed above, results from a combination of other factors which individually reflect either extrinsic or
intrinsic properties of a material which is subject to stress rupture. In this context V is given as

$$V = \beta\sigma_{th}/\sigma_{max}. \tag{4}$$

where β, with units of reciprocal stress, is a coefficient determining the stress dependence of the diffusion of vacancy defects causing
diffusional creep; σ_{th} is the cohesive strength of the grain boundaries of the material which part during high temperature stress rupture
and σ_{max} is the rupture stress of the material under sufficiently
fast loading that delayed failure effects are negligible. The factors β and σ_{th} are predominantly intrinsic characteristics of the material and, respectively, may be estimated from the hydrostatic pressure dependence of the controlling vacancy diffusion in the material
and the theoretical strength of the material as given by an appropriate fraction of the Young's Modulus. In practice, the product $\beta\sigma_{th}$
is of major interest and may be given by the factor $N = \beta\sigma_{th}$. The
instanteous strength, σ_{max}, is basically an extrinsic property of
the material and must be determined directly by experiment or by an
appropriate extrapolation of data. It is to be noted that N is a

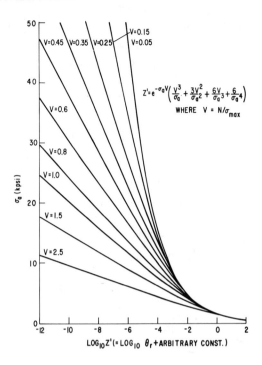

FIG. 1   A general stress rupture data matching plot for the determination of the parameter N/σ_{max} = V for various materials whose
rupture stress values would lie in the range 0 to 50 kpsi.

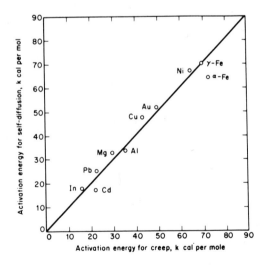

FIG. 2   Comparison of the activation energies for creep with the activation energies for self-diffusion in a number of metals.   (After
Dorn, J. E., ASM Seminar, Creep and Recovery, 1957, p 255)

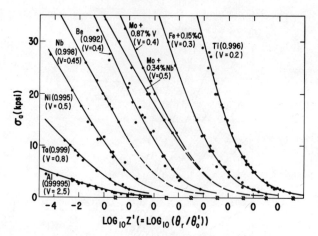

FIG. 3 Rupture stress versus the logarithm of the Z' rupture stress
function for some relatively pure metals. Points are experimental
values of the OSD parameter, $\Theta_r = t_f \exp(-\Delta H_c/RT)$, with t_f in hours,
T in °K. ΔH_c (kcals/mole) and $\log \Theta'_0$ values for the various metals
are, respectively, Al(36,-13.8); Ta(110-13.0); Ni(65,-9.0); Nb(75,-6.2);
Be(65,-12.4); Mo+0.35%Nb(120,-12.3); Mo+0.87%V(120,-12.4); Fe+0.15%C
(90,-16.9) and Ti(60,-12.2).

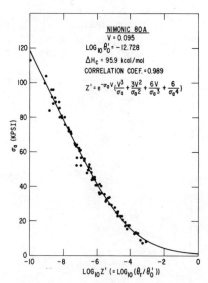

FIG. 4 Rupture stress vs. the logarithm of Z' rupture stress function (solid line; V=0.095) for the nickel base superalloy, Nimonic
80A. Points are experimental values of $\log(\Theta_r/\Theta_0')$ where Θ_r (OSD
parameters)$= t_f \exp(-\Delta H_c/RT)$, and $\log \Theta'_0 = -12.73$ (t_f in hours, T in °K
$\Delta H_c = 96$ kcal/mole).

dimensionless factor and to preserve dimensionless terms it is desirable to normalize the applied stress on a system with respect to the
instantaneous rupture stress such that σ_a/σ_{max} is given by another
factor, R. With these normalizations the Z' stress function of Eq.1
transforms to the following form given by Z

$$Z = mZ' = e^{-N(R-1)} \frac{(N^3/R + 3N^2/R^2 + 6N/R^3 + 6/R^4)}{(N^3 + 3N^2 + 6N + 6)} = \frac{\Theta_r}{\Theta_o} \qquad (5)$$

where m is an arbitrary constant dependent on a material. Figure 5
illustrates the general form of the Z function in terms of R, N and
the OSD parameter Θ_r.

     The utility of this more complicated form of the rupture stress
function of this work derives from the fact that the parameter N only
involves intrinsic factors relating to stress rupture and may be used
as a measure of the high temperature stress rupture sensitivity of a
material. It is thus useful for comparing different materials or for
comparing different modifications of a given material. In general,
the higher the value of N the less sensitive is the material to time
delayed failure and thus high N values are most desirable. For metals
and ceramics N may vary between about 5 and 50. Similar generaliza-
tions concerning the parameter V, discussed previously, cannot be made
since it involves extrinsic factors governing the basic strength of a
material as well as those reflecting diffusional properties of species
within the material and, additionally, it may exhibit a very large
range of values for different materials. Use of the N parameter does,
however, require further experimental data permitting estimates of
σ_{max} whereas such information for the V parameter is unnecessary.

                Applications of the Rupture Stress Function Z

A. Metals: Figure 6 illustrates the application of Eq. 5 to the
             stress rupture data of Simmons and Cross[18] for 304
stainless steel. Handbook data[19] allows an estimate of the instantaneous strength which is set equal to the strength of this alloy in
the recovery temperature range, 60 kspi. Using an activation energy for
creep, or stress rupture, of 80 kcals/mole the full stress rupture
curve may be prepared and, as shown in Figure 6, indicates that the
effective N value for this alloy equals 15. This single value of N
in Eq. 5 gives a Z function curve which closely describes the full
range of Θ_r parameters given by the experimental data. For comparison
we may estimate an instantaneous strength of 135 kpsi for the previously described alloy Nimonic 80A from the low temperature-short time
data of Betteridge[20]. The N value of this alloy, therefore, would
be indicated to be about 13 which is a value comparable to that determined for 304 SS.

B.   Ceramics: Polycrystalline ceramic materials would be expected

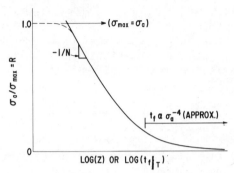

FIG. 5 Schematic illustration of the general features of the curve
corresponding to the stress rupture function of Eq. 5 showing the
principal slope -1/N. At small values of the stress ratio R (i.e.,
σ_a/σ_{max}) the curved portion of the solid line in the figure approximates an inverse fourth power relationship between R and the OSD
stress rupture parameter, Θ_r (or the failure time, t_f, at constant
temperature).

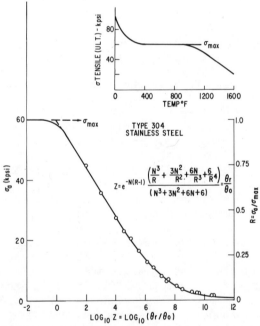

FIG. 6 Stress rupture plot for 18 Cr-8 Ni stainless steel showing
applied stress, σ_a, vs. the reduced OSD parameter $\log_{10}(\Theta_r/\Theta_o)$
$\Theta_r = t_f e^{-\Delta H/RT}$, $\log_{10}\Theta_o = -21.42$, $\Delta H = 80$ kcals/mole) for the data of
Simmons and Cross[18]. Full curve through the data corresponds to
the rupture stress function, Z, for N=15, σ_{max} = 60 kpsi. Inset illustrates the determination of σ_{max} from the recovery range of ultimate tensile stress data[19].

to exhibit high temperature stress rupture properties entirely analogous to those shown by metals. Indeed, for polycrystalline alumina
much of the creep behavior and grain boundary void and cavity formation processes under applied stress have been interpreted by models
similarly developed for metals[21].   Unfortunately, very little experimental data on stress rupture exists for polycrystalline ceramics
since, until recently, little interest has been shown in the use of
ceramics as high temperature, load bearing structural components.

Walles[22] has reported the results of a number of stress rupture
experiments at elevated temperatures on Al_2O_3 and SiC in fibre form.
Figure 7 reproduces these results as given in their original form by
the investigator and it is noted that these results bear a striking
similarity to typical stress rupture data for metals.

For the alumina, direct measurements of room temperature
strengths were reported thus, since a σ_{max} value may be closely estimated for this material as 28 kpsi, the analysis of the data may be
made in terms of the Z form of the rupture stress function given by
Eq. 5. In addition, a summary[23] of extensive work by many authors
on the sintering of alumina indicates that the activation energy for
this process and for diffusional creep would be about 115 kcals/mole.
Using the above two values, the experimental results are plotted in
Figure 8 according to Eq. 5. A Z function curve with N equal to about
16 fits the data reasonably well except for the low stress (low R)
regime. This is not a serious discrepancy since the data in this regime is complicated by the fact that the critical flaw sizes for rapid
crack propagation, as calculated by the Griffith theory, would have
been large fractions of the diameters (≈ 0.1 to 0.4) of the fibres used
in the experiments. The present analyses based on the Z and Z' stress
functions, demand that at some finite distance from the advancing crack
tip the concentrated stress falls to the macroscopic applied stress and
this requirement would not be met by the particular alumina samples in
Figure 9 which were tested at low stresses. It is of interest to note,
however, that the analytical procedure used does appear to correlate
data on polycrystalline alumina from two completely independent experimental sources - sintering phenomena and stress rupture phenomena.

With respect to Walles'[22] data on SiC, no room temperature or
very short term rupture data was reported. Consequently, the data is
best analyzed using the Z' formulation with a V parameter given in
Eq. 3. The data for each temperature was first plotted on a σ_a versus
$\log_{10} t_f$ basis and a best fit Z' function curve passed through each
series of points. This procedure served to identify an appropriate V
value of 0.04 and, by plotting horizontal intersections versus reciprocal temperature a ΔH_c value of 130,000 kcals/mole was obtained.
Figure 9 illustrates the final results plotted as σ_a versus $\log_{10}$
(Θ_r/Θ_o').   The corresponding log Z' curve fits the OSD parameter data
well and serves to rationalize the rather complicated data as given
in its original form in Figure 7. Analyzed in this fashion, however,
the data exhibits one or two somewhat unexpected features. On the one

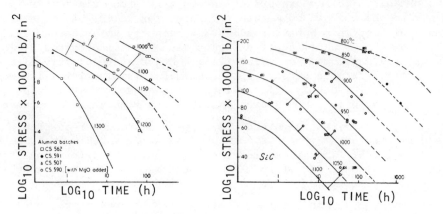

FIG. 7 Experimental stress rupture results on Al_2O_3 rods and SiC fibres as reported by Walles[22] (reproduction from the original).

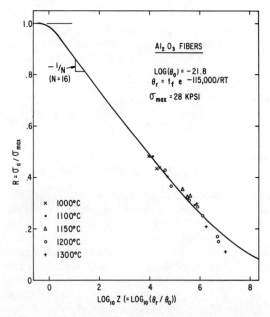

FIG. 8 Experimental stress rupture values of Walles[22] (points) for sintered alumina rods compared with the stress rupture curve (solid line) predicted by Eq. 5 utilizing the physical parameters listed in the figure.

hand the activation energy for creep of SiC, as reported by other investigators[24], would appear to be substantially higher than 130 kcal/mole and, secondly, the reduction in strength at moderate times at 1100°C is markedly greater than observed with other forms of SiC. Finally, if one makes a reasonable estimate of the quick time strength of the fibres as between 300 and 400 kpsi the experimental N value for the SiC would be 0.04 times these values or between 12 and 16. Such an N value is consistent with the behavior of alloys or metals, and as shown in a later section, is inconsistent with that indicated by limited data on other forms of SiC. While the fabrication procedure for the SiC fibres was not given by the investigator it is likely that the procedure utilized may have been the chemical vapor deposition of SiC on a refractory metal filament such as tungsten. The activation energy of self-diffusion in tungsten is indeed reported as about 130 kcal/mole and, consequently, if such was the origin of the experimental specimens, it may be suggested that the stress rupture behavior of the fibres was governed not by the SiC, per se, but the underlying refractory metal filament.

Trantina and Johnson[25] and Prochazka[26] have reported limited stress rupture data on a form of dense, pure phase sintered SiC described earlier by Prochazka[27]. The information consisted of both static fatigue (i.e., constant stress) and constant stress rate data. The former authors report their data over a limited range according to a semi-empirical relationship[28] derived for brittle materials on the basis of an early form of fracture mechanics theory and which may be given as follows:

$$\ln R = \ln(\sigma_a/\sigma_{max.}) = -\frac{1}{n}\ln t_f + K \tag{6}$$

where K and n are constants specific to the material. Charles[13] has shown that the slope parameter, N, utilized in the present work bears the following relationship to the constant, n, in Eq. 6 for relatively large values of R:

$$\frac{\partial \ln R}{\partial \ln t_f} = -\frac{1}{n} \simeq -\frac{1}{NR} \tag{7}$$

Thus, since the data of Trantina and Johnson were obtained for R values centering around about 0.84 and their value for n was given as about 33, the appropriate N value for the present analysis would be about 33/.84~39. Prochazka's creep rate studies[26] on this material indicate a temperature dependence of near steady state creep consistent with an activation energy of about 210-250 kcal/mole. Such a value is near the lower end of values determined by Goshtagori and Coble[24] for carbon diffusion in single crystal SiC (250-300 kcal/mole). Using a value for ΔH_c of 250 kcal/mole and N = 39, Eq. 5 may be applied to the combined stress rupture data (references 25 and 26) to obtain an estimation of the stress rupture performance of dense, sintered SiC over extended ranges of temperature and stress. Figure 10 shows the experimental data plotted as reduced Orr-Sherby-Dorn parameters (Θ_r/Θ_o where the appropriate $\log_{10}\Theta_o$ value for this material = -35.33) through which is passed a curve (solid line) corresponding to the Z stress rupture function for N = 39. Using this curve and a value of ΔH_c = 250 kcals/mole, rupture times for sintered silicon carbide may be estimated for any combination of stress or temperature.

In general, designers find stress rupture information to be most conveniently represented in a plot of applied stress versus temperature for various life-times (e.g., 100, 1000, 10,000 hours, etc.). Figure 11 is such a representation calculated for a 10,000 hour lifetime directly from Eq. 5 where $\log_{10}\Theta_o$ = -35.33 and ΔH_c = 250 kcals/mole. It is to be noted that the σ_{max} value utilized in these calculations was 340 MN/m^2 (50 kpsi) which represents a value shown to be appropriate by Johnson and Trantina for sintered SiC in pure tension in air. The pure tension σ_{max} value is an appropriate value for the direct comparison of stress rupture performance of ceramics versus alloys. Such a comparison is shown in Figure 11 wherein the previously described stress rupture data for Nimonic 80A superalloy is included. Also included in this figure are estimated stress rupture data on hot pressed forms of Si_3N_4 and SiC. These latter data are obtained from the original work of Lange[29] and were calculated using the same procedures as those outlined for the analysis of the data of Trantina and Johnson[25] on sintered SiC. In the case of the hot pressed materials, Lange reported constant stress rate data at several temperatures thus allowing a direct evaluation of the appropriate stress rupture temperature dependence parameters, ΔH_c. Finally in dashed lines, an estimated curve for an advanced superalloy, currently under development ($\gamma/\gamma'-\delta$)[30], is also included for comparison purposes.

There are several features of these comparisons that are of interest. For example, superalloys are superior in bulk strength at low temperatures and, in fact, strengths of 1000 MN/m^2 (150 kpsi) for these materials at low temperatures are common. There is, however, a cross-over at high temperatures where the stress rupture performances of the ceramics exceed the alloys. Another feature of interest is the fact that as longer and longer life-times are required of the materials, the de-rating in maximum allowable temperature, for a given applied stress, is much less for the ceramics than for the alloys. Also of interest is the indication that sensible strengths of one of the ceramics, sintered SiC, are predicted for temperatures well in excess of 1600°C.

Figure 12 shows a further comparison of stress rupture behavior of the above materials when such materials are used under inertial loading conditions where the strength to weight ratio is the dominant factor rather than simply strength. The low densities of the ceramics (about 1/3 of those of superalloys) moves the cross-over in performance for these materials to much higher levels compared to typical alloys.

Finally, Figure 13 displays much of the data previously described in this work but in terms of the normalized stress ratio, R = σ_a/σ_{max},

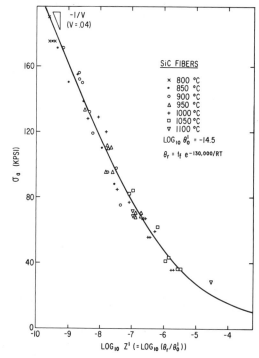

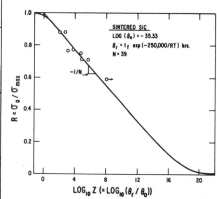

FIG. 10   Estimated stress rupture curve for sintered silicon
carbide defined by experimental
data (points) and Eq. 5 utilizing the physical parameters
listed in the figure.

FIG. 9   Experimental stress rupture
values of Walles[22] (points) for silicon carbide fibres compared with the
stress rupture curve (solid line) predicted by Eq. 3 utilizing the physical
parameters listed in the figure.

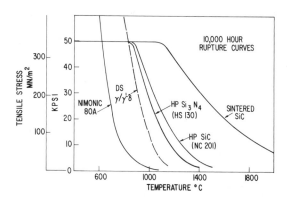

FIG. 11   10,000 hour stress rupture curves for several high temperature structural ceramics and for an early (Nimonic 80A) and an advanced (DS $\gamma/\gamma'-\delta$) superalloy as determined by Eq. 5 and available
experimental data.

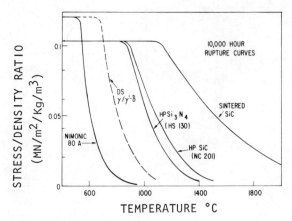

FIG. 12 Stress rupture curves as in Figure 11 corrected for alloy
or ceramic densities.

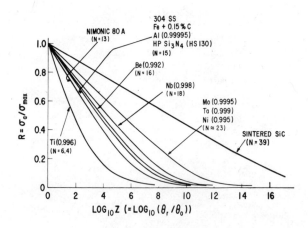

FIG. 13 Normalized rupture stress vs. the logarithm of the Z rupture
stress function for the metals in Figure 3 and the ceramics, sintered
SiC and hot pressed Si_3N_4 (Norton grade HS130). $\Theta_r = t_f \exp(-\Delta H_c/RT)$
with t_f in hours, T in °K. Log Θ_o and estimated σ_{max} (kpsi) and ΔH_c
(kcals/mole) values are, respectively, Ti(-18.3, 32, 60); Nimonic 80A
(-22, 135, 96); 304 SS (-21.3, 60, 80); Al(-18.4, 6, 36); Fe+0.15%C
(-27.7, 50, 90); HS130 Si_3N_4(-24.8, 50, 150); Be(-22.1, 37.5, 65);
Nb(-16.6, 40, 75); Nb(-16.6, 40, 75); Mo(-24.4, 45, 120); Ni(-21.5, 46,
65); Ta(-25.1, 30, 110); sintered SiC(-35.3, 50, 250).

and the log Z(=log(Θ_r/Θ_o)) function so that comparisons of intrinsic parameters, material to material, can be made. Values for σ_{max} for the metals and alloys were those reported along with the original stress rupture data[2,31,32] or estimates from general handbook data[19].

The materials exhibiting the lower slopes in the above figure (i.e., high N values) are inherently less sensitive to stress rupture than materials with higher slopes. In this regard the refractory metals Ta, Nb and Mo as well as Ni, exhibit a marked resistance (N=22 to 24) to stress rupture while Ti (N≈6.4) is highly sensitive to this process once stress rupture conditions are operative (i.e., $\Theta_r > \Theta_o$). It is noteworthy in this figure that the ceramic, sintered SiC, exhibits the highest estimated N value for the materials considered and thus shows the least sensitivity to stress rupture. It is evident that by the use of Eq. 5 and the data listed in the caption of the figure that it is a simple matter to prepare estimates of the lifetime-temperature-stress curves, customarily used for design purposes, for any of the materials listed.

One may further conclude from the above diagram that the marked life-time benefits of a few selected ceramics in high temperature load bearing applications might well exceed the disadvantages of design and construction attendant with the use of these fully brittle materials.

ACKNOWLEDGMENTS

The writer is indebted to S. Prochazka, R. A. Giddings and C. A. Johnson for assistance, provision of data, numerous discussions and critical comments.

REFERENCES

1. F. R. Larson and J. Miller: Trans. ASME, Vol 74, p 765 (1952).
2. R. L. Orr, O. D. Sherby and J. E. Dorn, Trans. ASM, Vol 46, p 113 (1954).
3. S. S. Manson and A. M. Haferd, NACA Tech. Note 2890, March (1953).
4. F. J. Clauss, Proc. ASTM, Vol 60, p 905 (1960).
5. H. Conrad, J. Basic Eng., Vol 81, p 617 (1959).
6. W. Kauzman, Trans. AIME, Vol 143, p 57 (1941).
7. R. N. Stevens and R. Dutton, Mater. Sci. Eng., Vol 8, pp 220-34 (1971).
8. D. P. H. Hasselman, Ultra-fine Grain Ceramics, pp 297-315, Proc. XVth Sagamore Army Materials Research Conference, Ed. J. J. Burke, N. L. Reed, V. Weiss, Syracuse Univ. Press, Syracuse, NY, 1970.
9. A. G. Evans, Journal of Materials Science, Vol 7, pp 1137-1146 (1972).
10. B. Burton and P. T. Heald, Phil. Mag., Vol 32, pp 1079-81 (1975).
11. R. P. Skelton, Phil. Mag., Vol 14, pp 563-571 (1966).
12. J. Weertman, Met. Trans., Vol 5, pp 1743-51 (1974).
13. R. J. Charles, Met. Trans. A, Vol 7A, pp 1081-1089 (1976).

14.   J. E. Dorn, ASM Seminar, Creep and Recovery, p 255 (1957).

15.   F. R. N. Nabarro, Report of a Conference on the Strength of
      Solids, p 75, Physcial Society, London, 1948.

16.   C. E. Inglis, Trans. Inst. Naval Arch., Vol 55, p 219 (1931).

17.   N. G. Needham, J. E. Wheatley and G. W. Greenwood, Acta. Met.,
      Vol 23, pp 23-27 (1975).

18.   W. F. Simmons and H. C. Cross, ASTM Spec. Tech. Publ. No. 124,
      p 5 (1952).

19.   Metals Handbook, 8th Ed., Vol 1, Published by ASM, Metals
      Park, Novelty, Ohio (1961).

20.   W. Betteridge, The Nimonic Alloys, p 187, Edward Arnold Pub.
      Ltd., London 1959.

21.   R. C. Folweiler, J.A.P., Vol 32 (5), pp

22.   K. F. A. Walles, Proc. Br. Cer. Soc., Vol 15, pp 157-171 (1970).

23.   J. H. Rosolowski and C. D. Greskovich, Jr. Am. Cer. Soc., Vol 5,
      p 177 (1975).

24.   R. N. Goshtagore and R. L. Coble, Phys. Rev., Vol 143, No. 2,
      pp 623-626 (1966).

25.   G. G. Trantina and C. A. Johnson, Jr. Am. Cer. Soc., Vol 58 (7-8)
      pp 344-345 (1975).

26.   S. Prochazka, Investigation of Ceramics for High-Temperature
      Turbine Vanes, p 26, Final Report Contract N00019-72-C-0129,
      Naval Air Systems Command, Dec. 1972.

27.   S. Prochazka, "Sintering of Silicon Carbide,: Proc. 2nd Army
      Materials Technical Conference (Nov. 1973), pp 239-251. Also
      see "Sintering of Silicon Carbide," Mass Transport Phenomena
      in Ceramics, Proc. IX University Conf. on Ceramic Sc., Case
      Western Reserve Univ., Cleveland, Ohio, June 3-5, 1974, Plenum
      Press, NY 1975.

28.   R. J. Charles, J.A.P., Vol 29, No. 11, pp 1554-1560 (1958).

29.   F. F. Lange, "High-Temperature Strength Behavior of Hot Pressed
      Si_3N_4: Evidence for Subcritical Crack Growth," Jr. Am. Cer. Soc.,
      Vol 57, No. 2, pp 84-87 (1974).

30.   A. F. McLean, E. A. Fisher and R. J. Bratton, Brittle Materials
      Design, High Temperature Gas Turbine, Interim Report No. 5,
      p 107, Contract No. DAAG 46-71-C-0162, Advanced Research Projects
      Agency, April 1974.

31.   A. Holden, et al.: Symposium on Newer Metals, ASTM Spec. Tech.
      Pub. 272, p 49 (1959).

32.   W. D. G. Bennett and G. Summer: The Metallurgy of Beryllium,
      pp 177-81, Inst. of Metals Monograph No. 28, Inst. of Metals,
      London 1963.

GROWTH OF CRACKS PARTLY FILLED WITH WATER

T. A. Michalske, J. R. Varner and V. D. Frechette

New York State College of Ceramics at Alfred University
and University of Erlangen-Nürnberg
Alfred, New York and Erlangen, West Germany

I. INTRODUCTION

Environmental effects on crack growth in soda-lime-silica glass have been the subject of several fracture mechanics studies. In particular the affect of water on slow crack growth has been studied by investigators including Schönert[1], Wiederhorn[2], and Kerkhof and Richter[3]. Their findings are in close agreement and are shown schematically in Fig. 1. From their findings it can be reasoned that the stress necessary to propagate a crack at a constant velocity of 10^{-2}mm/s will be greater in moist air than in liquid water and still greater in dry air than in moist air. Correspondingly at a constant velocity of 1mm/s the required stress in moist air and dry air will be equal and greater than that required in liquid water.

The effect of environment on slow crack growth in soda-lime-silica glass has also been studied by Varner and Frechette[4] and Quackenbush and Frechette[5]. Their work has shown that analysis of the fracture-generated surfaces may be used to interpret the significance of some fracture mechanics findings to specific effects at the crack front.

The present work is to study crack propagation in liquid water and moist air environments in the velocity range between 0.1 to 1mm/s. Methods of fracture surface analysis along with previous fracture mechanics findings are used to interpret crack growth in cases where the environment is nonuniform across the crack front.

639

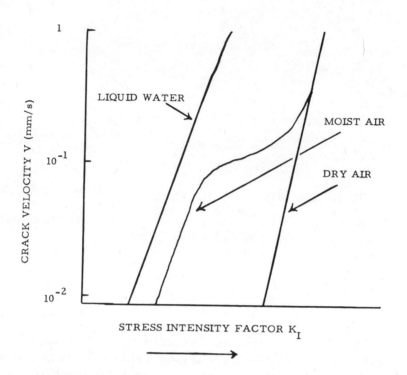

Fig. 1. Dependence of crack tip velocity V in the range between 0.01 to 1mm/s on the stress intensity factor K_I for cracks propagated in dry air, moist air and liquid water.

II. EXPERIMENTAL PROCEDURE

Single edge notched plates of soda-lime-silica glass measuring 3x2x1/8 inches were used in this study.

Starter cracks were extended using simplified double torsion and double cantilever techniques. Figure 2 shows the loading arrangements and crack front shapes developed in each technique.

A bordering reflected light technique described by Frechette[6] was used to monitor crack growth at 15X magnification. During crack growth some samples were mechanically impacted at a frequency of 2Hz. This technique is a modification of a method after Kerkhof[7] for measuring crack velocities. For the case of slow crack growth the technique also yields an accurate repre-

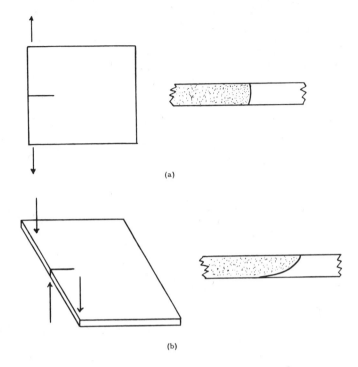

(a)

(b)

Fig. 2.   Loading geometry and corresponding crack front shapes for a) double cantilever and b) double torsion.

sentation of the crack front shape in the form of Wallner lines on the fracture generated surface.

After crack propagation was completed the fracture generated surfaces were examined at up to 500X under reflected light conditions using bordering dark field illumination.   Fracture generated surfaces were also studied with double and multiple beam interferometery.

Two methods were used to develop nonuniform environments across the crack front.   Some starting cracks were propagated in room air and then intersected a liquid water reservoir.   For other samples the starter cracks were allowed to fill with water and cracks were propagated until the supply of water to the crack front depleted.

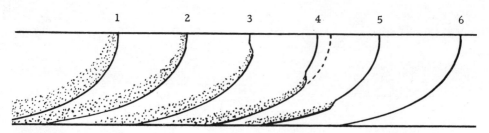

Fig. 3. Stages (1 to 6) in fracture front development from a wet
starter crack in a dry specimen under double torsion. Stippled
areas indicate liquid water trailing the crack front. The dotted
line (stage 4) emphasizes the extent of crack front distortion.

III. EXPERIMENTAL OBSERVATIONS

Cracks Depleted of Liquid Water

Figure 3 shows the sequence observed as a water-filled
starter crack propagated from left to right across a dry glass sheet
in the double torsion mode. Shaded areas behind the crack fronts
at times 1-5 indicate the presence of liquid water. At position 1
the typical double torsion crack front curvature is present and a
small volume of water is in continuous contact with all parts of the
front. At 2 only the amount of trailing water has changed. It is
reasoned that the decrease of water is due to some water left
clinging to newly generated surfaces. The note of interest is that
the trailing water has decreased preferentially along the leading
edge of the fracture front. Position 3 shows two simultaneous
phenomena. First, the trailing water has completely disappeared
from the upper portion of the crack front, and second, a discontinuity has developed in the crack front curvature. Views 4 and 5
show similar behavior. Each successive view shows further
decrease in trailing water and a matching change in the crack front
shape. The dotted line at 4 helps to show that the crack front
irregularity arises from lagging behind of the dry section. Finally
at 6 all trailing water is gone and the normal crack front shape
returns. Figure 4 is a photograph of the fracture generated surface
in the region just described. The smoothly undulating <u>scarp</u> from
upper left to lower right was created at the boundary between water-
contacted and water-free crack portions. The surface topography
of this marking can be seen in Fig. 5. The double beam interference fringes show that the surface marking left behind consists

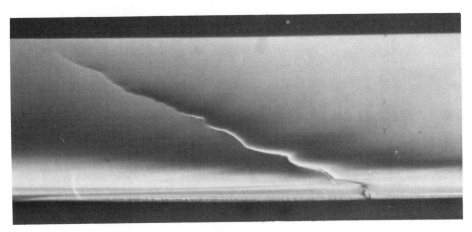

Fig. 4.   Fracture generated surface showing scarp marking
developed at water boundary.   Magnification 20X.

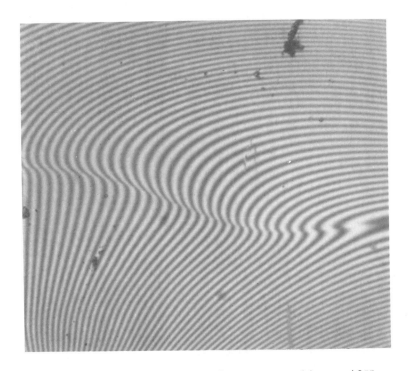

Fig. 5.   Interferogram of scarp marking.   40X.

of two separate planes of fracture generation separated by a steep
slope.   Because of its analogy to the geographical feature this
surface marking is called a <u>scarp</u> marking.   Figure 6 shows a
more accurate representation of the fracture front shape during
<u>scarp</u> formation.   The sample shown in Fig. 6 was prepared in the
same way as that of Fig. 4 and mechanical impulses were introduced during crack growth.   The event which is shown on the
fracture surface in Fig. 6 in similar to that described in Fig. 3.
The Wallner lines on the surface (dark curved lines) show the
crack front at a series of times as the crack ran from left to right.
The first two Wallner lines at the left show the normal crack front
shape associated with double torsion loading.   In this section the
crack front was continuously coated by trailing water.   At the
third and successive Wallner lines the trailing water was depleted
from the upper sections of the crack front.   The boundary of water-
contacted crack front is outlined by a discontinuity in each of these
Wallner lines.   The section of crack front which is water depleted
is strongly retarded from its normal position.   Although the crack
front in the water depleted area has fallen behind, spacings of the

Fig. 6.   Fracture-generated surface showing mechanically induced
Wallner lines in region of liquid water depletion.   50X.

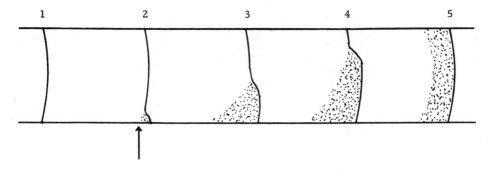

Fig. 7. Stages (1 to 5) in fracture front development from a dry
starter crack in a partly wetted specimen under double cantilever.
Stippled areas indicate liquid water trailing the crack front.
Arrow shows initial area of water contact.

Wallner lines show that the crack velocity remains relatively
constant (0.7mm/s). It is also noted that Wallner lines occurring
in the water depleted region show greater width than the corresponding regions of water contacted crack front.

Cracks Enriched with Liquid Water

Figure 7 shows the sequence observed as a crack was propagated from left to right in the double cantilever mode in a dry
specimen. At position 1 the normal crack front shape is present.
At position 2 liquid water is introduced to the running crack. The
spread of water into the crack front at positions 3, 4, and 5 is
shown by the stippled area behind the crack front representations.
Upon water contact the reaction of the crack front is similar to
that described in the previous section. In regions where only part
of the crack front is in water contact those sections in contact are
advanced with respect to their counterparts in the water-free
regions. Figure 8 is a photograph of a fracture-generated surface
prepared in the previously described manner. A strong <u>scarp</u>
marking can be seen following the boundary of liquid water penetration.

IV. DISCUSSION

Cracks Depleted of Liquid Water

Propagation of partly water-filled cracks in the velocity range
10^{-1} - 1mm/s display three interesting effects: 1. As the water

Fig. 8. Fracture generated surface showing <u>scarp</u> marking developed at water boundary. 30X.

supply is depleted, the water-starved portion begins to lag behind the rest of the crack. This causes the shape of the crack front to become irregular at the water-air boundary. 2. Wallner lines generated from mechanical impulses during crack propagation appear broader along water-depleted areas. 3. The partly water-filled crack fronts propagate along two different planes which connect at a steep slope or <u>scarp</u>.

To interpret these findings, it will be recalled from Fig. 1 that cracks traveling in the velocity range 0.1 to 1mm/s require more stress to propagate in air than in water. From this it may be reasoned that a crack which is only partly water-filled will require some overall stress level which is actually an <u>average</u> taken between the local water-free and water-aided stress requirements. In Fig. 3, as the crack loses trailing water the section left dry requires more than the <u>average</u> stress and the water-filled section requires less than this <u>average</u> stress value to maintain constant crack velocity. This increased stress requirement drags on the dry section and causes it to fall behind the water-filled crack front which has sufficient stress for its

velocity. Accordingly, a redistribution of stress occurs along
the crack front and more stress is delivered to the lagging, dry
section. This keeps the local fracture velocities nearly equal
as shown by the regular spacing between Wallner lines in Fig. 6.
The unusual character of those Wallner lines is probably due to
the higher slope in the V vs. K_I diagram for dry crack fronts.
Thus Wallner lines appear broader in the dry section of the crack
front. Since the crack fronts to either side of the <u>scarp</u> are
propagating under slightly different stress configurations, it is
reasonable to imagine the two sides being somewhat out of plane
with one another and that the convergence of these different planes
forms a steep declivity or <u>scarp</u> on the fracture surface.
Figure 9 is a higher magnification photograph of the intersection
between a Wallner line (upper left-lower right) and a <u>scarp</u>
(lower left-upper right). Figure 10 is a sketch of the same view
with arrows indicating the directions of local fracture travel.
They are normal to the Wallner line and show that the crack
fronts on either side of the <u>scarp</u> are convergent.

Fig. 9. Intersection of Wallner line and <u>scarp</u> on fracture
generated surface. 90X.

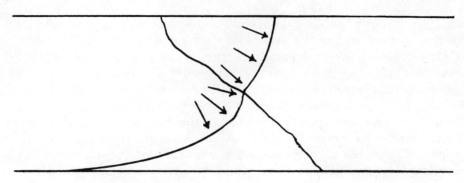

Fig. 10.   Sketch of fracture surface from Fig. 9 showing local direction of fracture travel.

Cracks Enriched with Liquid Water

Results shown in Figs. 7 and 8 show that behavior of water-enriched cracks may be treated exactly as the water-depleted case. As water is drawn into the dry crack the average stress to maintain constant velocity decreases with increasing percentage of water-aided area. The crack front shape and scarp formation may be interpreted identically.

V.   CONCLUSIONS

The previously unreported scarp marking noted on fracture-generated surfaces of cracks having limited access to water and traveling in the velocity range of 0.1 to 1mm/s was found to be the locus of intersection of the water-assisted and water-starved regions of the crack front.

REFERENCES

1. K. Schönert, H. Umhauer and W. Klemm, pp. 474 Fracture 1969, Proceedings of the Second International Conference on Fracture, Brighton, April 1969, P. L. Pratt (Ed. -in-Chief), Chapman and Hall Ltd. London, 1969.
2. S. M. Wiederhorn, J. Am. Ceram. Soc., 50 407 (1967).
3. F. Kerkhof and H. Richter, Glastech. Ber., 42 129.
4. J. R. Varner and V. D. Frechette, J. Appl. Phys., 42 1983.
5. C. L. Quackenbush and V. D. Frechette, to be published.
6. V. D. Frechette, pp. 433 Introduction to Glass Science, L. D. Pye, H. J. Stevens and W. C. LaCourse, eds., Plenum Press, New York (1972).
7. F. Kerkhof, Bruchuorgänge in Gläsern, Deutsche Glastech. Gesell., Frankfurt (1970).

ELECTROLYTIC DEGRADATION OF LITHIA-STABILIZED POLYCRYSTALLINE β"-ALUMINA

D. K. Shetty, A. V. Virkar and R. S. Gordon

Department of Materials Science and Engineering

University of Utah

I. INTRODUCTION

Certain compositions of β"-alumina ceramics possessing high ionic conductivity and low electronic conductivity have been developed for use as solid electrolyte membranes in sodium-sulfur batteries.[1,2] These electrolytes, however, do undergo deterioration during service in Na-S cells. This deterioration is manifested both as a loss of mechanical integrity[3] as well as development of electronic conductivity.[4]

The mechanical degradation of the ceramic electrolyte was investigated in a study by Tennenhouse and coworkers.[3] The progress of degradation of the electrolyte was followed by measuring the fracture strength of specimens in four point bending after they were subjected to electrolysis. This study revealed several interesting features of ceramic degradation. Simple immersion of the ceramic electrolytes in liquid sodium or liquid polysulfides causes no deterioration. Ceramic deterioration occurs only during the $Na^+ \rightarrow Na$ electrolysis, i.e., during the cell charging cycle and is observed only on the ion neutralization surface. This is also evidenced by the formation of a network pattern of cracks filled with sodium. Armstrong et.al.[4] used measurements of cell resistance for monitoring the progress of the "breakdown" of ceramic electrolytes. The cell resistance decreased during continued electrolysis and sodium filaments were observed to grow in the electrolyte matrix.

Richman and Tennenhouse[5] recently proposed a stress corrosion model for crack growth in ceramic electrolytes. In this model crack growth occurs in a subcritical manner by the stress assisted

dissolution of the ceramic in liquid sodium. Armstrong et.al.[4]
qualitatively suggested a mechanism where crack growth was assumed
to occur in a critical manner when the stress intensity reached
the critical stress intensity, K_{IC}. But their mathematical development for the growth of sodium dendrites was empirical in nature
and did not include this critical fracture concept. Both the
treatments utilized two common ideas; a) the inherent surface flaws
on the sodium reservoir side are filled with sodium and these act
as current concentrators, b) the Poiseuille pressure developed in
the cracks due to the flow of sodium causes the cracks to grow.

The models presented by both Richman and Tennenhouse[5] and
Armstrong et.al[4] assume a constant crack width with arbitrary
adjustable value. Such an assumption is incorrect since the shape
of the crack and the pressure distribution (due to the fluid flow)
are interdependent. The strong dependence of the Poiseuille pressure
on the crack width makes any quantitative predictions of these
models questionable. In the present paper a quantitative model
based on the critical fracture concept is developed. A first
order approximation of the crack shape – Poiseuille pressure
interrelationship is presented in which the crack opening displacement is allowed to be determined by the internal pressure.

II. THEORETICAL MODELS FOR CRACK GROWTH IN CERAMIC ELECTROLYTES

The theoretical development of the proposed critical fracture
model is presented in this section. It is, however, important to
emphasize the fundamental difference between the approaches taken
by Richman and Tennenhouse[5] and the present model. In the
stress corrosion model crack growth occurs at stress intensity
values well below the critical stress intensity, K_{IC}. It is thus
a subcritical crack growth model. The crack growth model presented
in this paper, on the other hand, is based on critical and yet
stable (not catastrophic) fracture of the ceramic. Before we
present the formulation of this model a brief review of the
Richman-Tennenhouse model (hereafter abbreviated as R-T model) is
presented. The predictions of this stress corrosion model will
also be compared with the subcritical crack growth data obtained
for β''-alumina ceramics in liquid sodium environments.

A. Stress Corrosion Model

In the R-T model the dissolution of the ceramic in liquid Na
is preferrentially enhanced at the crack tip by two processes.
The Na^+ flow in the vicinity of the crack is "focussed" to the tip
of the crack. This sets up a flux of the solute away from the
crack tip, J_E. Secondly, the Poiseuille pressure developed
inside the crack by the flow of Na develops a stress intensity
at the crack tip and hence an excess chemical potential which

leads to the second flux component away from the crack tip, J_σ. These two flux components are opposed by the flux due to capillarity force, J_γ, that tends to blunt the crack. The crack growth rate is determined by the net flux, J_T.

$$J_T = J_\gamma + J_E + J_\sigma \tag{1}$$

$$J_T = \frac{C_o D \Omega \gamma}{kTr^2} - \frac{6.14 i v_m \ell C}{F \Pi r} - \frac{C_o D \Omega \ell \sigma_a^2}{EkTr^2} \tag{2}$$

and

$$\sigma_a = \frac{6.14 \eta i v_m \ell^2}{Fr^3} \tag{3}$$

Crack propagation rate: $\quad d\ell/dt = - \Omega J_T \tag{4}$

where C_o : Equilibrium concentration of solute in liquid Na.
$\quad\quad\; C^o$: Concentration of solute near the interface.
$\quad\quad\; D$: Diffusion coefficient of solute in liquid Na.
$\quad\quad\; \gamma$: Solid-liquid interfacial energy.
$\quad\quad\; \ell$: Crack length
$\quad\quad\; r$: Crack tip radius
$\quad\quad\; i$: Current density
$\quad\quad\; \eta$: Viscosity of liquid Na
$\quad\quad\; v_m$: Molar volume of liquid Na
$\quad\quad\; F$: Faraday constant
$\quad\quad\; \Omega$: Atomic volume of the solid
$\quad\quad\; E$: Young's modulus of the solid
$\quad\quad\; \sigma_a$: Poiseuille pressure on the crack faces

For net zero flux the model predicts a critical current density, i_{cr}, only above which crack growth should be observed. Fracture strength measurements in four point bending after 10 minute electrolysis at various current densities qualitatively support this prediction. For finite crack velocities the characteristics of crack growth are determined by the flux component that is dominant; i.e. either J_E or J_σ. Based on a comparison of solubilities, C_o, calculated for both flux components to account for the same strength degradation, the model suggests that the stress induced flux term, J_σ, is the dominant one. In other words, stress-aided dissolution should be more likely mechanism of electrolytic degradation.

B. Critical Fracture Model

Unlike the stress corrosion model, in the critical fracture approach the crack extension occurs when the stress intensity at

the crack tip attains the critical value, K_{IC}. As demonstrated
in the appendix, the fracture initiated by an internal liquid
pressure is not catastrophic. Indeed if the fluid amount is
fixed, the crack arrests immediately. In order for the crack to
grow continuously, more and more fluid must be brought into the
crack to maintain criticality. Such a situation exists in a
ceramic electrolyte during battery charging. Na^+ ions traversing
through the electrolyte in the vicinity of a crack filled with
sodium are "focussed" toward the crack in the general manner as
described by Richman and Tennenhouse.[5] A flow of Na is maintained in the crack channel at a rate that is determined by
the current density and the crack length. The flow of liquid Na
in the crack leads to a pressure head development over the crack
surfaces and this is given by the Poiseuille law. When this
pressure reaches a value such that stress intensity reaches
critical, crack extends a small amount, the pressure head drops
and the crack arrests. But then more Na enters the crack and
the process of crack extension and arrest repeats. This sequence
can occur continuously and the rate of crack growth at any instant
is determined by the rate of Na flow into the crack. In essence
this model is based on critical and yet slow fracture and hence
may give rise to crack growth kinetics somewhat similar to the
conventional subcritical crack growth.

C.   Crack Shape and Pressure Distribution

A rigorous mathematical formulation of the above ideas is a
complex problem. This complexity is due to the fact that the
Poiseuille pressure distribution within a crack and the shape of
the crack under that pressure distribution must be self consistent
since one determines the other. Thus the problem must be solved
iteratively whereby successive shape and pressure distribution are
calculated until no significant change in either takes place. In
this paper we will present a first order approximation of this
iterative problem.

We consider an edge crack of length ℓ in a semi-infinite plate
We will assume that the Poiseuille pressure distribution along
the crack length is a linear one as indicated in figure 1a.

$$p = p_o \cdot \frac{x}{\ell} \qquad\qquad (5)$$

It may be noted that this pressure distribution would be consistent
with a constant width channel if a constant volume flow rate is
maintained throughout the length of the channel.

The shape of an edge crack under linearly increasing pressure
distribution is calculated by the method indicated by Sneddon and
Das.[6] The normal displacement of the crack face is given by the
equation

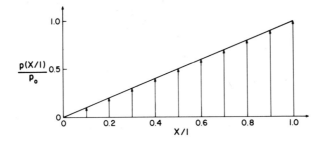

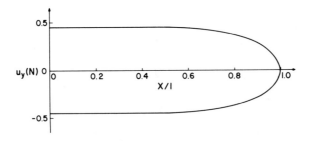

Figure 1. (a) A plot of assumed pressure distribution on the crack faces. (b) The shape of an edge crack in a semi-infinite plate (in plane strain.) $u_y(N)$ is the normalized displacement, $u_y(N) = \varepsilon u_y(x,o)/b$.

$$u_y(x,o) = \frac{2(1-\nu^2)p_o\ell}{E} \int_x^1 \frac{ta(t)dt}{\sqrt{(t^2-x^2)}}, \qquad o \leq x \leq 1 \qquad (6)$$

where the function $a(t)$ satisfies the integral equation

$$a(t) - \int_o^1 a(\tau)L(t,\tau)d\tau = \phi(t), \qquad o \leq t \leq 1 \qquad (7)$$

where the kernel $L(t,\tau)$ is given by

$$L(t,\tau) = \frac{16t\tau^2}{\Pi^2} \int_o^\infty \frac{u^3du}{(u^2+t^2)^2(u^2+\tau^2)^2} \qquad (8)$$

and the function $\phi(t)$ is determined by the pressure distribution on the crack faces, $f(s)$

$$\phi(t) = \frac{2}{\pi} \int_{o}^{t} \frac{f(s)ds}{\sqrt{(t^2-s^2)}} \tag{9}$$

One of the methods suggested by Sneddon and Das[6] for solving the integral equation (7) is to replace the integral by a finite sum using the formula for Gaussian quadrature. This leads to a set of n algebraic equations

$$a(x_i) - \sum_{j=1}^{n} w_j L(x_i, x_j) a(x_j) = \phi(x_i), \quad (i = 1, 2, \ldots n) \tag{10}$$

This procedure was followed for the case n = 3. The values of x_i and the weighting factors, w_j, were obtained from reference (7).

The shape of the crack thus generated is indicated in figure 1b. The accuracy of this numerical procedure was checked by comparing the strain energy and crack opening calculations with the more rigorous calculations of Stallybrass.[8] These, strain energy W and half the crack end displacement, b, are given as

$$W = 0.1831 \left[\frac{2(1-\nu^2)p_o^2 \ell^2}{E} \right] \tag{11}$$

$$b = \varepsilon \frac{(1-\nu^2)p_o \ell}{E} \qquad \text{where } \varepsilon = 0.8854 \tag{12}$$

The constants appearing in equations (11) and (12), evaluated by the approximate method were 0.18325 and 0.8982 respectively, thus showing excellent agreement.

The crack shape indicated in figure 1b shows that crack face displacement is nearly constant for about half the length of the crack from the open end. Beyond this region the crack width decreases. If we assume constant flow rate all along the channel, and this would imply point current focussing at the crack tip, the pressure distribution would deviate from linearity toward higher pressures at about half the crack length. But a more realistic current focussing should be as indicated schematically in figure 2. The volume flow rate decreases near the crack tip and compensates for the decreasing crack width. In this sense the triangular pressure distribution due to fluid flow may not be inconsistent with the shape of the crack calculated at least in a first order approximation. A more exact calculation of current focussing near the crack of shape shown in figure 1b and its effect on Poiseuille pressure calculation will be reported at a later date.

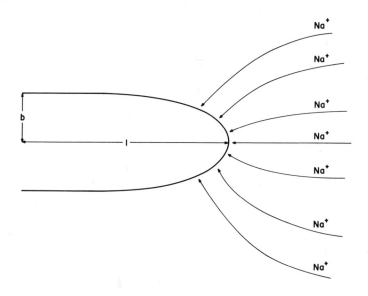

Figure 2. A schematic of ion focussing at the crack tip.

D. Crack Growth Kinetics during Critical Fracture

The Griffith criterion for critical fracture of an edge crack under a linearly increasing pressure profile can be written as[6,8]

$$P_o = \left[\frac{E\gamma_{eff}}{\alpha (1 - \nu^2)\ell} \right]^{1/2} \tag{13}$$

where $\alpha = 0.3662$, γ_{eff}: fracture surface energy. This critical pressure simultaneously satisfies the Poiseuille pressure equation. In our first order approximation this can be written as

$$P_o = \frac{12\eta\ell}{4b^2} \left[\frac{V_{avg}}{2b} \right] \tag{14}$$

Where V_{avg} is the average exit volume flow rate of Na at the open end. During continued fracture this is given as the total inflow of Na into the crack less the increase in (critical) volume of the crack. For unit thickness, volume of the crack of a shape indicated in figure 1b may be written as

$$v_c = \frac{\beta(1-\nu^2)P_o\ell^2}{E} \tag{15}$$

where β is a constant that can be numerically evaluated from the knowledge of the exact shape of the crack such as indicated in figure 1b. For this case $\beta = 1.6083$. Now the average exit flow rate of Na can be written as

$$V_{avg} = V_{Na} - \frac{dv_c}{dt} \tag{16}$$

The instantaneous flow rate of Na into the crack, V_{Na}, is a function of the current density, i, and the effective area of current focussing, A_{eff}

$$V_{Na} = \frac{iv_m A_{eff}}{F} \tag{17}$$

Making an approximation similar to R-T model, we will assume that effective area of current focussing is determined by the instantaneous crack length. For unit crack thickness then

$$A_{eff} = 2\ell \tag{18}$$

Substitution of equations (15) through (18) in equation (14) and equating to equation (13) leads to the crack growth rate equation

$$\frac{d\ell}{dt} = A\ell^{1/2} - B\ell^{-1/2} \tag{19}$$

where

$$A = \frac{4iv_m}{3F\beta} \left[\frac{\alpha E}{(1-\nu^2)\gamma_{eff}} \right]^{1/2}$$

$$B = \frac{4\epsilon^3(1-\nu^2)\gamma_{eff}^2}{9\beta\eta E\alpha^2} \left[\frac{\alpha E}{(1-\nu^2)\gamma_{eff}} \right]^{1/2}$$

On integration equation (19) yields

$$t = \frac{2}{A}(\ell^{1/2} - \ell_o^{1/2}) + \frac{B^{1/2}}{A^{3/2}} \ell n \left[\frac{\sqrt{A\ell} - \sqrt{B}}{\sqrt{A\ell} + \sqrt{B}} \Big/ \frac{\sqrt{A\ell_o} - \sqrt{B}}{\sqrt{A\ell_o} - \sqrt{B}} \right] \tag{20}$$

For net zero crack growth rate, $\frac{d\ell}{dt} = o$, this model also predicts a critical current density, i_{cr}. Equation (19) is applicable only for the current regime

$$i > i_{cr}$$

III.   EXPERIMENTAL RESULTS AND THEORETICAL PREDICTIONS

A.   Subcritical Crack Growth Data for β"-Alumina in
Liquid Na

One important conclusion of the R-T model[5] was that the
stress-aided flux (J_σ) controls the rate of deterioration. It
should therefore, be possible to critically test this model by
mechanically stressing the ceramic in liquid Na to evaluate subcritical crack growth that is characteristic to stress corrosion.
Some of the results of such tests are discussed here.

Subcritical crack growth in β"-alumina in liquid Na was
investigated by a load relaxation technique using double cantilever
beam specimen geometry. The details of this technique have been
discussed before.[9] Specimens of two different compositions
(8.8 wt.% Na_2O - 0.75 wt.% Li_2O - 90.45 wt.% Al_2O_3 and 10.0 wt.%
Na_2O - 1.2 wt.% Li_2O - 88.8 wt.% Al_2O_3) were fabricated by
isostatic pressing (50,000 psi) into rectangular bars and sintering
at 1580°C for 20 minutes in packed β"-alumina powder. Postsinter
annealing was carried out at 1450°C for 6 hours. The high Na_2O
and Li_2O composition was chosen for comparison purposes since it
exhibited low critical current density in electrolytic degradation
tests.[3] The load relaxation tests were carried out at 300°C
subsequent to a wetting excursion at 375°C for 1 hour. The subcritical crack growth data for the two compositions plotted in
the conventional form of Log V vs. Log K_I plots are shown in
figures 3 and 4 respectively. The linear plots seen in these
figures conform to the empirical equation

$$V = A K_I^N \qquad\qquad (21)$$

The large values of N observed (N = 562 and 355) discount any
serious stress corrosion in the 8.8 - 0.75 ceramic. For typical
stress-corroding ceramics N values are generally below 50. The
high Na_2O and Li_2O composition shows slightly increased susceptibility to stress corrosion as reflected in the lower N values. In
this respect, the poor performance of the high Na_2O - high Li_2O
ceramic in stress corrosion tests is qualitatively consistent
with the electrolytic degradation test results.[3] It is
interesting to note that R-T model predicts a value of N = 2 for
the situation when stress-induced flux term is dominating.

B.   Prediction of Critical Current Densities

Both the stress corrosion model (equations (1) through (4)
as well as the critical fracture model (equation 19) predict a
critical or threshold current density when the crack growth rate
is set equal to zero. Estimation of critical current density
based on the R-T model is not possible because the crack tip

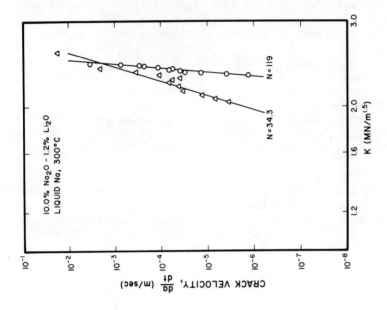

Figure 4. Crack velocity (V) – stress intensity factor (K_I) diagrams for β″-alumina of composition 10% Na_2O – 1.2% Li_2O in liquid sodium at 300°C.

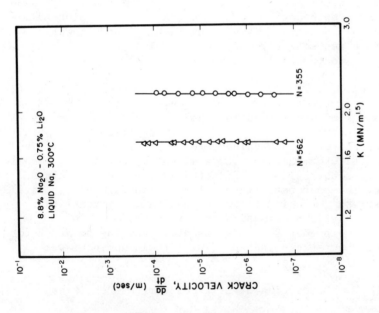

Figure 3. Crack velocity (V) – stress intensity factor (K_I) diagrams for β″-alumina of composition 8.8% Na_2O – 0.75% Li_2O in liquid sodium at 300°C.

radius, r, is an undetermined arbitrary parameter. Assigning an
arbitrary value to r is not meaningful since critical current
density is a very sensitive function of r (i_{cr} α r^3).

The calculation of critical current density, i_{cr}, based on
the critical fracture model (equation 14) indicates several
interesting results. In addition to the physical properties of
the β"-alumina ceramic (E, ν) and liquid Na (η, v_m), the critical
current density is a function of the fracture energy of the
ceramic and the initial size of the flaw at the surface. Polycrystalline β"-alumina has a typical fracture energy of 20 J/m^2.[10]
For an initial crack size of 25 μm this would correspond to a
critical current density of 1500 amp/cm^2. This is clearly too
high as compared to the critical current densities observed in
practice in the operation of Na-S cells[3] which are typically of
the order of 1-2 amp/cm^2. For a fracture energy of 1 J/m^2, however,
a value that would be typical of single crystal β"-alumina, the
predicted critical current density is 3.8 $amps/cm^2$. The formulation
developed in this paper with the idealized crack shape and geometry
is clearly more applicable to single crystal β"-alumina which has a
well defined cleavage plane (basal). This is also the plane of
highest ionic conductivity. In a polycrystalline ceramic the
crack is likely to take a tortuous path along grain boundaries.
These disturbances in the flow path of liquid Na are likely to
enhance the Poiseuille pressures that are generated at the crack
tip. Presently, calculations are underway which attempt to
include the tortuosity of the crack path in the calculation of
Poiseuille pressure. This also allows one to include microstructural
effects such as grain size in the crack growth calculations.

IV. DISCUSSION

A logical development of a model that describes the pseudo
steady state crack growth kinetics in a ceramic electrolyte during
charging has been presented in this paper. The complexity of the
problem necessitated several assumptions. These simplifying
assumptions must be clearly pointed out and their implications
elucidated. The calculation of the exact crack shape and the
Poiseuille pressure generated is an integral part of the development of the model. The first order approximation of a triangular
loading of the edge crack and the calculation of the crack shape
under that loading, has been justified on qualitative arguments.
A second order improvement would be to evaluate the new Poiseuille
pressure distribution corresponding to the above shape. This
requires a more rigorous treatment of current focussing near the
crack tip by the solution of the field equations. Such calculations
of pressure distribution and crack shape must be repeated until
they are self consistent. The criterion of zero crack growth rate
in the crack growth rate equation (equation 19) automatically
leads to a critical current density in agreement with experimental

observations. But the quantitative value predicted is substantially
higher than experimental values particularly for polycrystalline
ceramics. From purely mechanical consideration, the model should
be more applicable to single crystal β''-alumina since it has a
well defined cleavage plane (basal plane) and the idealized crack
shape may be achieved. But in the calculation of critical current
density for the case of a single crystal with a crack in the
basal plane, it was assumed that the ion focussing is similar to
that in the polycrystalline material. In view of the large
anisotropy in conductivities in the basal plane and c-direction
(the conductivity in the basal plane is about 10^7 times larger than
the c-direction), such an assumption cannot be true. In fact the
ion focussing should be insignificant suggesting that the actual
i_{cr} could be much larger than the one calculated here. Consideration
of crack tortuosity in polycrystal matrix will enhance the Poiseuille
pressure head calculated due to pressure losses at crack bends.
How this will affect the crack opening displacement, 2b, and the
critical current density is difficult to evaluate at this stage.
Finally the crack growth equations developed in the present model
reflect a pseudo steady state situation. It is implicitly assumed
in the treatment that current focussing and ion flow to the crack
tip are continuous and rapid and are not the rate determining
steps. Transient situations are not treated in this formulation.
These include initial transient ion flow when the charging cycle
is started. Also if crack blunting occurs due to plastic deformation
or other reasons, crack growth may occur in discrete jumps. In
such a situation adjustment of current focussing and ion flow to
the successive crack configurations may become the rate controlling
step.

The large slopes observed in the log velocity - log stress
intensity plots obtained for the β''-alumina ceramics suggest that
simple mechanical stress corrosion is not the primary mechanism
for electrolytic degradation. This does not, however, rule out
the applicability of the R-T model. It does indicate that mechanical
stressing of the electrolyte ceramic in liquid sodium does not
simulate the conditions during electrolysis. The application of
the electric field during electrolysis and the resultant "focussed"
ions flow has a more significant role in electrolytic
degradation. Thus for example the electrolytic flux (J_E) could
be important in overcoming such rate limiting phenomena as non-
wetting of the ceramic by liquid sodium or elimination of super-
saturation of liquid sodium near the crack tip by the flow of
fresh sodium from the electrolyte matrix. While the basic ideas
in the R-T model may well be applicable to electrolytic degradation,
their formulation based on arbitrary adjustable crack width makes
any quantitative prediction such as critical current density
questionable. This is particularly true in the present situation
where crack is loaded by an internal pressure which is a sensitive
function of crack opening, 2b ($p \propto 1/b^3$). In a situation where the

loading is external the constant crack width assumption would not be too serious. Secondly in the R-T model, the stress flux term, J_σ, is dominant at small crack opening displacements, $2r$, ($r = b$) while the electrical flux term, J_E, becomes significant at large values of r. For reasonable values of pressure, p_o = 30,000 psi and crack length, ℓ = 25 μm, the crack opening displacement, $2b$, calculated from equation (12) turns out to be about 460°A. In their analysis this value is large enough for the electrical flux term to make substantial contribution.

It is evident that refinement of the theoretical formulations of both the models is necessary to obtain better quantitative predictions. The suitability of either of the models to explain electrolytic degradation can, however, be tested only with detailed experiments. One of the key experiments to achieve this purpose would be temperature dependence of degradation. In the critical fracture model the temperature dependence is mainly due to the temperature dependence of the viscosity, η. It, therefore, predicts decreasing crack velocities at elevated temperatures for the same current density. Or conversely temperature dependence of the critical current density (viz. the activation energy) would be identical to that of the viscosity of liquid sodium. In the R-T model, on the other hand, temperature dependence is due to the temperature dependencies of the diffusivity of the solute in liquid sodium D, solubility of the solute, C, in liquid sodium as well as the viscosity of liquid sodium. At a given current density, i, ($i > i_{cr}$), the crack velocity may increase or decrease with increasing temperatures depending upon the relative magnitudes of the terms in their flux equation.

APPENDIX

Criterion for brittle fracture when a Griffith crack is opened by internal pressure of a fluid of compressibility 'k'.

It has been shown[11] that for a Griffith crack of length '2ℓ' in an infinite plate, stressed internally by a uniform fluid pressure, p, the strain energy stored is given by

$$W = \frac{\Pi(1-\nu^2)p^2\ell^2}{E} \tag{A1}$$

Under uniform pressure, the crack assumes the shape of an ellipse with the semimajor axis ℓ and the semiminor axis, b, given by

$$b = \frac{2(1-\nu^2)p\ell}{E} \tag{A2}$$

The volume of the crack per unit thickness is given as

$$v = \frac{2\Pi(1-\nu^2)p\ell^2}{E} \tag{A3}$$

Now consider this crack volume to be occupied by a fixed amount of a fluid of compressibility, $k = -\frac{1}{v}\frac{dv}{dp}$, whose relaxed volume, v_o, is given by

$$v = v_o \cdot \exp(-kp) \tag{A4}$$

From (A3) and (A4)

$$v_o = \frac{2\Pi(1-\nu^2)p\ell^2}{E} \cdot \exp(kp) \tag{A5}$$

It is also shown that[11] the Griffith criterion for crack initiation under internal pressure, p, can be written as

$$p = \left[\frac{2E\gamma_{eff}}{\Pi(1-\nu^2)\ell}\right]^{1/2} \tag{A6}$$

or $p^2\ell$ = constant

The condition for continued fracture by satisfying the above criterion is

$$\frac{d}{d\ell}(p^2\ell) > o$$

or

$$p^2 + 2p\ell \cdot \frac{dp}{d\ell} > o \tag{A7}$$

But from equation (A5), from the constancy of the relaxed volume, v_o, it can be shown that

$$\frac{dp}{d\ell} = -\frac{2p}{\ell(1+kp)} \tag{A8}$$

Hence the criterion for continued fracture is

$$\frac{d}{d\ell}(p^2\ell) = \frac{p^2(kp-3)}{1+kp} > o \tag{A9}$$

For incompressible fluids k = o and

$$\frac{d}{d\ell}(p^2\ell) = -3p^2 < o \tag{A10}$$

Even for an ideally compressible fluid such as ideal gas kp = 1 and hence $\frac{d}{d\ell}$ (p$^2\ell$) < o. Hence for all real fluids fracture will not be catastrophic since a small increment in crack length leads to a drop in the stress intensity below the critical value and crack comes to arrest.

ACKNOWLEDGEMENT

    This work was supported by the National Science Foundation (RANN) under Contract No. NSF-C805.

REFERENCES

1. J. T. Kummer and N. Weber, Trans. Soc. Auto. Engrs., 76, p. 1003-1007 (1968).

2. "Research on Electrode and the Electrolytes for the Ford Sodium-Sulphur Battery", Annual Report for period June 30, 1975 to June 29, 1976. NSF (RANN) Contract No. NSF-C805, July 1976.

3. G. J. Tennenhouse, R. C. Ku, R. H. Richman and T. J. Whalen, Bull. Amer. Ceram. Soc., 54, No. 5, p. 523 (1975).

4. R. D. Armstrong, T. Dickinson and J. Turner, Electrochimica Acta, 19, p. 187 (1974).

5. R. H. Richman and G. J. Tennenhouse, J. Amer. Ceram. Soc., 58, No. 1-2, p. 63 (1975).

6. I. N. Sneddon and S. C. Das, Int. J. Engg. Sci., 9, p. 25-36 (1971).

7. M. Abramowitz and I. A. Stegun, Handbook of Mathematical Functions, National Bureau of Standards, App. Math. Series, No. 55, Washington, D. C. (1964).

8. M. P. Stallybrass, Int. J. Engg. Sic., 8, p. 351-362 (1970).

9. A. V. Virkar and R. S. Gordon, J. Amer. Ceram. Soc., 59, No. 1-2, p. 68 (1976).

10. A. V. Virkar and R. S. Gordon, J. Amer. Ceram. Soc., 60, No. 1-2, p. 58 (1977).

11. I. N. Sneddon and M. Lowengrub, "Crack Problems in the Classical Theory of Elasticity", John Wiley & Sons, Inc., (1969).

ENGINEERING DESIGN AND FATIGUE FAILURE OF

BRITTLE MATERIALS

John E. Ritter, Jr.

Mechanical Engineering Department

University of Massachusetts, Amherst, MA 01003

ABSTRACT

Fracture mechanics provides the foundation necessary for making failure predictions to assure the reliability of ceramic materials under various loading conditions and before and after proof testing. The parameters necessary for making these predictions can be obtained directly from crack velocity measurements or, indirectly, from static or dynamic fatigue strength experiments. Experimental confirmation of the fracture mechanics theory was obtained using soda-lime glass in water as the model material/environment system. The application of this theory to engineering design and fatigue failure is discussed.

1.0   INTRODUCTION

The key to the utilization of ceramic materials in structural applications is to assure their mechanical reliability and safety. Because ceramics are brittle, their strength is not a well-defined quantity but can vary by as much as 20% from the average measured strength. To complicate matters even more, strength is generally found to be dependent on the time of loading. Ceramics loaded slowly or forced to support a load for a long time are found relatively weak; whereas, they are relatively strong if loaded rapidly or have to support a load for only a short time. To account for this variability and time dependency of strength, critical ceramic components in the past have been designed to support maximum tensile stress no greater than about 10-20% of the average strength.

A sound, fundamental theory has been recently developed for

making failure predictions for ceramic materials based on fracture mechanics principles [1-3]. It is believed that this fracture mechanics theory will not only increase the reliability of structural ceramics but also result in a reduction of the large safety factors associated with these materials. This paper will review the fracture mechanics background for making lifetime predictions, give experimental confirmation of this theory, and discuss the applicability of the theory in the engineering design of ceramics.

2.0   FRACTURE MECHANICS FUNDAMENTALS

2.1   Fatigue Failure

Fracture mechanics theory is based on the reasonable assumption that fatigue failure of ceramics occurs from stress-dependent growth of preexisting flaws to dimensions critical for spontaneous crack propagation. The time to failure (t_f) under a constant applied tensile stress (σ_a) is derived from the definition of the stress intensity factor (K_I):

$$K_I = \sigma_a Y \sqrt{a} \tag{1}$$

where Y = constant that depends on flaw geometry and a = flaw size. For surface flaws, which are usually the most critical flaw in ceramics, $Y = \sqrt{\pi}$. On taking the derivative of Eq.(1) with respect to time:

$$\frac{dK_I}{dt} = (\frac{\sigma_a^2 Y^2}{2K_I}) V \tag{2}$$

where V = crack velocity. By rearranging and integrating Eq.(2), the time to failure under constant stress is:

$$t_f = (\frac{2}{\sigma_a^2 Y^2}) \int_{K_{Ii}}^{K_{IC}} (K_I/V)\ dK_I \tag{3}$$

where K_{Ii} = initial stress intensity factor at the most serious flaw and K_{IC} = critical intensity factor of this flaw.

For most glasses, subcritical crack velocity can be expressed as a power function of the stress intensity factor so that [1,2]:

$$V = A \ K_I^N \tag{4}$$

where A and N are constants that depend on the environment and material composition. On substituting Eq.(4) into Eq.(3), Eq.(3) can be integrated to give:

$$t_f = [\frac{2}{AY^2 \ (N-2) \ \sigma_a^2}] \ [K_{Ii}^{2-N} - K_{IC}^{2-N}] \tag{5}$$

Generally, $K_{IC}^{2-N} << K_{Ii}^{2-N}$ since for ceramics N > 10 and for the usual range of service stresses $K_{Ii} < 0.9 \ K_{IC}$. The initial stress intensity factor can be expressed as:

$$K_{Ii} = (\sigma_a/S_i) \ K_{IC} \tag{6}$$

where S_i = fracture strength in an inert environment where no subcritical crack growth occurs prior to fracture. By substituting Eq.(6) into Eq.(5) and neglecting K_{IC}^{2-N} with respect to K_{Ii}^{2-N}, the failure time becomes:

$$t_f = [\frac{2}{AY^2 \ (N-2) \ K_{IC}^{N-2}}] \ S_i^{N-2} \ \sigma_a^{-N} = B \ S_i^{N-2} \ \sigma_a^{-N} \tag{7}$$

where $B = [\frac{2}{AY^2 \ (N-2) \ K_{IC}^{N-2}}]$ = a constant for a given material and

environment. In Eq.(7) t_f simply represents the time required for a flaw to grow from an initial, subcritical size to dimensions critical for catastrophic propagation and B and N are the constants that characterize this subcritical crack growth. The initial flaw size is characterized in Eq.(7) by the fracture strength in an inert environment. From Eq.(7) it is seen that the time to failure decreases with increasing stress. This delayed failure behavior is commonly known as static fatigue, i.e. fatigue under a static stress.

It is also possible to derive the fracture strength (S) of ceramics loaded at a constant stressing rate ($\dot{\sigma}$) by combining Eq.(4) with Eq.(2) and integrating. In this case flaws grow from subcritical to critical size under a time dependent stress. The result of this analysis is [4]:

$$S^{N+1} = B \ (N+1) \ S_i^{N-2} \ \dot{\sigma} \tag{8}$$

where B and N are the same fatigue constants as in Eq.(7). From Eq.(8) it is seen that fracture strength decreases with decreasing stressing rate since the flaws are given more time to grow. This behavior is known as <u>dynamic fatigue</u>, i.e. fatigue under dynamic stressing conditions.

In addition to constant stress and stressing rate loading conditions, cyclic loading must also be considered. It is generally assumed for ceramic materials that crack propagation rates under cyclic loading can be predicted from crack growth parameters determined under static load experiments, i.e. there is no significant enhanced effect of cycling. This has been verified in measurements of slow crack growth rates under static and cyclic loading for a number of ceramics [5,6]. From this assumption it can then be derived that the failure time under cyclic stress (t_c) is directly proportional to the failure time under constant stress [5]:

$$t_c = g^{-1} t_s \qquad\qquad (9)$$

where t_s = static time to failure at the average cyclic stress and g^{-1} = a proportionality factor that depends on the type of stress cycle, the amplitude of the cycle, and N. The factor g^{-1} can be evaluated by numerical integration for any periodic load cycle. For square wave, sinusoidal, or saw-tooth type loading, values of g^{-1} have been evaluated and are available in diagrams that express g^{-1} as a function of N and the ratio of the stress amplitude to the average applied stress [5].

Many cyclic loading conditions can be treated as a simple extension of constant stress and constant stress rate loading [7]. Fig. 1a and 1b show that for a fracture stress of σ_f the failure time under constant stress is $\tau/(N+1)$ and is τ for constant stress rate. This result can be derived from Eq.(7) and (8). In Fig. 1c, a general linear cycle is considered where the stress varies linearly from 0 to σ_f to 0 in time τ. With this cyclic loading condition, the failure time is the same as for a constant stress rate loading up to the fracture stress. Extending this to any number of repetitive linear cycles (Fig.1d) results in the same relationship where τ is now the number of cycles multiplied by the period of the cycle. Random linear cycles (Fig. 1e) also results in the constant stress rate relationship where τ is now the sum of the periods of each of the cycles. Since a sinusoidal cycle lies between a linear, constant stress rate cycle and a constant stress cycle (Fig. 1f), lifetimes under sinusoidal loading conditions fall between constant stress and constant stress rate lifetimes. Thus, for ceramic materials <u>cyclic fatigue</u>, i.e. fatigue under cyclic stressing conditions, is just an extension of static and dynamic fatigue.

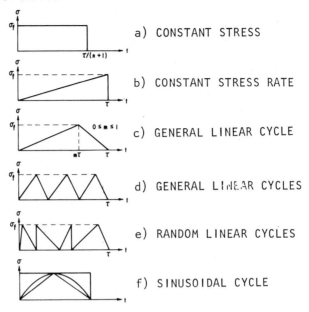

a) CONSTANT STRESS

b) CONSTANT STRESS RATE

c) GENERAL LINEAR CYCLE

d) GENERAL LINEAR CYCLES

e) RANDOM LINEAR CYCLES

f) SINUSOIDAL CYCLE

Figure 1.   Possible loading conditions
(after Ref. 7).

2.2   Probability of Failure

Since the distribution of initial flaw sizes in ceramic material
is directly related to the distribution of inert strengths (Eq.(6)),
the probability of failure (F) in service for a given t_f and σ_a can
be obtained from Eq.(7) by expressing the inert strength (S_i) of the
components in terms of their failure probability distribution.   By
so doing, it is assumed that the origin of fracture is the same for
both fatigue and inert failures and that the sample with the shortest
fatigue life has the lowest inert strength (largest initial flaw)
and the sample with the median life has the median inert strength,
etc.

Generally, the inert strength distribution of ceramics can be
approximated by the Weibull relationship [7]:

$$\ln \ln \frac{1}{1-F} = m \ln \left(\frac{S_i}{S_o}\right) \qquad (10)$$

where m and S_o = constants.   By substituting Eq.(10) into (7), the
time to failure as a function of failure probability is obtained:

$$\ln t_f = \frac{(N-2)}{m} \ln \ln \left(\frac{1}{1-F}\right) - N \ln \sigma_a + (N-2) \ln S_o + \ln B \qquad (11)$$

Similarly, the strength distribution at a fixed stressing rate and time to failure distribution under cyclic loading can be obtained by substituting Eq.(10) into Eq.(8) and (9), respectively.   It should be noted that other statistical distributions for the inert strength could be used if they are shown to be more applicable. This would be the case, for example, if the distribution was bimodal.

2.3   Proof Testing

In proof testing, components are subjected to stresses that are greater than those expected in service in order to assure that all weak samples are eliminated.   In this manner, the proof test technique attempts to assure that every component surviving the proof test will have a minimum lifetime.   To insure the validity of proof testing, certain precautions must be followed [8].   For the proof test to be most effective, it should be conducted in a relatively inert environment and rapid unloading rates should be used.   Also, the proof test must duplicate in the component the actual state of stress expected in service and the components after proof testing must be protected from any subsequent structural damage.   If these precautions are followed, proof testing can be a practical method for assuring the mechanical reliability of ceramics [8].

From a fracture mechanics viewpoint, the value of proof testing is that it characterizes the largest effective flaw possible in the tested components since any larger flaw would have caused failure during the proof test.   The minimum time to failure after proof testing is the time it takes for this maximum flaw to grow to critical dimensions for spontaneous fracture and is given by [1]:

$$t_{min} = B \, \sigma_p^{N-2} \, \sigma_a^N \qquad\qquad (12)$$

where σ_p = proof test stress.   By comparing Eq.(12) with Eq.(7) it is seen that σ_p simply represents the minimum inert strength of the components after proof testing.

The minimum lifetime predictions given by Eq.(12) are only valid when no crack growth occurs on unloading during the proof test.   Crack growth during loading and at the proof stress does not present a problem since the maximum flaw size can still be characterized.   However, if flaw growth occurs on unloading, then the maximum flaw size cannot be defined and no assurance of a minimum service life can be given.   Before proof testing can be used to assure a minimum lifetime, the effectiveness of the proof test in truncating the inert strength at σ_p should be experimentally determined.

Evans and Wiederhorn [1] have shown that the inert strength after proof testing (S_i'), considering crack growth during loading but not unloading, is:

$$\left(\frac{S_i'}{S_i}\right)^{N_p-2} = 1 - \left(\frac{\sigma_p^*}{S_i}\right)^{N_p-2} \left[\ 1 - \left(\frac{\sigma_p}{\sigma_p^*}\right)^{N_p-2}\ \right] \tag{13}$$

where N is the crack propagation parameter appropriate for the proof test conditions and σ_p^* is the equivalent proof stress for inert conditions. Assuming that the initial inert strength distribution is given by the Weibull relationship, σ_p^* is determined from:

$$\ln \ln \left(\frac{1}{1-F_p}\right) = m \ln \left(\frac{\sigma_p^*}{S_o}\right) \tag{14}$$

where F_p = failure probability during proof testing, and m and S_o are the initial Weibull inert strength constants. The failure probability after proof testing (F_a) is related to the initial failure probability before proof testing by [1]:

$$F_a = \frac{F-F_p}{1-F_p} \tag{15}$$

Substituting Eq.(14) and (15) into Eq.(13) gives:

$$\left(\frac{S_i'}{S_o}\right)^{N_p-2} = \left(\ln \frac{1}{1-F_a} + \ln \frac{1}{1-F_p}\right)^{\frac{N_p-2}{m}} - \left(\ln \frac{1}{1-F_p}\right)^{\frac{N_p-2}{m}} + \left(\frac{\sigma_p}{S_o}\right)^{N_p-2} \tag{16}$$

From Eq.(16) it is seen that the inert strength distribution after proof testing depends upon the initial inert strength distribution parameters m and S_o, the crack propagation parameter N_p, and the failure probability during proof F_p. It is also significant to note that from Eq.(16) the inert strength distribution after proof testing, is truncated at σ_p and will be greater than the initial inert strength at all levels of probability if $m < N_p-2$.

The minimum lifetime predicted by Eq.(12) will generally be substantially smaller than the observed failure time for most of the samples that pass the proof test because of the distribution in the after-proof strengths, hence, failure times. The time to failure after proof testing (t_f') can be calculated by using Eq.(7) and (12) to give:

$$\frac{t'_f}{t_{min}} = \left(\frac{S'_i}{\sigma_p}\right)^{N-2} \qquad\qquad (17)$$

By expressing S'_i in terms of its failure probability distribution (Eq.(16)), t'_f at a given level of probability can be calculated.

2.4 Experimental Techniques for Obtaining Parameters Necessary for Failure Prediction

Eq.(7),(8),(9),(12), and (17) summarize failure prediction for ceramic materials under various loading conditions and before and after proof testing. These predictions are dependent on the inert strength distribution of the components and the fatigue parameters B and N. The inert strength of the components can be determined directly or estimated from measurements of the inert strength of laboratory samples using statistical scaling laws [9,10]. The fatigue parameters must be determined under the service environent and can be obtained from one of three types of experiments: crack velocity, static fatigue, or dynamic fatigue.

Crack velocity can be measured directly as a function of the stress intensity factor on specimens specifically designed for fracture mechanics experiments[2]. These specimens contain a macroscopic crack that allow accurate measurement of the crack velocity and stress intensity factor. A regression analysis of this data using Eq.(4) gives A and N. By determining K_{IC} in a separate measurement on identical samples, B can be determined from its definition, see Eq.(7). Crack velocity measurements have been widely used in determining the fatigue characteristics of ceramic materials and have yielded much fundamental information regarding the details of subcritical critical crack growth in ceramics [2]. Unfortunately, for purposes of failure prediction crack velocity data is not as reliable as that from static or dynamic fatigue strength experiments because data from large, preformed cracks is not necessarily relevant to the propagation of the microscopic cracks present in structural ceramics [11,12]. Also lifetime predictions based on crack velocity data generally involve greater extrapolation of the data than do those based on static or dynamic fatigue data [13].

Static fatigue data is generally obtained by measuring the time to failure of a large number of samples at several constant applied stresses [14]. From this data the median value of t_f can be determined as a function of σ_a. By measuring the median $\tilde{S}_i$ on a group of statistically identical samples, B and N can be determined from Eq.(7). Equation (7) can be rewritten in terms of the median values of t_f and S_i as:

$$\ln \overline{t}_f = \ln B = (N-2) \ln \overline{S}_i - N \ln \sigma_a \tag{18}$$

where $\overline{t}_f$ and $\overline{S}_i$ = median values of t_f and S_i. From a regression analysis of $\ln t_f$ vs $\ln \sigma_a$, N and B are determined from:

$$\text{slope} = N \tag{19}$$

$$\text{intercept} = \ln B + (N-2) \ln \overline{S}_i$$

The median values of t_f and S_i are used since these values are at equal failure probabilities and since it is the median value that is best estimated by the sample distribution. Although the "median value" technique for analyzing static fatigue data is quite straight forward to apply, it makes inefficient use of the data since only the median t_f values are used in the regression analysis; thus, the uncertainty $\ln B$ and N can be large.

Recognizing the need for more efficient utilization of fatigue data, several researchers [15,16] have suggested a method of data reduction that is based on a homologous stress ratio (σ_{HS}), defined as

$$\sigma_{HS} = \frac{\sigma_a}{S_i} \tag{20}$$

To obtain fatigue data in terms of σ_{HS}, t_f values for each applied stress are ranked and then paired with equal ranked S_i values, i.e. the shortest t_f value is paired with the lowest S_i, the next shortest t_f with the next lowest S_i, and so on. Since σ_a is fixed for a given ranking of fatigue lives, the relationship between σ_{HS}, S_i, and t_f is established. In terms of the homologous stress ratio, Eq.(7) can be rewritten

$$\ln (t_f S_i^2) = \ln B - N \ln \sigma_{HS} \tag{21}$$

From a regression analysis of $\ln (t_f S_i^2)$ vs $\ln \sigma_{HS}$, the parameters N and B are determined from the slope and intercept, respectively. Since the data for all samples that fail are used in the regression analysis, the homologous stress method for analyzing static fatigue data greatly increases the confidence in N and B as compared to the median value technique.

Dynamic fatigue data is generated by measuring fracture strength as a function of stressing rate [14] and Eq.(8) is used to analyze the data. Since Eq. (8) is of the same form as Eq.(7), the analysis of dynamic fatigue data can be carried out in a manner similar to that of static fatigue data. For example, Eq.(8) can be rewritten in terms of median values of S and S_i as

$$\ln \bar{S} = \frac{1}{N+1} [\ln B + \ln (N+1) + (N-2) \ln \bar{S}_i + \ln \dot{\sigma}] \qquad (22)$$

From a regression analysis of $\ln \bar{S}$ vs $\ln \dot{\sigma}$, N and B are determined from

$$\text{slope} = \frac{1}{N+1}$$
$$\qquad \qquad \qquad \qquad \qquad \qquad \qquad \qquad \qquad \qquad (23)$$
$$\text{intercept} = \frac{1}{N+1} [\ln B + \ln (N+1) + (N-2) \ln \bar{S}_i]$$

Likewise, by defining a homologous stress ratio to be

$$\sigma_{HD} = \frac{S}{S_i} \qquad (24)$$

Eq. (8) can be rewritten

$$\ln \sigma_{HD} = \frac{1}{N+1} [\ln B + \ln (N+1) + \ln (\frac{\dot{\sigma}}{S_i^3})] \qquad (25)$$

By ranking the S data for a given $\dot{\sigma}$ and then pairing this data with equal ranked S_i data, a regression analysis of $\ln \sigma_{HD}$ vs $\ln (\dot{\sigma}/S_i^3)$ will give N and B from the slope and intercept, respectively.

It is important to realize that there is an uncertainty in failure predictions due to the experimental uncertainty associated with determining the fatigue parameters and the inert strength. Statistical techniques have been derived to estimate this uncertainty in the failure predictions [17,18] and the analysis of typical fatigue data show that this uncertainty is quite sensitive to the experimental error in the fatigue parameters. Fortunately, much of this uncertainty in the failure predictions can be eliminated either by increasing the required proof stress or by decreasing the allowable stress in service by a small increment [17,18].

2.5 Design Diagrams

Eq. (7) and (12) are most easily understood by expressing them in terms of a design diagram [1,8]. These diagrams can be developed for a particular component in the service environment and give the probability of failure in service for a given lifetime and applied stress, as well as, the proof test stress necessary to assure a minimum lifetime in service. Design diagrams are most useful to the engineer in deciding if a proof test is necessary and, if necessary, in determining the proof test required to insure a given lifetime.

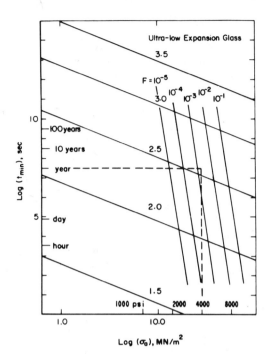

Figure 2.   Design diagram for the windows of the space
            shuttle [8].   The lines that run from left to
            right relate the minimum failure time after
            proof testing to the applied stress (Eq.(12))
            where the numbers over each line give the ratio
            σ_p/σ_a; the lines that give the probability without
            proof testing are included for comparison (Eq.(11)).

Figure 2 gives the design diagram developed for the windows of
the space shuttle [8].   For a minimum lifetime of 1 year at an
applied stress of 4000 psi, a proof test ratio of 2.6, corresponding to a proof stress of about 10,500 psi, will be necessary for
these windows.   By comparison, the failure probability is 2×10^{-3}
for the same service conditions without proof testing.   It was estimated that about one window in ten would be broken during the
proof test.   Thus, by breaking about 10% of the windows, a minimum
lifetime performance of the ramaining windows is assured.

3.0   EXPERIMENTAL VERIFICATION

The fracture mechanics principles described in Section 2.0
have been verified using as a model material soda-lime glass
microscope slides.   The fatigue parameters in water were determined

by crack velocity, dynamic fatigue, and static fatigue experiments and are summarized in Table I. It should be noted that the fatigue strength data was analyzed using both the median value and homologous stress techniques. To determine if the differences in the fatigue parameters in Table I lead to significant differences in failure predictions, the allowable applied stress for a lifetime of 10^5 sec. was calculated. The σ_a based on proof testing was calculated from Eq.(12) choosing σ_p = 79.3 MN/m^2 and that based on no proof testing from Eq.(7) choosing F = 10^{-3} and using the inert (liquid nitrogen) strength distribution of the slides as given by:

$$\ln \ln \frac{1}{1-F} = 8.19 \ln S_i/137.41 \ (MN/m^2) \qquad (26)$$

Table II summarizes these predictions and it is seen that all the predictions are within about 10% of each other and are not sig-nificantly different. Thus, for soda-lime glass in water it appears that the accelerated tests (crack velocity and dynamic fatigue) can be used to develop data for long term failure predictions.

It is also important to point out in Table I the larger un-certainties in the fatigue parameters based on the median fatigue data. This illustrates the advantage of analyzing the fatigue data by the homologous stress techniques where data for each sample that fails is included in the regression analysis.

TABLE I. Summary of Fatigue Parameters for Soda-Lime Glass in Water

Type of Test Data	N	$\ln B \ [(MN/m^2)^2 \cdot s]$
Static Fatigue (Median)	19.32(3.10)*	-1.66(3.11)
Static Fatigue (Homologous)	15.12(.38)	1.82(.36)
Dynamic Fatigue (Median)	18.42(.79)	-1.70(.59)
Dynamic Fatigue (Homologous)	17.96(.39)	-1.78(.28)
Crack Velocity[+]	16.03(.40)	-.889(.37)

*The numbers in parentheses represent the standard deviation of the fitted parameters.

[+]Data was obtained by S.M. Wiederhorn.

TABLE II. Values of Allowable Stress for a Lifetime of 10^5 sec. with and without Proof Testing using Fatigue Parameters Given in Table I.

Type of Data	With Proof Testing $\sigma_p = 79.3$ MN/m^2 Allowable σ_a (MN/m^2)	Without Proof Testing F = .001 Allowable σ_a (MN/m^2)
Static Fatigue (Median)	25.50	19.61
Static Fatigue (Homologous)	23.42	18.15
Dynamic Fatigue (Median)	24.08	18.54
Dynamic Fagitue (Homologous)	23.25	17.91
Crack Velocity	21.20	16.39

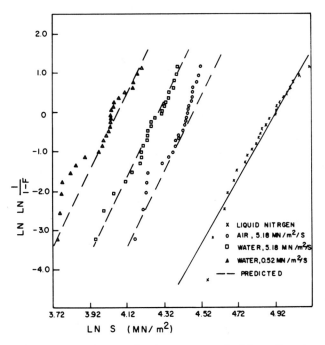

Figure 3. Comparison of the strength distribution data for soda-lime glass in water to that predicted using Eq.(8) coupled with Eq.(26).

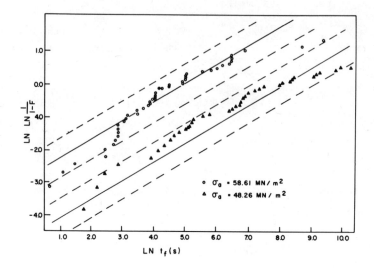

Figure 4.   Comparison of the failure time distributions
             for soda-lime glass in water to that predicted
             using Eq.(7) coupled with Eq.(26). The dashed
             lines indicate the limits of ± a standard deviation.

     The applicability of fracture mechanics theory in describing
the fatigue test data can be examined by comparing the predicted
failure time and strength distribution to that actual measured.
The predicted distributions are given by Eq.(7) and (8) using the
average of fatigue parameters given in Table I and the inert
strength distribution given in Eq.(26). Figures 3 and 4 show
these results and the good correlation between the experimental
and predicted distributions is evident. These results largely
confirm that fatigue failure is caused by slow crack growth of
preexisting flaws and that fracture mechanics theory can be used
in making failure predictions.

     The inert strength distribution after proof testing in air is
compared in Fig. 5 to the initial distribution (Eq.(26)) and to
that predicted theoretically from Eq.(16). Excellent agreement is
obtained between theory and experiment; thus, it is believed that
in this case proof testing was effective in truncating the after-
proof, inert strength at σ_p. In addition, the minimum lifetime in
water after proof testing was measured at various applied stresses.
The experimentally measured minimum time to failure is compared in

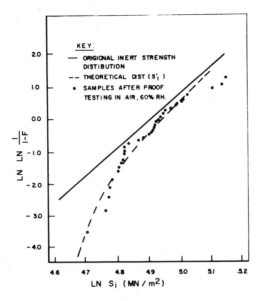

Figure 5.   Inert strength distribution of soda-lime glass before and after proof testing in air compared to the theoretical, after-proof distribution given by Eq.(16)   (σ_p = 79.3 MN/m^2, F_p = 0.33, N_p = 18.42).

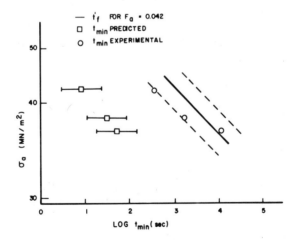

Figure 6.   Comparison of Experimentally determined minimum failure times of soda-lime glass in water after proof testing to that predicted from Eq.(12) and Eq.(17).   The proof test conditions were the same as in Fig. 4.   The indicated brackets for t_{min} and the dashed lines about the t_f prediction correspond to $\pm$ a standard deviation.

Fig. 6 to that predicted from Eq.(12) using the average of the
fatigue parameters in Table I. In all cases, failure occured at
times greater than the predicted minimum lifetime after proof testing. A more precise test of the analytical predictions is provided
by estimating failure times after proof testing on a probability
basis using Eq.(17) in conjunction with Eq.(16). Since a total of
23 samples were tested at each of the 3 stress levels in Fig. 5,
the weakest specimen had a failure probability of 0.042. Thus t_f
from Eq.(17) is calculated for F_a = 0.042 and is plotted in Fig. 5.
The measured failure time after proof testing of the weakest specimen at each stress level is in close agreement with the predicted
failure time, showing that the analytical predictions are of reasonable accuracy and are not excessively conservative.

4.0   ENGINEERING DESIGN

Based on the test results for soda-lime glass presented in
the previous section, it is believed that fracture mechanics
theory provides an excellent foundation for making failure predictions for brittle materials. This theory has been successfully
used in engineering design for assuring the mechanical reliability
of brittle materials in a number of applications such as: space-
craft windows [19,20], porcelain insulators [20], vitrified grinding wheels [22], optical glass fibers [23], and ceramic turbine
components [7]. In each of these applications, the fatigue param-
eters(B and N) were measured by at least one of the three experimental techniques: crack velocity, static fatigue, and dynamic fatigue.
These parameters were then used to predict a proof test for assuring a minimum lifetime in service of the ceramic component in question or to predict the probability of failure for a given lifetime.
However, before these fracture mechanics principles can be assumed
to be generally valid for all ceramic/environment systems, several
very important questions must be resolved.

First the applicability of the two major assumptions inherent
in fracture mechanics theory, namely failure occurs by subcritical
growth of preexisting flaws and the fracture origin is the same for
failure under both inert and fatigue conditions, must be demonstrated. It is quite possible that in the long-term service use of
ceramics, alternate failure mechanisms may become important and
control the lifetime of material. For example with Si_3N_4 at elevated temperatures, oxidation effects could overshadow the stress
corrosion growth of preexisting flaws by either initiating new
flaws and/or healing preexisting ones. Also, with glass fibers,
the long-term lifetime may be controlled by the breakdown of the
coating rather than the slow growth of flaws. Related to the
second assumption, recent work [24,25] has indicated that fracture
origins in some ceramics tend to be internal under inert conditions
and at the surface under fatigue conditions.

The implications of these results for lifetime predictions is quite serious since the inert strength cannot be used as a measure of the initial flaw size and an alternate means of characterizing the initial flaw size distribution must be found. Without this characterization, we cannot derive the lifetime of sample in service from fracture mechanics theory. To assure that the two key assumptions of fracture mechanics theory are valid, one should experimentally demonstrate that the inert strength distribution, coupled with the fatigue parameters B and N, can be used to predict the failure time and strength distributions (Eq.(7) and (8)). In addition, SEM photographs of the fracture origin under inert and fatigue conditions can be obtained.

Second, the effectiveness of proof testing in truncating the strength and thereby assuring a minimum lifetime in service must be proven. For example, recent work with Si_3N_4 [26] has indicated that the flaws being eliminated by proof testing under inert conditions are not the flaws controlling strength at elevated temperatures; hence, proof testing was not effective in improving the high temperature strength. On the other hand, with proof testing at high temperature, flaw growth on unloading became possible and proof testing was not effective in truncating the strengths. Thus, experimental confirmation of the effectiveness of proof testing should be obtained before it is assumed that proof testing can assure the reliability of structural ceramics. However, it should be realized that even if proof testing is not effective in truncating the strength distribution due to crack growth on unloading, it may still be beneficial in eliminating components with manufacturing defects.

Third, before crack velocity experiments can be used to make long-term failure predictions, agreement between these experiments and fatigue strength data must be demonstrated since recent papers on Si_3N_4 [11] and fused silica fibers [12] show that crack velocity experiments do not agree with the fatigue strength experiments. One explanation of these results is that the kinetics of slow growth of microscopic flaws as present on strength samples is different from macroscopic flaws as present in crack velocity samples. This could be due to differences in crack tip environments and/or differences in how large and small cracks interact with the microstructure.

5.0 SUMMARY

This paper has presented a review of the techniques that have been developed recently to assure the reliability of ceramic components in structural applications. Although fracture mechanics forms the basis of the failure predictions methods, the fatigue paramaters necessary for these predictions can be obtained from static and

dynamic fatigue strength experiments. The validity of the theory
has been tested experimentally and agreement between theory and
experiment is good. However, additional experimentation is needed
to fully evaluate the limits of the theory. Nevertheless, within
its range of applicability, the theory provides a rational basis
for design and is being used successfully in a number of applications.

ACKNOWLEDGEMENT

This work was supported by National Science Foundation.

REFERENCES

1. A.G. Evans and S.M. Wiederhorn, "Proof Testing of Ceramic
    Materials-An Analytical Basis for Failure Prediction," Int.
    J. Fract., 10, 379-92 (1974).

2. S.M. Wiederhorn, "Subcritical Crack Growth in Ceramics,"
    pp. 613-46 in Fracture Mechanics of Ceramics, Vol. 2, ed.
    by R.C. Bradt, D.P.H. Hasselman, and F.F. Lange, Plenum Press,
    New York (1974).

3. J.E. Ritter, Jr. and J.A. Meisel, "Strength and Failure Pre-
    dictions for Glass and Ceramics," J. Am. Ceram. Soc., 59,
    478-81 (1976).

4. A.G. Evans, "Slow Crack Growth in Brittle Materials Under
    Dynamic Loading Conditions," Int. J. Frac., 10, 251-59 (1974).

5. A.G. Evans and E.R. Fuller, "Crack Propagation in Ceramic
    Materials Under Cyclic Loading Conditions," Met. Trans.,5,
    27-33 (1974).

6. A.G. Evans, L.R. Russell, and D.W. Richerson, "Slow Crack
    Growth in Ceramic Materials at Elevated Temperatures," Met.
    Trans., 6A, 707-16 (1975).

7. G.G. Trantina, "Brittle Fracture and Subcritical Crack Growth
    in a Ceramic Structure," pp. 921-27 in Fracture 1977, Vol. 3,
    ed. by D.M.R. Taplin, Univ. of Waterloo Press, Waterloo,
    Canada (1977).

8. S.M. Wiederhorn, "Reliability, Life Prediction, and Proof Test-
    ing of Ceramics," pp. 635-65 in Ceramics for High Performance
    Applications, ed. by J.J. Burke, A.E. Gorum, and R.N. Katz,
    Brook Hill Pub. Co., Chestnut Hill, MA (1974).

9.   F.A. McClintock and A.S. Argon, <u>Mechanical Behavior of Materi-</u>
     <u>als</u>, pp. 504-08, Addison-Wesley, Reading, MA (1966).

10. F.A. McClintock, "Statistics of Brittle Fracture," pp. 93-116
     in <u>Fracture Mechanics of Ceramics</u>, <u>Vol. I</u>, ed. by R.C. Bradt,
     D.P.H. Hasselman, and F.F. Lange, Plenum Press, New York
     (1974).

11. J.E. Ritter, Jr., "Assuring Mechanical Reliability of Ceramics,"
     to be published in the proceedings of NATO Advanced Study
     Institute on Nitrogen Ceramics, University of Kent August 1976.

12. J.E. Ritter, Jr. and K. Jakus, "Applicability of Crack Velocity
     Data to Lifetime Predictions for Fused Silica Fibers," J. Am.
     Ceram. Soc., $\underline{60}$, 171,(1977).

13. S.M. Wiederhorn, "A Critical Analysis of Failure Prediction
     Techniques," presented at 1977 Annual Meeting of American
     Ceramic Society.

14. J.E. Ritter, Jr. and C.L. Sherburne, "Dynamic and Static
     Fatigue of Silicate Glasses, "J. Am. Ceram. Soc., $\underline{54}$, 601-05
     (1971).

15. B.J.S. Wilkins, "Engineering Design and the Probability of
     Fatigue Failure of Ceramic Materials," pp. 875-82 in <u>Fracture</u>
     <u>Mechanics of Ceramics</u>, <u>Vol. 2</u>, ed. by R.C. Bradt, D.P.H.
     Hasselman, and F.F. Lange, Plenum Press, New York (1974).

16. J E. Burke, et al., "Static Fatigue of Glasses and Alumina,"
     pp. 435-39 in <u>Ceramics in Severe Environments</u>, ed. by W.W.
     Kriegel and Hayne Palmour III, Plenum Press, New York (1971).

17. D.F. Jacobs and J.E. Ritter, Jr., "Uncertainty in Minimum
     Lifetime Predictions," J. Am. Ceram. Soc., $\underline{59}$, 481-87 (1976).

18. S.M. Wiederhorn, et al., "An Error Analysis of Failure Pre-
     diction Techniques Derived from Fracture Mechanics," J. Am.
     Ceram. Soc., $\underline{59}$, 403-11 (1976).

19. S.M. Wiederhorn, et al., "A Fracture Mechanics Study of the
     Skylab Windows," pp. 829-42 in <u>Fracture Mechanics of Ceramics</u>,
     <u>Vol. 2</u>, ed. by R.C. Bradt, D.P.H. Hasselman, and F.F. Lange,
     Plenum Press, New York (1974).

20. S.M. Wiederhorn, et al., "Application of Fracture Mechanics to
     Space-Shuttle Windows," J. Am. Ceram. Soc., $\underline{57}$, 319-23 (1974).

21. A.G. Evans, et al., "Proof Testing of Porcelain Insulators

and Application of Acoustic Emission," Am. Ceram. Soc. Bull., 54, 576-81 (1975).

22. J.E. Ritter and S.A. Wulf, "Designing a Proof Test to Assure Against Delayed Failure," presented at 1977 Annual Meeting of American Ceramic Society, to be published.

23. B.K. Tariyal, et al., "Proof Testing of Long Length Optical Fibers for a Communication Cable," Bull. Am. Ceram. Soc., 56, 204-05 (1977).

24. G.K. Bansal, W. Duckworth, and D.E. Niesz, "Strength-Size Relationships in Ceramic Materials: Investigation of a Commercial Glass-Ceramic," Bull. Am. Ceram. Soc., 55, 289-92 and 307 (1976).

25. R.M. Gruver, W.A. Sotter, H.P. Kirchner, "Variation of Fracture Stress with Flaw Character in 96% Al_2O_3." Bull. Am. Ceram. Soc., 55, 198-201 and 204 (1976).

26. N.J. Tighe, et al., "Proof Testing of Si_3N_4," presented at the Fall Meeting of Basic Science Division, American Ceramic Society, San Francisco (1976).

SUBCRITICAL CRACK GROWTH IN PZT[*]

J. G. Bruce, W. W. Gerberich[†] and B. G. Koepke
Honeywell Corporate Material Sciences Center
Bloomington, Minnesota 55420

[†]Department of Chemical Engineering and Materials
Science
University of Minnesota, Minneapolis, Minnesota 55455

ABSTRACT

Subcritical crack propagation in the transducer ceramic, PZT, has been studied using the double torsion technique. The effects of testing environment and temperature as well as the state of poling in the material have been characterized in detail. Tests run in water and in environments inert with respect to water such as toluene, mineral oil and Freon, a corona suppressant, have established that water enhances slow crack propagation in PZT. Fracture has also been found to depend sensitively on the state of poling. Crack propagation is hindered if poling is perpendicular to the crack plane but is hardly affected if the material is poled parallel to the crack. These results can be explained more in terms of the residual stresses introduced by poling than in terms of the microstructural (i.e., domain structure) changes accompanying poling. A thermal activation analysis carried out on crack-propagation data measured in water on unpoled PZT yielded a stress free activation energy of 100 kcal/mol.

INTRODUCTION

The propagation of surface and subsurface flaws under subcritical[†] loads in a ceramic high-drive sonar transducer can be detrimental

---

[*]The majority of this work was supported by the Office of Naval Research under contract No. N00014-76-C-0625.

[†]A subcritical load is any load less than that necessary to catastrophically propagate the flaw.

to both the mechanical strength and the electrical characteristics of the device. The mechanical properties degrade since the fracture stress, σ_F, of a brittle material depends sensitively on the initial flaw size, a, through the Griffith equation $\sigma_F = AK_{IC}a^{-\frac{1}{2}}$ where A is a geometrical term on the order of unity, and K_{IC} is the critical stress intensity factor. The electrical characteristics can be altered by flaw growth since flaws are sites of corona discharge. Electrical discharge contributes to noise in the device. If the device output is increased or if the unit is subjected to shock loading, both these concerns are amplified.

The piezoelectric properties of sonar transducer ceramics such as barium titanate and solid solutions of lead zirconate and lead titanate (PZT) have been intensively studied and improved through the addition of a host of elements (many proprietary)[1]. The mechanical properties, particularly the fracture characteristics, on the other hand, have only recently attracted attention[2,3]. Recent work at the Naval Research Laboratory has identified predominant fracture origins in PZT as pores, pore clusters, large grains and machining flaws[3,4]. Origins of surface fracture have also been observed at the edges of electrodes on transducer elements in regions of high electric field gradients[5]. Pohanka, et. al.[6] have further shown that internal stresses resulting from the paraelectric → ferroelectric phase transformation occurring in PZT upon cooling through the Curie temperature can measurably decrease the fracture resistance of the material.

The phenomenon of subcritical crack growth has been studied in many ceramic materials[7] but has only recently been investigated in transducer ceramics. Freimen et. al.[8] published data showing that water enhances slow crack growth in PZT 5800 (a Navy Type I material), and Caldwell and Bradt[9] established that slow crack growth in PZT could be detected using dynamic fatigue tests in both 3-point bend and in compressive tests. These studies have established that subcritical crack growth occurs in PZT and the effect is sensitive to the chemical environment. More work is needed, however, to determine in greater detail the effects of chemical environment on flaw growth in a transducer ceramic, particularly with respect to the effects of temperature and to the effects of realistic transducer operating environments such as corona suppressants. In addition, it is not known how the microstructural changes introduced by poling effect crack growth. In $BaTiO_3$, Freiman et. al.[8] observed an enhancement of crack propagation in samples poled parallel to the crack plane. The effects of poling perpendicular to the crack plane have not been reported.

In this paper we present the results of an extensive series of

measurements in which slow crack growth in PZT was examined in a
number of testing environments. The effects of temperature on
crack propagation in one environment (water) are reported for the
first time. Finally the anisotropic effects of poling on slow
crack growth are demonstrated by testing specimens poled both
parallel and perpendicular to the crack plane.

EXPERIMENTAL PROCEDURE

Slow crack growth in PZT was studied using the double torsion
(DT) technique popularized by Evans et. al.[10,11]. A schematic
showing the specimen and loading geometry is shown in Figure 1.
The stress intensity for this configuration is independent of
crack length and is given by

$$K_I = PW_m \left[\frac{3(1+\nu)}{Wd^3 d_n} \right]^{\frac{1}{2}} = AP \qquad (1)$$

where P is the load, ν is Poisson's ratio and the other terms are
defined on the Figure.

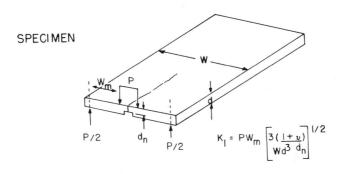

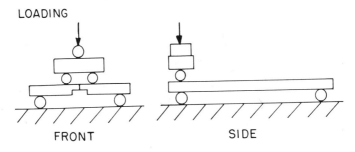

Fig. 1. Schematic showing specimen and loading geometry for
double torsion tests.

Slow crack growth was studied using the load relaxation technique. In a load relaxation test a precracked specimen is
rapidly loaded to a load, P, somewhat less than the critical load,
P_{IC}, necessary to initiate fast fracture of the specimen, and the
crosshead of the testing machine is arrested. If slow crack
growth occurs the load will decay as a function of time since the
compliance of the system is an increasing function of crack
length. Providing the compliance is a linear function of crack
length the crack velocity is related to the instantaneous load and
the rate of decay of the load according to[11].

$$V = \frac{-(P_{i,f})(a_{i,f})}{P^2} \left(\frac{dP}{dt}\right) \qquad\qquad (2)$$

where $P_{i,f}$ and $a_{i,f}$ are the initial (or final) load and crack
length respectively. Thus, in a single load relaxation test V is
measured over a range of K_I. The results are usually plotted as
log V vs log K_I since log V is experimentally found to be a linear
function of log K_I for many ceramics[7,12].

The samples were PZT plates nominally 2.54 cm (1 inch) wide,
7.62 cm (3 inch) long and had thicknesses ranging from 15 mm (.060)
to 23 mm (.090 inch). A side groove was cut in each specimen to a
depth of about one-half the specimen thickness and to a width of
approximately 3 mm. We will comment on the side groove width in
the results. A specimen is loaded through four ball bearings and
is supported on the base of the jig by four ball bearings as shown
in Figure 1. Before testing, a thin starter notch is cut in the
end of the specimen to a depth of 3 mm, and a short scratch diamond
scribed on the bottom of the specimen at the base of the notch to
insure that a crack will initiate down the center of the side
groove.

In a typical test a specimen is tested with the side groove up
(i.e. opposite to that shown in Figure 1) to facilitate the determination of the final crack length with dye penetrant. The specimens were loaded in an Instron TM testing machine at a crosshead
speed of 5×10^{-3} mm/min (2×10^{-4} inch/min) until a crack popped in
as noted by a rapid decrease in the load. The specimens were then
unloaded and reloaded at a crosshead speed of 0.25 mm/min (10^{-2}
inch.min) until the crack just started running at which time the
crosshead was arrested. The load vs time curves during relaxation
were recorded on teletype tape to facilitate computer analysis of
the data. The duration of most tests was 15 to 20 min. since under
ordinary testing conditions slight oscillations in the load due to
small thermally induced deformations in the loading train obscure
the data at the extremely low relaxation rates[10]. In computing

the V-K_I curves, provisions were made in the program to include machine relaxations during the test. Accordingly a background relaxation tape was recorded at a load of approximately ½ to 3/4 P_{IC} before a test. This relaxation was then subtracted from the total relaxation. After a relaxation test K_{IC} was measured by reloading the specimen to failure at a crosshead speed of 0.25 mm/min (10^{-2} inch/min).

Material and Testing Environment

The PZT specimens tested in this study were cut from trans-ducer tubes manufactured by the Honeywell ceramics Center. The PZT is a Navy Type III high drive sonar ceramic with a nominal composition of $Pb_{0.94}Sr_{0.06}Ti_{0.47}Zr_{0.53}O_3$ plus proprietary additions. This composition is tetragonal below the Curie point and is dielectrically "hard". The tubes were manufactured by cold isostatic pressing and sintering. All samples were cut from un-poled tubes with a 100 grit diamond cut-off wheel and subsequently surface ground to shape with a 100 grit diamond wheel. All grind-ing was carried out wet.

Samples were tested in liquid environments by immersing the DT loading jig in the fluid in a covered stainless steel chamber. The liquids used were water, toluene, Freon and mineral oil. The latter three are considered inert with respect to water. Freon, in addition, is a known corona suppressant. The stainless steel chamber had resistance heating elements attached to it to enable tests to be run at elevated temperatures.

Poled Specimens

The effects of poling on crack propagation were studied using the composite DT specimen shown in Figure 2. The specimen was pro-duced by first poling a 5 mm wide PZT beam and then cementing it between two wider unpoled PZT beams with high strength epoxy. A side groove was then machined in the poled section and the surfaces trued before testing. A narrow center section is necessary when testing poled specimens because the electrode spacing must be small to produce the high field. For poling a 35 KV/cm field was applied at 150°C for 2.5 min. The poling field was not applied for longer times due to the danger of microcracking the samples[2]. Composite DT specimens were fabricated from the poled material with the direction of polarization both parallel to and perpendicular to the crack plane as shown in Figure 2.

RESULTS

Compliance Measurements

The elastic analysis for a DT specimen is predicated by the

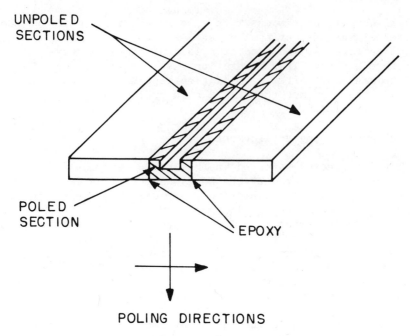

UNPOLED
SECTIONS

POLED
SECTION

EPOXY

POLING DIRECTIONS

Fig. 2. Schematic showing composite PZT double torsion specimen with poled center section cemented between two unpoled sections.

condition that the compliance of the system is a linear function of crack length. To this end the compliances of a number of uncut and composite specimens were measured as a function of crack length by introducing simulated cracks with a thin diamond slicing wheel. The results are shown in Figure 3. Since specimens had varying thicknesses, t, the compliances plotted in Figure 3 have been normalized by t^3 according to Equation 3. The normalized compliance of both the uncut and composite specimens is seen to be a linear function of the crack length, a. It is significant to note that cutting and rejoining the specimens did not alter the compliance calibration.

The theoretical compliance of a DT specimen is given by[11]:

$$C = \frac{y}{p} \approx \frac{3W_m^2 a}{Wt^3 G} \tag{3}$$

where Y is the deflection, P is the load and G is the shear modulus. The other terms are defined in Figure 1. If we let $G = C_{44}^E =$

$2.56 \times 10^4 \text{MNm}^{-2}$, i.e. the short circuit value of the shear modulus of Sr modified PZT[1] then

$$t^3 \frac{dc}{da} = 8.76 \times 10^{-5} \frac{m^3}{MN} \left(5.1 \times 10^{-8} \frac{inch^3}{16}\right).$$

This is nearly identical to the experimental slope of the line shown in the Figure indicating that the elastic analysis[11] can be applied with confidence in this case to both the uncut and the composite specimens.

Specimen Geometry

    Before the crack propagation data are presented a comment concerning the geometry of the DT specimen is in order. Other experimenters have noted that in some cases the load relaxation during a DT test can be discontinuous resulting in discontinuous load-time curves[13,14]. Early in this study we found that samples with relatively narrow (i.e. .75mm) side grooves also exhibited discontinuous load relaxation curves. Furthermore, it was found first that the load during relaxation would frequently arrest at values higher than those expected from previous results and secondly K_{IC} measured by retesting these specimens to failure was higher than expected. Examination of the specimens exhibiting these effects indicated that during relaxation the crack had wandered to the corner of the side groove where it had been held up. Discontinuous relaxation curves resulted because the crack would sometimes break away and start running again. If the crack trapped itself at the corner of the side groove, it would fan out and increase in area. Presumably this accounts for the anomalously high measured values of K_{IC}.

    These problems have been solved by increasing the width of the side groove from 0.75 mm to 3.25 mm. With the wider side groove the crack can wander off the centerline of the specimen but will turn in the stress field back towards the centerline before it intersects the side of the groove. As a result the crack remains perpendicular to the bottom of the specimen and propagation occurs in a geometrically stable manner resulting in smooth load relaxation curves.

    In a DT specimen the crack should be at least as long as one-half the specimen width to avoid end effects[11,14,15]. In the measurements reported here it was generally observed that after pop-in the initial cracks were about 1.3 cm (i.e. 0.5 W) long.

Environmental effects on slow crack growth in unpoled PZT

    Slow crack growth curves for PZT tested in a number of environments are shown in Figure 4. In the Figure the logarithm of the crack velocity, V, is plotted as a function of the logarithm of the applied stress intensity, K_I. The individual points on the Figure

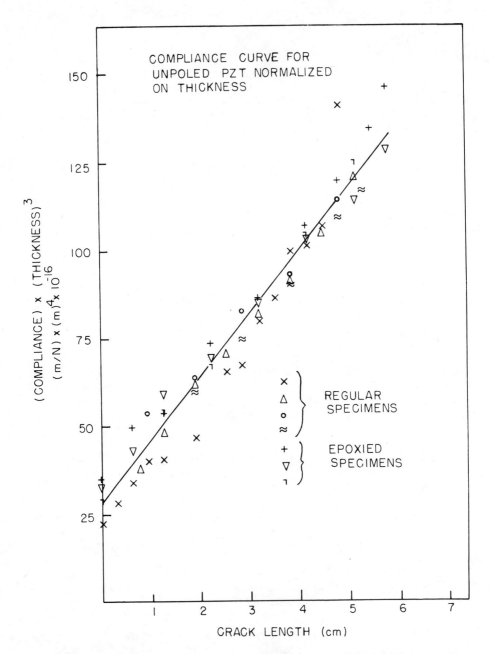

Fig. 3. Compliance of regular and composite PZT double torsion
specimens plotted as a function of crack length. The compliance
has been normalized by t^3 where t is the specimen thickness.

are plotted at intervals dictated by the computer program used to
analyze the load relaxation curves and do not represent actual data
points. By plotting the data in this manner scatter in the data
due to minor irregularities in the load relaxation curves is
readily visualized. The reproducibility of the V-K_I curves shown
in Figure 4 is quite good. Data from 5 tests in water, 5 tests in
toluene, 3 tests in Freon and 2 tests in mineral oil are plotted.[8]
Also included are V-K_I data for PZT 5800 taken by Freiman et. al.
using the constant moment technique[16]. The fracture toughness,
K_{IC}, of this material is low with respect to that of other oxide
ceramics such as alumina and is about the same as that for soda-
lime-silicate glass (i.e., .75 $MNm^{-3.2}$)[7].

The data can be conveniently discussed in terms of the three-
stage V-K_I curves exhibited by many materials[7,12]. In PZT water
is seen to measurably enhance subcritical crack propagation as it
does in other crystalline ceramics and glasses[7,8,10,12,16,17].
The data for water are in Stage I of the V-K_I curve and can be
represented by $V = AK_I^n$. The line drawn on Figure 4 through the
water data shows n is on the order of 55. The water data agree
reasonably well with that of Freiman et. al.[8]. Water free
environments such as toluene, Freon and mineral oil act to retard
slow crack growth. The toluene data exhibit all three stages of
the V-K_I curve while only Stage III crack propagation is observed
in the velocity range examined when tests were run in Freon and
mineral oil. The slope of the V-K_I curve in Stage III is about
130.

Slow crack growth in PZT can also be material dependent. Data
taken on PZT from another batch of the same material showed similar
effects of environment (i.e. the same n values and relative positions
of the V-K_I waves) but the data were shifted by about 0.5 $MNm^{-3/2}$
to higher values of K_I. Thus when the sometimes subtle effects of
testing environment on slow crack growth are examined it is important to use material from the same batch.

Effects of temperature on crack propagation

In order to examine the effects of temperature on slow crack
propagation in PZT and to determine if the kinetics of crack growth
could be described by some thermally activated process, V-K_I curves
were determined on unpoled samples tested in distilled water at 0°,
25°, 50° and $75^\circ C$. All the samples for each series of tests were
cut from the same transducer tube to reduce sample to sample variation. V-K_I curves for the samples tested in water are shown in
Figure 5. At least two runs at each temperature are plotted. The
data for $V > 10^{-6}$ m/s were fitted to $V = AK_I^n$ with the least squares
approach and the values of the exponents are indicated on the
Figure. We note that the crack velocity increases with increasing

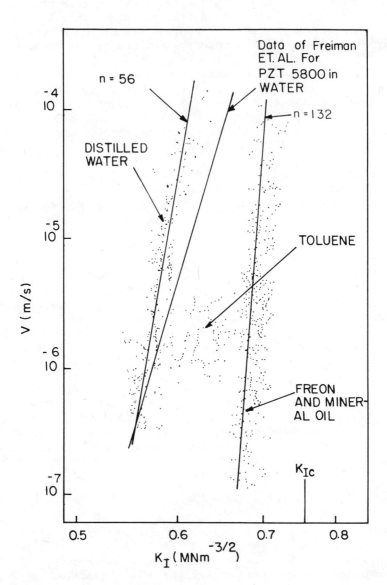

Fig. 4. Crack velocity in unpoled PZT measured as a function of
applied stress intensity in a number of environments at room
temperature. The fracture toughness, K_{IC}, is also indicated on
the Figure.

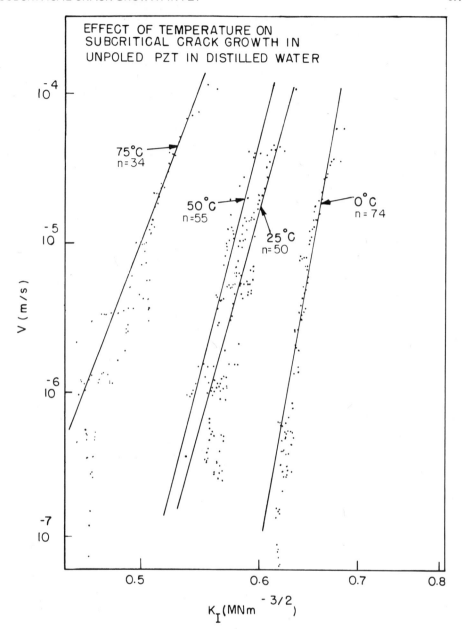

Fig. 5. Effect of temperature on crack propagation in unpoled PZT measured in distilled water.

temperature. If slow crack growth in PZT is thermally activated and if a single process is rate controlling, the crack velocity can be

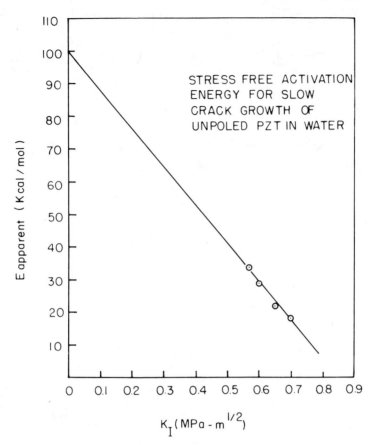

Fig. 6. Apparent activation energy for crack propagation in un-poled PZT in water plotted as a function of applied stress intensity.

written as

$$V = V_o \exp(-E* + \alpha K_I)/RT$$

where $E*-\alpha K_I$ is the apparent activation energy for fracture. $E*$ is the stress free activation energy and V_o and α are constants[19]. The activation volume is contained in α. The apparent activation energy calculated from the data of Figure 5 is plotted as a function of the applied stress intensity in Figure 6. $E*$ is approximately 100 kcal/mole.

Effects of poling on crack propagation in PZT

The effects of testing environment on slow crack growth of PZT poled perpendicular to the crack (see Figure 2) are shown in Figure 7. Included on the Figure are data taken in distilled water (4 tests), toluene (3 tests) and mineral oil (2 tests). Comparing

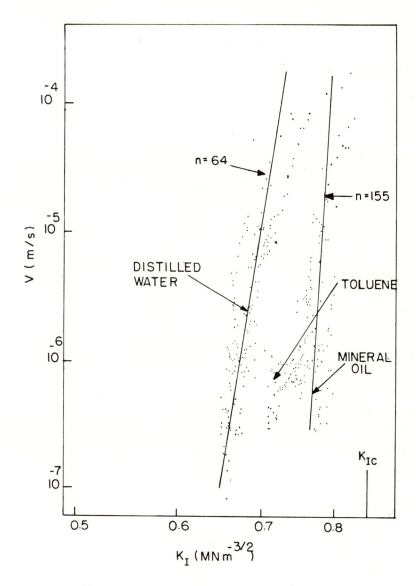

Fig. 7. The effect of environment on crack propagation in PZT poled perpendicular to the crack. Poling has shifted the curves to higher values of K_I.

Figures 4 and 7 we note that the major effect of poling on environmentally sensitive slow crack growth in PZT is to displace the V-K_I
curves by about 0.1 MNm$^{-3/2}$ to higher K_I values. The slopes of the
curves in Stage I (i.e. water) and in Stage III (i.e. mineral oil
and toluene) are only slightly higher in the poled material. Samples
poled parallel to the crack plane were only tested in distilled
water. The range of these data along with those for other material
tested in water at ambient temperature are shown in Figure 8. Note
that poling parallel to the crack had little effect on the V-K_I
curves.

Crack profiles, fracture surfaces and microstructures

As a result of the loading geometry in a double torsion test the
crack profiles are curved as shown schematically in Figure 9[7,10,11].
If crack propagation is assumed to be orthogonal to the crack front
(this has not been experimentally verified for PZT) the crack velocity given by Equation 2 is too high but can be approximately
adjusted by the correction factor $\phi = d_n/(c^2+d_n^2)^{\frac{1}{2}}$ [10,11]. Photographs of two crack profiles in PZT are also shown in Figure 9b and
9c. The shape of the arrested crack can be seen following a high
loading rate test to failure. A dotted line has been drawn on the
photos of the top specimen half in Figures 9b and 9c to better reveal the crack profile. Based on these observations and others
$\phi \approx 0.45$. The data presented in this paper have all been corrected
by that amount.

Close examination of some of the crack profiles showed them to
meet the bottom surface at an angle greater than 90°. An example
is shown in Figure 9c. This is attributed to the presence of
grinding induced compressive stresses in the near surface layers of
the specimen. Nadeau[20] has observed similar profiles in vitreous
carbon.

A scanning electron micrograph of the fracture surface of an
unpoled PZT specimen is shown in Figure 10a. With exception to
regions where the crack has intersected pores and pore clusters,
crack propagation has occurred predominantly by transgranular
fracture. Micrographs taken in regions of both slow and fast
fracture revealed no difference in fracture appearance. The fracture
surface of a specimen poled perpendicular to the crack is shown in
Figure 10b. Fracture is again transgranular and, in general, the
surface is quite similar to that of the unpoled specimen. The only
notable difference is the appearance of some parallel arrays of
markings visible on the surface. Presumably the markings are due
to the intersection of the crack with domain boundaries introduced
by poling. Similar features were observed on the fracture surfaces
of samples poled both parallel to (see Figure 10c) and perpendicular to the crack plane. No obvious differences were noted on the
fracture surfaces that could account for the shift in the V-K_I
curves to higher K_I when poling was perpendicular to the crack.

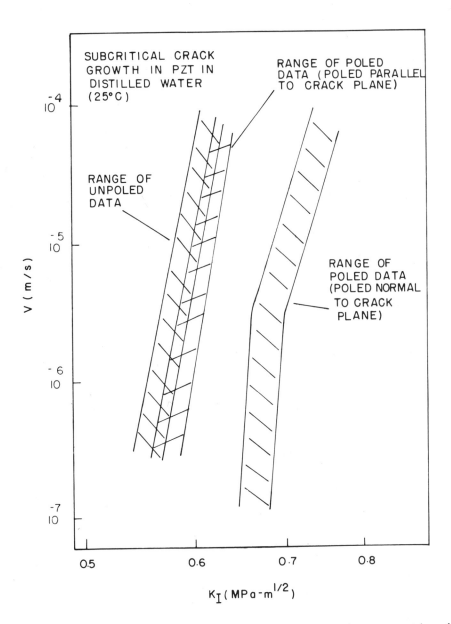

Fig. 8. Plot showing the effect of poling on crack propagation in
PZT in distilled water at 25°C. Note that poling parallel to the
crack has little effect on crack propagation while poling perpendicular to the crack measurably retards crack propagation.

DISCUSSION OF RESULTS

Specimen Design

It is significant to note that problems due to intermittent crack propagation during a DT relaxation test can be essentially eliminated by using a specimen with a wider side groove. By widening the groove the stress concentrations at the corners are removed to a distance from the propagating crack where they do not interact with the stress field of the crack. The crack geometry is them dominated by the outer fiber tensile stresses in the specimen and the crack, even though it may wander from the specimen center, is turned back towards the center and remains perpendicular to the specimen surface throughout a test. We suggest that this aspect of the DT specimen geometry can increase the "yield" of valid DT tests and should be considered by others using the test. Our experimental results and the analytical results of Trantina[15] further show that the starter crack should be at least as long as the specimen half width, $W/2$; otherwise erroneously high values of K_I will be measured. If possible specimens should be used that are longer than the ones employed in this study (i.e. $3W$) to enable multiple relaxations to be performed. Multiple relaxations must give repeatable data if confidence in the results is to be expected.

Environmental effects on crack propagation in unpoled PZT

The data shown in this report and by others[8,9] on the effects of environment on crack propagation in PZT conclusively show that the material is susceptible to environmentally enhanced slow crack growth. Water appears to be the active medium. As shown in Figure 4, the V-K_I curves[7] of PZT exhibit the three stages characteristic of other ceramics. When the data are fitted to $V = AK_I^n$, n is about 55 in Stage I and about 130 in Stage III. The n values are slightly higher than those determined by Freiman, et. al.[8] for a PZT 5800 and are essentially identical to those found by Caldwell and Bradt[9] on another commercial PZT tested in air. This is an interesting point since the fracture mode was intergranular in Caldwell and Bradt's material and was transgranular in ours and in that of Freiman et. al. From the V-K_I data shown in Figures 4 and 7 we further note that the operating lifetime of a component can be significantly extended if the service environment is a medium that is inert with respect to water. Stage I crack growth was essentially eliminated when tests were run in mineral oil and Freon.

Recent work has shown that the lifetimes of structural ceramic components can generally be predicted from V-K_I data using proof testing techniques[7,12]. In the case of transducer ceramics, however, a note of caution is in order. Transducers generally operate under cyclic loading conditions and data such as that shown in Figures 4 and 6 were obtained under "static" conditions. Evans

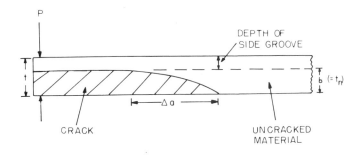

(a)

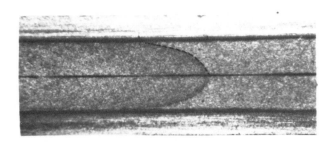

(b)

(c)

Fig. 9. The crack profile in PZT double torsion specimens. (a)
schematic (b) normal crack profile (c) crack front intersecting
the bottom surface at angle greater than 90°.

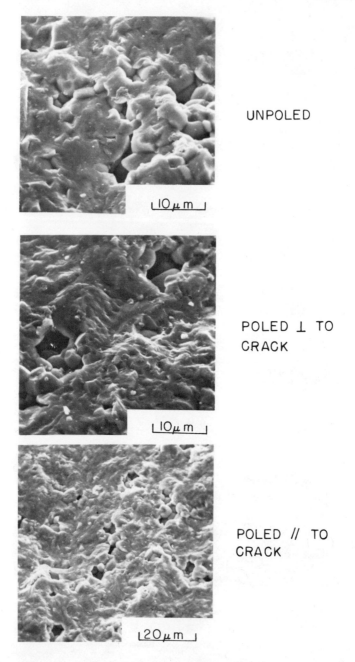

UNPOLED

POLED ⊥ TO
CRACK

POLED // TO
CRACK

Fig. 10.   Scanning electron micrographs of the fracture surfaces
of (a) unpoled PZT (b) PZT poled perpendicular to crack (c) PZT
poled parallel to crack.   Fracture is predominantly transgranular in all cases.

and Fuller[21] have shown that $V-K_I$ curves under cyclic loading can
be predicted from static $V-K_I$ data providing the characteristics of
slow crack growth are not altered by the loading. It is not clear
if this is the case in a transducer component. In an operating
transducer, other effects such as that of a high applied field must
also be considered. Preliminary studies[22] have shown, in fact,
that $V-K_I$ data obtained with a.c. fields applied to a crack propa-
ting in PZT are not the same as that predicted from the static $V-K_I$
curves using the analysis of Evans and Fuller[21].

Crack Propagation in Poled PZT

Poling caused a measurable increase in the stress to propagate a
crack when the poling direction was perpendicular to the crack plane.
Only slight shifts in the $V-K_I$ curves were noted when the poling
direction was parallel to the crack plane. This behavior can be qualitatively accounted for by residual stresses introduced during poling.
Since the PZT used in this study is predominantly tetragonal below the
Curie temperature, polarization results from the switching of 90° and
180° domains[1]. Only 90° domains, however, contribute to the strain
accompanying polarization. When polycrystalline PZT is poled, 90°
domain rotation tends to deform each grain in the polarization
direction. The deformation is, in turn, restricted by the surrounding
grains and intergranular residual stresses result. One contribution
to aging* in piezoelectric ceramics is, in fact, back switching of the
90° domains due to residual poling stresses[1]. The residual stresses
(and remanent strains) are highest in the poling direction. The stresses
are compressive and hinder crack propagation when the poling direction
is perpendicular to the crack. A compressive stress parallel to the
crack is not expected to interact strongly with the stress field at the
crack tip providing a maximum principal stress fracture criterion is
controlling. Thus, cracks propagating in samples poled parallel to
the crack plane show $V-K_I$ behavior similar to unpoled samples.

The magnitude of the residual stress introduced by poling can be
estimated from the shift, ΔK_I, in the $V-K_I$ curves by considering how
the residual stress alters the stress field of the crack. The maximum
tensile stress in front of a Mode I crack is

$$\sigma_{yy} = \frac{K_I}{\sqrt{2\pi r}}$$

where r is the distance from the crack tip[23]. The stress in the

---

*Aging is the time dependent degradation of the dielectric properties of
a ferroelectric material following poling, thermal treatments or stress
applications.

presence of a residual stress, σ_r, is

$$\sigma_{yy}' = \frac{K'_I}{\sqrt{2\pi r}} - \sigma_r .$$

Fracture occurs in both cases when $\sigma_{yy} = \sigma_{yy}' = \sigma_F$ where σ_F is the fracture stress.   Thus

$$\sigma_r = \frac{K'_I - K_I}{\sqrt{2\pi r}} = \frac{\Delta K}{\sqrt{2\pi r}} .$$

At distances from the crack tip corresponding to realistic flaw sizes in PZT eg. 10 to 50 μm, the residual stress is about 6 to 13 MNm^{-2} which is reasonable.

An alternate explanation of the effects of poling on crack propagation is based on the interaction of a crack with the differ-ent microstructures of poled and unpoled material. Gerson[24] has shown, for instance, that the density of both 90° and 180° domain boundaries in PZT decrease drastically after poling. The resulting "coarser" microstructure could lead to higher fracture energies. The removal of domain boundaries upon poling does not, however, account for the anisotropy observed in crack propagation rates with respect to poling direction. The fact that poled and unpoled material exhibited V-K$_I$ curves with similar shapes* when tested in a number of environments also indicates that the presence (or absence) of domain boundaries plays a minor role in the fracture process in PZT[25].

Fracture Mechanism

As shown in Figure 6, the apparent activation energy for slow crack propagation in unpoled PZT in water is a linear function of K$_I$ and extrapolates to a stress free activation energy of about 100 kcal/mol. If the bond energy is defined as the oxide dissocia-tion energy divided by the coordination in the Perovskite structure[26] the following Table indicates that the measured value is about 30 percent higher than either the Zr-0 or Ti-0 bond energies which, in PZT, are presumably rate controlling.

Other rate controlling mechanisms might be associated with the stress-enhanced diffusion of some species near the crack tip. How-

---

*but were shifted to higher values of K$_I$ when poling was perpen-dicular to the crack.

Cation	Coordination Number in PZT	Oxide Dissociation Energy Kcal/mol	Bond Energy Kcal/mol
Pb	12	145	12
Ti	6	435	72
Zr	6	485	81

ever, the observed stress free activation energy is too high for reasonable candidates. Furthermore, if stress enhanced diffusion were important, one would expect the samples poled parallel to the crack to exhibit $V-K_I$ behavior different from that of unpoled material. That is, since V can be related to the pressure tensor gradient driving the "embrittling species" by[27]

$$V \alpha \frac{d\sigma_{ii}}{dx} \text{ and } \sigma_{ii} = \sigma_x + \sigma_y + \sigma_z \qquad (4)$$

if σ'_x is decreased as a result of the poling induced residual compressive stress then the velocity should decrease. Since this is not the case, a normal stress criterion for the cohesive bond is much more realistic in the present case.

Clearly more activation energy measurements are needed both in inert environments and in poled material before fracture in PZT is better understood. A major experimental problem in studying thermally activated crack propagation in PZT however is that, while the test temperatures are low with respect to the melting point, they are high with respect to the Curie temperature and close to the morphotropic boundary. Any stress and/or temperature induced structural changes complicate the application of a simple thermal activation analysis.

Two other aspects of fracture in PZT deserve mention. These are the possible existence of a deformed anelastic zone at the crack tip resulting from the stress induced migration of 90° domain boundaries[1,28] and the resulting electrostrictive charge buildup. Both will tend to hinder crack propagation but, since they are stress dependent, should not be reflected in the stress free activation energy. If the dependence of these effects on stress is not linear, however, a linear extrapolation to determine the activation energy is not justified.

ACKNOWLEDGEMENTS

The authors are indebted to Drs. R. J. Stokes and R. G. Johnson for helpful and critical discussion and to S. J. Tibbetts for experimental assistance during the course of this study. The continued interest of Dr. A. N. Diness, Office of Naval Research, is gratefully acknowledged.

REFERENCES

1. B. Jaffe, W. Cook Jr. and H. Jaffe, Piezoelectric Ceramics, Academic Press, New York (1971).

2. B. K. Molnar and R. W. Rice, Bull. Am. Ceram. Soc., 52, 505 (1973).

3. R. C. Pohanka, et. al., Proc. Workshop on Sonar Transducer Mater., P. L. Smith and R. C. Pohanka, ed., Naval Res. Lab (1976), p. 205.

4. R. W. Rice in Fracture Mechanics of Ceramics, Vol. 1, R. C. Bradt, D. P. H. Hasselman and F. F. Lange, ed., Plenum Publishing Co., New York (1974) p. 323.

5. R. G. Johnson, Honeywell Corporate Material Sciences Center, private communication (1977).

6. R. C. Pohanka, S. W. Freiman and B. E. Walker, Bull. Am. Ceram. Soc., 56, 291 (1977) (abstract only).

7. A. G. Evans and T. G. Langdon, Prog. Mat. Sci., 21, 171 (1976).

8. S. W. Freiman, K. R. McKinney and H. L. Smith, in Fracture Mechanics of Ceramics, Vol. 2, R. C. Bradt, D. P. H. Hasselman and F. F. Lange, ed., Plenum Publishing Co., New York, (1974) p. 659.

9. R. F. Caldwell and R. C. Bradt, J. Am. Ceram. Soc., 60, 168 (1977).

10. A. G. Evans, J. Mater. Sci., 7, 1137 (1972).

11. D. P. Williams and A. G. Evans, J. Testing and Eval., 1, 264 (1973).

12. S. M. Wiederhorn, p. 613 in reference 7.

13. P. H. Hodkinson and J. S. Nadeau, J. Mater. Sci., 10, 846 (1975).

14. B. J. Pletka, National Bureau of Standards, private communication (1977).

15. G. G. Trantina, to be published in J. Am. Ceram. Soc. (1977).

16. S. W. Freiman, D. R. Mulville, and P. W. Mast, J. Mater. Sci., 8, 1527 (1973).

17. R. W. Adams and P. W. McMillan, J. Mater. Sci., 12, 643 (1977).

18. A. G. Evans and E. A. Charles, J. Am. Ceram. Soc., 59, 371, (1976).

19. W. W. Gerberich and M. Stout, J. Am. Ceram. Soc., 59, 222 (1976).

20. J. S. Nadeau, J. Am. Ceram. Soc., 57, 303 (1974).

21. A. G. Evans and E. R. Fuller, Met. Trans., 5, 27 (1974).

22. R. G. Johnson, S. J. Tibbetts, J. G. Bruce and B. G. Koepke, Bull. Am. Ceram. Soc., 56 302 (1977) (abstract only).

23. G. R. Irwin and P. C. Paris in Fracture an Advanced Treatise, Vol. III, H. Liebowitz ed., Academic Press, New York (1971) p. 1.

24. R. Gerson, J. Appl. Phys., 31, 188 (1960).

25. R. C. Pohanka, Naval Res. Lab., private communication (1977).

26. K. H. Sun, J. Am. Ceram. Soc., 30, 277 (1947).

27. W. W. Gerberich, Y. T. Chen and C. St. John, Met. Trans., 6A, 1485 (1975).

28. E. C. Subbarao, M. C. McQuarrie and W. R. Buessem, J. Appl. Phys., 28, 1194 (1957).

SUBCRITICAL CRACK GROWTH IN ELECTRICAL PORCELAINS

M. Matsui, T. Soma and I. Oda

Research and Development Laboratory

NGK INSULATORS, Ltd., Nagoya, Japan

ABSTRACT

The subcritical crack growth in porcelains was studied in various environments by using the double torsion technique at constant load. The obtained K_I-V data exhibited typical three stage curves, consisting of regions I, II, and III. The K_I-V relation in region I was not extended to the low stress region, where the crack growth rate was below 10^{-6} m/sec and the crack decelerated and stopped with time. This phenomenon was interpreted in terms of the stress corrosion limit and the heterogeneous microstructure which pins up the crack. The lifetime prediction of porcelains was discussed on the basis of the obtained data, and the resistance to delayed failure of porcelains was compared with that of soda-lime silicate glass.

INTRODUCTION

The subcritical crack growth in ceramics has been extensively studied, and the lifetime prediction of ceramic components has been discussed on the basis of the relation between the stress intensity factor (K_I) and the crack growth rate (V) [1]. This approach will be useful for the design, material selection and non-destructive inspection of porcelain insulators, which are required to endure high mechanical tension for a long time, about 20 years.

Slow crack growth data on the opaque materials such as porcelain have been collected mainly by using the constant displacement technique of DT (Double Torsion) method [2], where the measuring system is rather simple but the crack growth behavior under the constant K_I condition cannot be observed because the K_I at the crack tip changes with crack growth.

In this study, the K_I-V relations of the typical high strength porcelains were measured in various environments to investigate the precise behavior of the crack growth in porcelains having complicated microstructure by using the constant load technique of DT method, where the K_I at the crack tip is independent of crack length. The obtained data were used for the lifetime prediction of porcelains.

EXPERIMENTAL

Materials and Environments

Testing was conducted on three typical porcelains and a commercial soda-lime silicate glass in air with relative humidity of 60 to 80 %, in distilled water, in kerosene with water content of 0.004 wt%, in aqueous NaOH solution of PH 13, and in aqueous H_2SO_4 solution of PH 1. In porcelain A, mullite, quartz and corundum particles were dispersed in glass matrix and the total content of crystalline phases was about 30 wt%. Porcelains B and C contained cristobalite in addition to the above three kinds of crystal and the total contents of crystalline phases were about 50 and 60 wt%, respectively. The particle size in each porcelain was smaller than 30 μm.

The specimen was machined to a rectangular plate 4 mm thick, 22 mm wide and 120 mm long. Grooves, 0.1 mm wide and 0.3 mm deep, were cut along the length on both the upper and lower surfaces so carefully that the positions of both grooves coincided with a discrepancy less than 50 μm. The discrepancy between them was one of the principal causes of the scatter of the data. A notch was cut into one end of the specimen, so that the crack could easily start.

Apparatus

The schematic of the constant load DT apparatus used in this study is shown in Figs. 1-(a) and (b). Fig. 1-(a) shows the side view of the apparatus. Fig. 1-(b) shows the magnified front view of the parts for distorting the specimen and measuring the distortion. A weight of 15 to 25 kg was hung from the loading frame and the small load change required for starting the crack and controlling the growth rate was accomplished through spring by moving the crosshead of the testing machine. The load was changed continuously without giving shock or vibration to the crack tip in

the specimen and the change of the load due to the crack growth in
the specimen or due to the change of the room temperature was
negligibly small in this system. This was significant when very
slow crack growth rate ($V<10^{-6}$ m/sec) was measured. The load
supplied to the specimen was measured using a loadcell located
between the crosshead and the fixer for the loading spheres. The
crack growth rate was determined by measuring the deflection of the
specimen by using a displacement transducer of unbonded type strain

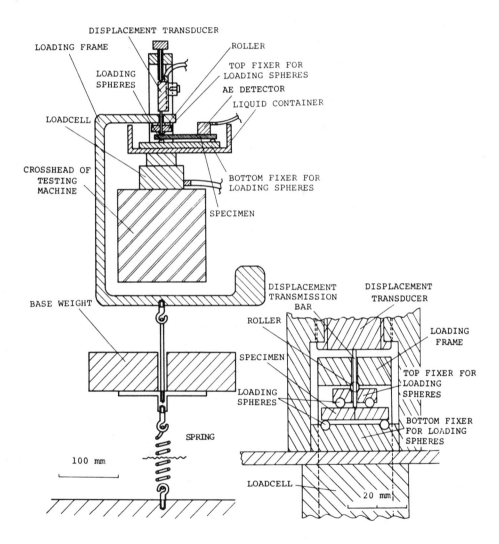

Fig. 1. Schematics of constant load double torsion apparatus
(a) side view. (b) magnified front view of the parts for distorting the specimen and measuring the distortion.

gauge which was attached to the specimen at the middle point between the upper loading spheres. The sensitivity of the displacement tranducer was 3/100 μm. The deflection of the specimen due to the crack growth was relatively small and the change of the room temperature caused considerable drift of the transducer signal. This drift resulted from the differences in thermal expansion between some mechanical parts in the measuring system. To diminish the drift, a dummy equipment with the same structure as shown in Fig. 1-(b) minus the loading system was placed near the main equipment and the difference in signals between the main and dummy transducers was recorded on a strip chart recorder. Room temperature was controlled at 25 ± 0.5 °C. Acoustic emission was measured during the crack growth rate measurement by attaching a detector to the end of the DT specimen as shown in Fig. 1-(a).

The stress intensity factor, K_I, and the crack velocity, V, were calculated by the following equations [3]:

$$K_I = PW_m [\frac{3(1+\nu)}{Wd^3 d_n (1-1.26\frac{d}{W})}]^{\frac{1}{2}} \quad (1)$$

$$V = \frac{\Delta y}{\Delta t} \cdot \frac{Wd^3 G (1-1.26\frac{d}{W})}{3PW_m \cdot W_n} \quad (2)$$

where P is the applied load, ν Poisson's ratio, G the shear modulus , $\Delta y/\Delta t$ the average displacement rate of the point where the transducer was attached, and the other symbols are given in Fig. 2. The crack growth rate in the porcelain fluctuated in spite of the constant load condition because of its heterogeneous microstructure. Therefore, the crack growth rate was defined as the average rate of the crack growth in 500 μm in this study.

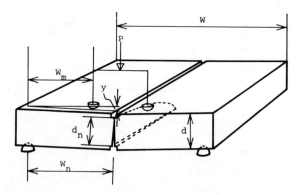

Fig. 2. Double tortion specimen

RESULTS

The K_I-V relations obtained for the porcelains in air, in distilled water and in kerosene are shown in Fig. 3. Each curve was determined from the data points of 50 to 200, some of which are shown in the figure to demonstrate the scatter of the data. These K_I-V relations exhibit the well-known three stage curves, consisting of regions I, II and III, and differ depending on the water

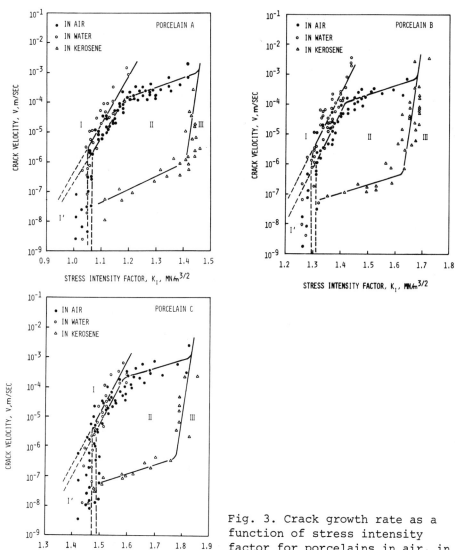

Fig. 3. Crack growth rate as a function of stress intensity factor for porcelains in air, in distilled water and in kerosene.

content in each environment. Time dependent crack growth behavior
was observed in the very slow crack growth region as shown between
two broken lines for each condition in Fig. 3, which is termed
region I'. The features of the crack growth behavior in region I'
were as follows:
1)   When K_I was decreased to a lower value in region I' (after the
crack propagated some length under the condition in region I), the
crack growth rate, at first, coincided to the value obtained by
extraporating the K_I-V relation in region I to region I'. The rate
then decreased gradually until the crack stopped. Some typical
examples of the change of the crack growth rate in region I' for
porcelain A are shown in Fig. 4. Below the critical stress (about
80 % of the K_I where the crack growth rate is 10^{-4} m/sec), the crack
growth under constant load was not observed. In the case of soda-
lime silicate glass, the change of the crack growth rate with time
was also not observed under constant load within the crack growth
rate measured in this study.
2)   When the crack was arrested in region I' and kept for a while,
the required load to restart the crack was larger than that
corresponding to the K_I at the boundary between regions I and I'.
The restarted crack gradually accelerated until reaching the value
given by the K_I-V relation in region I.
3)   Under the condition in region II in kerosene, the crack propagated at a constant rate of 10^{-8} to 10^{-6} m/sec, where the crack
stopped before propagating by about 1 mm in the environment of air
or water.

The K_I-V relation in region I could be fitted to the following
exponential or power functions:

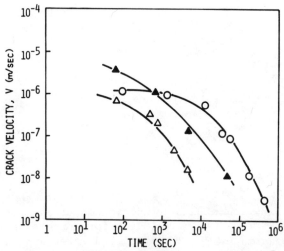

Fig. 4. Typical examples of the change of the crack growth rate
with time in region I' for porcelain A.

$$V = \nu_0 \exp \beta K_I \qquad (3)$$

$$V = AK_I^N \qquad (4)$$

where ν_0, β, A and N are constants. Those for the environment of air are shown in Table 1. The crack growth rate in distilled water is about 4 times higher than in air at the same K_I. Region II in the environment of water, which has been reported for porcelain by Evans and Linzer [4], was not observed in this study. The crack growth behaviors in aqueous NaOH solution and in aqueous H_2SO_4 solution are almost equivalent to that in distilled water within experimental uncertainty. The K_I-V relation curve for porcelains in each environment falls at a higher stress region as the content of crystalline phase in the body increases.

Table 1. Crack propagation data in region I in air (mks units)

porcelains	$V = \nu_0 \exp \beta K_I$		$V = AK_I^N$	
	ν_0	β	N	log A
A	1.6×10^{-22}	35 ± 5	40 ± 10	-247
B	3.7×10^{-26}	35 ± 5	50 ± 10	-311
C	1.4×10^{-28}	35 ± 5	60 ± 10	-375

Fig. 5 shows the fracture surface of the DT specimen for porcelains and the roughness of the fracture surface. The fracture surface for porcelain A is relatively smooth and that for porcelain B is relatively rough. Porcelain A has a high content of glass phase and porcelain B has a high content of crystalline phase, particularly cristobalite which induces localized residual thermal stress in the porcelain body.

Fig. 6 shows the crack growth behavior in region I in air for porcelains and for soda-lime silicate glass under the constant load condition with the result of acoustic emission measurements. Acoustic emission was not observed in soda-lime silicate glass during crack growth. The degree of the fluctuation of the crack growth rate corresponds to roughness of the fracture surface. The acoustic emission rate is almost in proportion to the crack growth rate.

Fig. 7 shows the comparison of the crack growth behavior in regions I, II and III in air. The crack growth curve in each region exhibits a characteristic feature almost independent of the experimental environment and the growth rate. In region II the crack growth curve is very smooth, in region III the crack growth exhibits sudden acceleration and deceleration, and in region I the fluctuation degree of the crack growth rate is between in regions II and III. This phenomenon can be related to the heterogeneity of the porcelain microstructure and the dependence of the crack growth

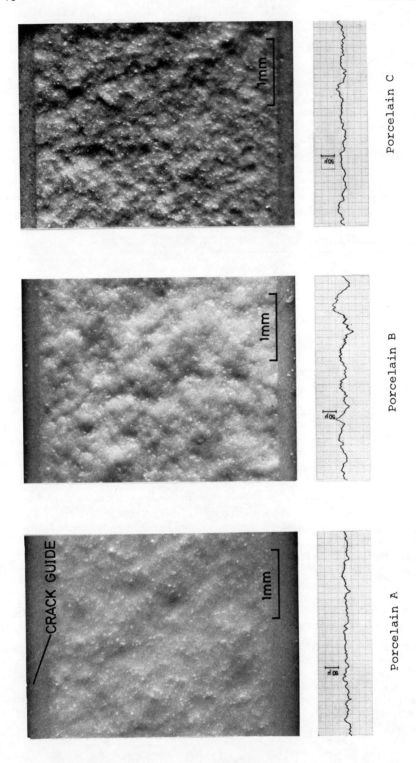

Fig. 5. Photographs and roughnesses of fracture surface for porcelains

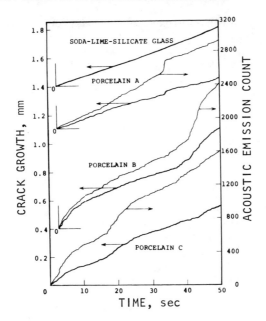

Fig. 6. Crack growth behavior and acoustic emission in region I for pocelains and soda-lime silicate glass.

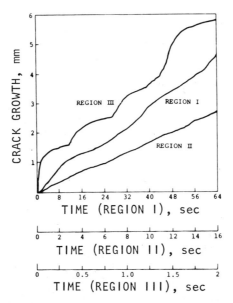

Fig. 7. Comparison of crack growth behavior in regions I, II and III in air for porcelain C.

rate on K_I. The former causes the fluctuation of the stress concentration at the crack tip. The latter is large in region III and small in region II as compared with that in region I, as shown in Fig. 3.

DISCUSSION

Stress Corrosion Limit

The crack growth behavior, observed in region I', is considered to indicate that the stress relaxation at the crack tip occurs with time when the crack growth rate is low. It is remarkable that the critical crack growth rate, below which the crack deceleration phenomenon occurs, is relatively high about 10^{-6} m/sec in porcelains, while in soda-lime silicate glass the stress corrosion limit appears at the crack growth rate about 3×10^{-10} m/sec [5] and no dependence of the crack growth rate on time is observed.

The stress relaxation at the crack tip in porcelain material may be caused by the stress corrosion limit and the heterogeneous microstructure which pins up the crack. The stress at the microscopic crack tip portion is largely distributed along the crack front because of the heterogeneity in the porcelain body as suggested by the roughness of the fracture surface in Fig. 5. The microscopic crack tip is considered to move in incremental bursts between succesive blocking positions influencing each other with adjacent crack tip portions as is indicated by the isolated event of acoustic emission from porcelain material [4]. The crack front will be blunted by mechanism of the stress corrosion limit [6] in the presence of water at some crack tip portions where the distributed stress is lower than that of the stress corrosion limit. Once the crack tip blunting is initiated, it proceeds with time and blocks the motion of the whole of the crack front by pinning up more strongly unless the blunted microscopic crack tip is resharpened by the driving force supplied by adjacent microscopic crack tip motions. Therefore, the crack front will be gradually decelerated when its average growth rate is low, while the crack front will propagate at a constant rate when the average growth rate is high enough not to allow crack tip blunting.

It is easily understood that additional load is required to restart the crack arrested by the above mechanism because the crack tip is rounded and its stress concentration is relaxed. The fact that the crack grows steadily at the rate from 10^{-8} to 10^{-6} m/sec in kerosene can be accounted for in terms of the crack growth mechanism in region II as follows. Even if the stress concentration at some crack tip portion is decreased by the heterogeneous microstructure of the porcelain body, the microscopic crack tip velocity does not

decreased below the critical value where the crack tip blunting
is initiated because the crack growth rate in region II is
principally determined by the diffusion rate of water [1].

E ffect of Environment

The experimental result in distilled water indicates that the
crack growth is accelerated about 4 times higher than that in air.
However, the difference in the stress intensity factor at the
equivalent crack growth rate in air and in distilled water is only
about 5 % of the instantaneous fracture stress. The decrease in
the mechanical strength of porcelain components to this extent is
not serious in practical use. On the other hand, the experimental
result in kerosene indicates that the slow crack growth at the rate
of 10^{-8} to 10^{-6} m/sec occurs in such an environment with very little
water.

The experimental results in aqueous NaOH solution of PH 13 and
in aqueous H_2SO_4 solution of PH 1 indicate that the PH of the
solution does not have a great effect on the mechanical properties
related to the subcritical crack growth in the velocity range
measured in this study.

Lifetime Prediction

Assuming that the K_I-V relation in region I is applicable in
the lower stress region and no subcritical crack growth occurs
during proof testing in the porcelain component which survives the
proof test, the minimum lifetime of the component after proof testing can be calculated to construct a design diagram using experimental data [7]. The minimum lifetime derived from equation (3) is
more conservative and suitable for safe design diagraming than that
derived from equation (4). The minimum lifetime, t_f, after proof
testing at stress, σ_p, is calculated by using equation (3) as
follows [7]:

$$t_f = \frac{2K_{IC}}{Y^2 v_0 \beta (\sigma_a)^2 x \exp(\frac{\beta K_{IC}}{x})} \tag{5}$$

where σ_a is the applied stress in service, K_{IC} is the critical
stress intensity factor and x is the proof test ratio, σ_p/σ_a.
Logarithm of the minimum lifetime after proof testing at various
proof test ratios for porcelains A, B and C versus applied stress
is plotted in Fig. 8. For example, the proof test ratio is determined as 2.0 from the lifetime diagram to assure the lifetime of
20 years when porcelain A is used under the applied stress of 100
kg/cm^2. Theoretically, the larger the proof test ratio, the larger
the assured minimum lifetime. However, with the excessive overload,
there is the danger that new flaws are introduced in the ceramic

components and remain after the proof test without leading the
components to complete fracture [8]. The slippage at the porcelain
/cement/metal interfaces is especially serious when the assembled
porcelain insulator is loaded. To avoid this trouble, it is
desirable that the proof test is conducted on each component
individually before assembling them, if not, the crack propagation
and slippage at the interface will be detected by the acoustic
emission measurment or another inspection technique.

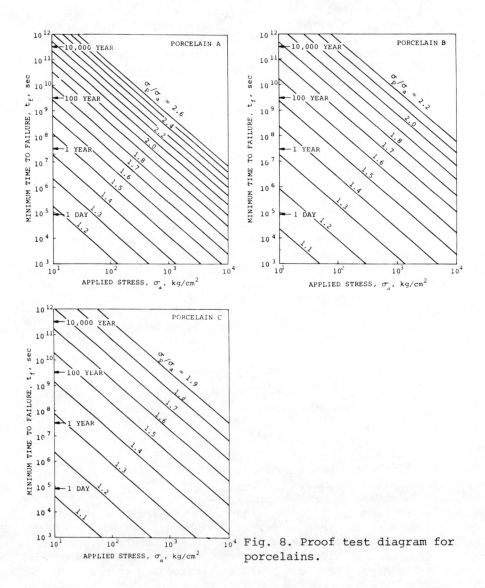

Fig. 8. Proof test diagram for porcelains.

It must be emphasized here that the propagating crack decelerates and stops with time under the slow crack growth condition and the stress corrosion limit further exists in porcelain. Therefore, the minimum lifetime in actual porcelain components could be much longer than that estimated on the assumption that the K_I-V relation in region I can be extended to the lower crack velocity region. The initial crack growth rate is estimated as an order of 10^{-15} m/sec for the delayed failure of 20 years when the lifetime is calculated by using equation (5). The direct measurement of the crack growth rate in this order is impossible to conduct by using the DT method. Another problem in predicting the lifetime of the ceramic components lies in the differences between the inherent cracks in the real porcelain component and the artificial cracks in the DT specimen. For example, the Griffith flaw in porcelains is estimated to be 50 to 200 µm in length and the crack growth behavior is influenced strongly by the heterogeneous microstructure of the body, in comparson to the DT specimen. In conclusion, the K_I-V relation determined by the DT expeiment should be examined by experimenting on the fracture initiated from the inherent crack and the information on the very slow crack growth (to about 10^{-15} m/sec) should be collected for the precise lifetime prediction of the porcelain component.

Comparison of Resistance to Delayed Failure between Glass and Porcelain

In Fig. 9. the characteristcs of the subcritical crack growth in water for porcelains are compared with that for soda-lime silicate glass which is the typical material used for the ceramic components. The danger of a delayed failure exists in the stress range between the stress corrosion limit and the instantaneous fracture stress, which is defined here as the stress producing the crack growth rate of 10^{-4} m/sec. The dangerous stress range of

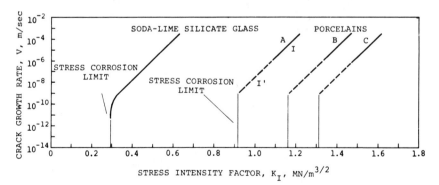

Fig. 9. Comparison of the resistance to delayed failure between soda-lime silicate glass and porcelains.

delayed failure for porcelains is above 80 % of the instantaneous
fracture stress even if in region I' delayed failure might occur.
On the other hand, the range for soda-lime silicate glass is above
50 % of the instantaneous fracture stress. Therefore, porcelain is
superior to soda-lime silicate glass in the reliability on the
resistance against delayed failure.

CONCLUSIONS

1)   The K_I-V relation in region I for porcelains is not extended
to the low stress region, where the crack growth rate is below
10^{-6} m/sec and the crack decelerates and stops with time. This
phenomena may be attributed to the stress corrosion limit and the
heterogeneous microstructure which pins up the crack.
2)   A conservative failure prediction is posible on the basis of
the K_I-V relation (10^{-9}<V<10^{-2} m/sec) obtained by the DT experiment.
However, to predict the precise failure time and to determine the
practical safe working stress of the porcelain components which
are required to endure high mechanical tension for a long time
about 20 years, the behavior of the inherent crack with a very
slow crack growth rate of about 10^{-15} m/sec should be further
investigated.
3)   Porcelain was cofirmed to have the stress corrosion limit at
a high stress level of more than 80 % of the instantaneous fracture
stress and to be superior to soda-lime silicate glass in the
reliability on the resistance against delayed failure.

REFERENCES

1.   S. M. Wiederhorn, pp 613-46 in Fracture Mechanics of Ceramics
     Vol. 2. Edited by R. C. Bradt, D. P. H. Hasselman and F. F.
     Lange, Plenum, New York, 1974.
2.   D. P. Williams and A. G. Evans, J. Test. Eval., 1, 4, 264 (1973)
3.   K. R. Mckinney and H. L. Smith, J. Amer. Ceram. Soc., 56, 1,
     30 (1973)
4.   A. G. Evans and M. Linzer, J. Amer. Ceram. Soc., 53, 10, 543
     (1973)
5.   S. M. Wiederhorn and L. H. Boltz, J. Amer. Ceram. Soc., 53, 10,
     543 (1970)
6.   W. B. Hilling and R. J. Charles, pp. 682-705 in High Strength
     Materials. Edited by V. F. Zackey. John Wiley & Sons Inc.,
     New York, 1965.
7.   S. M. Wiederhorn, J. Amer. Ceram. Soc., 56, 4, 227 (1973)
8.   A. G. Evans, S. M. Wiederhorn, M. Linzer and E. R. Fuller, Jr.
     Ceramic Bulletin, 54, 6, 576 (1975)

FRACTURE MECHANICS OF ALUMINA IN A SIMULATED BIOLOGICAL
ENVIRONMENT

E. M. Rockar and B. J. Pletka

Institute for Materials Research
National Bureau of Standards
Washington, D. C. 20234

ABSTRACT

Dense, polycrystalline alumina, because of its high compressive strength and excellent biocompatibility, will soon be used in human prosthetic devices. However, the environmental effects of the biological system on the long term strength and wear of alumina have not been evaluated. As a preliminary investigation of this problem, crack propagation data on nine types of highly dense alumina were obtained in environments of distilled water, and Krebs-Ringer solution. The data were obtained primarily by strength measurements using a biaxial tension test. Variations in crack propagation data were observed among the different materials. The causes of these variations are discussed in terms of microstructure and environment.

1. INTRODUCTION

High purity, fine grained, low porosity alumina is now being used as component material in artificial hip joints. As the "ball and socket" part of a total hip prosthesis, the alumina can endure the high compressive loads encountered in the hip joint (1). Alumina has approximately 2 to 3 times the compresssive strength of living bone. Because of its excellent biocompatibility, alumina is also being considered as the material for ankle, shoulder, elbow, and finger joint prostheses (2). In these applications the alumina could experience tensile stresses up to $\sim$200 MPa as well as cyclic loads for extended periods. Unlike natural bone, however, which is self-healing, alumina under stress, will undergo slow crack growth.

725

The goal in prosthetic surgery is to leave these devices implanted for 20 or more years. Therefore, careful design criteria and material characterization must be developed before these devices can be implanted in humans. It is especially necessary to assess the importance of slow crack growth and the critical flaw size as a function of microstructure in these materials, and from this information, to develop proof testing criteria.

Wiederhorn, Tighe, and Evans have reviewed the techniques which have recently been developed to assure the reliability of ceramic components in structural applications (3). Time-to-failure design diagrams can be derived for a given material based on either fracture mechanics or strength techniques, once the strength of the component in an inert environment is known. Time-to-failure in a static fatigue test is given by

$$t = A \, \sigma^{-n} \qquad (\sigma \text{ constant}). \qquad (1)$$

The strength in a dynamic fatigue test is likewise given by

$$\sigma = B \, \dot{\sigma}^{\,1/(n+1)} \qquad (\dot{\sigma} \text{ constant}). \qquad (2)$$

For a given material and environment, n is a material constant which is independent of the surface treatment of the specimen or the specimen history. The larger the n-value, the greater the resistance to crack growth. On the other hand, A and B depend on the specimen history and surface conditions. A and B are determined when the strengths in an inert environment of both laboratory specimens and components are known. Time to failure diagrams can be used to determine the proof test ratio which will guarantee a minimum lifetime for a given service stress (3). Applying these techniques, Greenspan, et al., have compared alumina with bioglass coated alumina, another material which may be used for prosthetic implants (4).

This paper compares values of n ("the crack susceptibility") which were determined from breaking strength as a function of stressing rate for nine samples of high density alumina. These nine samples represent a wide range of grain and pore size. Since the service environment for prosthetic implants might affect the crack growth rate, three of the samples were tested in simulated biological environment. Krebs-Ringer solution (a buffered salt solution, ph=7) and bovine serum were used at room temperature (∼23°C) to simulate body fluids.

2. MATERIALS

The alumina samples used in this study were obtained from six suppliers and represent nine different sample materials. The characteristics and method of preparation of the seven samples labelled M1-M6 and M8 are described in a paper by Wachtman et al., (5). The samples M9 and M10 were obtained from a seventh supplier, whose method of preparation was not specified. The densities of all samples was 3.70 gm/cm^2 or greater.

All of the specimens were disk shaped, approximately 1 mm thick and 32 mm in diameter. The surface characteristics of the disks varied between sample sets. Some of the disks were punched out, some cut from rods, with no additional surface finishing. Other samples were ground to a given thickness after being cut into disks. For this study the samples were not categorized according to their surface characteristics. In one case, M5, one side of each specimen had been polished. The polished sides of specimens M5 were subjected to tensile loading in the test jig for this study.

Figure 1 contains photomicrographs of four of the nine sample materials studied. The average grain size of M8 and M9 were determined by the linear intercept method, which gave a value of 3.5 μm for M8 and 19 μm for M9. These specimens represent the extremes in grain size for the materials of this study. No attempt was made to quantify pore size although the larger grain material appeared to contain larger pores than the smaller grain material.

3. EXPERIMENTAL PROCEDURES

Strength as a function of stressing rate was determined for 50 specimens of each sample material. The disk-shaped specimens were mounted in a test fixture which provided maximum tensile stress on the lower face of the disk. The upper surface of the specimens was loaded with a small piston, while the disk was supported on three spheres placed equidistant on a 12.7 mm radius. The test fixture and its dimensions are described in detail by Wachtman et al. (5). The tensile stress in the center of the tensile surface of the specimens was calculated from an equation that was derived by Kirstein and Woolley (6). Ten specimens were broken at each of five stressing rates an order of magnitude apart.

All nine sample materials were loaded to failure in distilled water at room temperature. In addition M4, M5, and M8 were loaded to failure in Krebs-Ringer solution. M8 was also tested in bovine serum. All tests were conducted at room temperature.

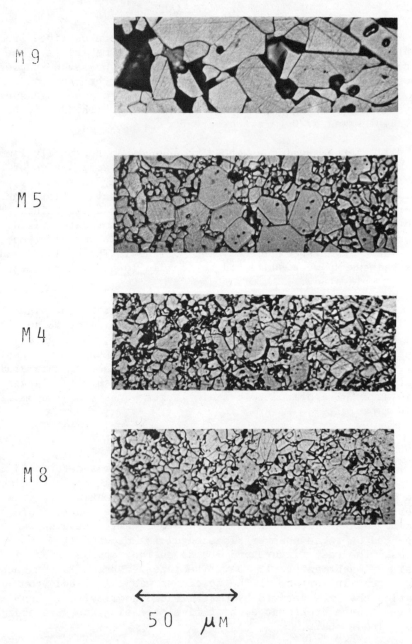

Figure 1 - Optical micrographs of four alumina sample materials.

4. DATA ANALYSIS

To extract a value of n from breaking strength vs. stressing
rate data Equation (2) was rewritten in the form,

$$\ln \sigma = \ln B + \frac{1}{n+1} \ln \dot{\sigma}, \qquad (4)$$

Two assumptions are made. The first assumption is that for the
range of stressing rates studied the crack velocity is in region
I of the K_I-V plot for this material. This assumption is
necessary since Equation 4 is valid when only one mechanism for
crack growth is dominant (7). The plot of ln σ vs. ln $\dot{\sigma}$ should
curve upward at higher loading rates as velocities approach
region II of the K-V curve (7).

The second assumption is that the specimens broken at each
loading rate have equal flaw size distributions. This means that
the distribution functions which characterize the breaking strengths
at each loading rate are the same. Ordering the breaking strengths
at each loading rate gives a means for plotting Eqn. (4) at equal
probabilities for failure. The median values therefore represent
a measure of the 50% failure probability level. When the number
of specimens broken is small and the breaking strength distribution
is not known *a priori*, the median is a better representation of
the 50% failure probability level than is the mean (8). To use
all of the data points, sets of equal probability breaking strengths
can be formed from the ordered data set.

Therefore, in addition to using the median values at each
loading rate to determine n from the slope, the slope can be
determined from each of the ordered sets. Values of ln σ and ln
$\dot{\sigma}/\dot{\sigma}_o$ were plotted for each set of data, where $\dot{\sigma}_o$ is the middle
stressing rate. A straight line was then fitted to each set using
a linear regression analysis. In this analysis it was assumed
that there was no error in the values for stressing rate.

Values of n were determined from the slope of the straight
lines. The correlation coefficient of each fit was also determined. Values of correlation coefficient close to 1.0 indicate
that the points are a "good" fit to a straight line. A composite
value of n for each sample was determined by calculating a weighted
average of the 10 values of n. The weights were determined by
the correlation coefficients of the fits using standard weighing
techniques (8).

5. RESULTS AND DISCUSSION

Table I lists values of n for the nine sample materials.
Column A lists values based on the median breaking strengths at

TABLE I

| Sample | A | | B | |
	n-value	Correlation coefficient	n-value	Correlation coefficient
M9	35	0.96	35 $\pm$ 1	0.97
M2	24	0.95	26 $\pm$ 2	0.92
M3	38	0.97	36 $\pm$ 5	0.91
M5	35	0.98	39 $\pm$ 6	0.91
M10	40	0.95	34 $\pm$ 6	0.90
M6	45	0.90	44 $\pm$ 6	0.83
M1	40	0.75	37 $\pm$ 4	0.82
M4	49	0.7	61 $\pm$ 9	0.6
M8	118	0.4	68 $\pm$ 9	0.6

each of the five loading rates. Column B lists values based on a weighted average of 10 determinations of n for each sample. These weighted averages were determined from the ordered breaking strengths at each loading rate as discussed above. It is noted that if the scatter in the data is small (i.e. the breaking strengths vs stressing rates gives a good fit to a straight line) then the median values (Column A) give as good a determination of n as is obtained by using weighted averages (Column B). However, when the median values do not give a good fit, the determination of n is not substantially improved by using the method of weighted averages. The samples are listed in Table I in order of the correlation coefficients of the fits. When listed in this manner the samples are also approximately in order of grain size. M9 has the largest grain size whereas, M8 has the smallest grain size.

    The dynamic fatigue data for M8 and M9 are given in figure 2. We note that although the same number of specimens were broken in both sets of tests, the scatter for M8 is considerably greater than for M9. This dependence of scatter on grain-size does not seem to hold for the other samples in this study. The data from sample M5 for example, also exhibited large scatter even though M5 contained some large grains. Therefore we cannot generalize to conclude that average grain-size alone is correlated

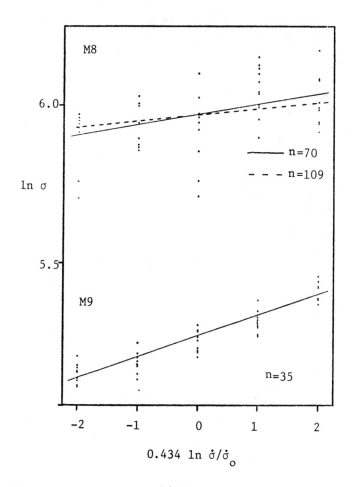

Figure 2 – Ln σ vs 0.434 ln $\dot{\sigma}/\dot{\sigma}_0$ where σ_0 is the middle stressing
          rate. Points represent data values. For sample M8,
          the dashed line is a best fit to the median values;
          the solid line is an average of ten linear regressions.
          For M9 the solid line is an average of ten linear
          regression using all the data points.

with the scatter in the dynamic fatigue data. The results
presented here do indicate, however that the microstructure has
an influence on the scatter in the data.

    Of the materials studied, samples M4 and M8 had the largest
n-values and the poorest correlation coefficients. All of the

other samples had approximately the same n-values and correlation
coefficients. Again, it is tenuous to conclude that small grained
materials have the larger n-values, especially, when their correlation coefficients are so poor. However, some microstructural
characteristic that is related to the grain size or the grain size
distribution appears to give rise to the scatter and the variation
in the n-values.

The generally poor fit of some of the data to Eqn. 4 can be
accounted for in several ways. First, the assumptions upon which
the derivation of Eqn. 4 is based may not hold for these materials.
We may not be in region I of the K_I-V plot and more than one
failure mechanism may be operating (intergranular and transgranular
fracture, for example). Each mechanism could have different
fracture kinetics, responding perhaps to a different flaw population. Two overlapping flaw distributions contributing to two n-
values would complicate the analysis.

The fit of the data depends on the amount of scatter. Scatter
can come from different flaw densities caused by differences in
surface preparation or from variations in microstructure. While
the surface treatment should not affect the n-value, it could
affect the scatter in the breaking strength data leading to a
statistical uncertainty in the value of n. Similarly, departures
from grain size uniformity influence the size of the critical
flaw, an effect which is more pronounced in a smaller grain size
material and which causes more scatter in its strength data than
in a large grain size material. In either case, more specimens
must be broken to reduce the statistical scatter.

A more complete investigation of these and other similar
materials first requires a careful characterization of the underlying flaw population which is contributing to failure. Furthermore, this paper shows that the statistical nature of each of the
fracture mechanics parameters determined in any experiment must be
clearly accounted for before any conclusions can be drawn.
Although it is indecisive, the present data does suggest that some
microstructural parameter other than grain size alone is responsible for the differences in n-values calculated for these materials.

Figures 3 through 6 compare the breaking strength vs loading
rate for specimens broken in water with those broken in Krebs-
Ringer solution. For M4, M5, and M8, there does not seem to be a
change in n-values, M4 has a larger n-value when broken in Krebs-
Ringer solution than it does when broken in water. For sample M5
the reverse is true.

Table II compares the median strength values at two loading
rates for M8 in water, Krebs-Ringer solution, and bovine serum.

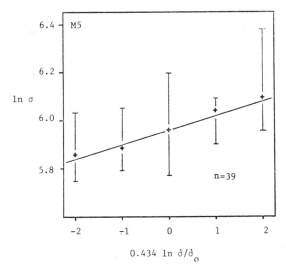

Figure 3 - Ln σ vs 0.434 ln $\dot{\sigma}/\dot{\sigma}_o$ for sample material M5 broken in
distilled water. Crosses represent the median values,
error bars indicate the range of the data. Line is a
linear regression using all of the data points, giving
n-equal to 39.

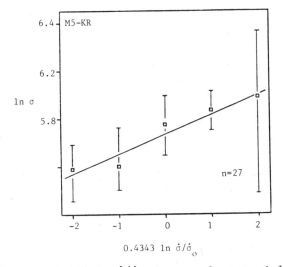

Figure 4 - Ln σ vs 0.434 ln $\dot{\sigma}/\dot{\sigma}_o$ for sample material M5 broken in
Krebs-Ringer solution. Boxes represent median values,
error bars indicate the range of the data. The line
is from a linear regression using all of the data points
giving n equal to 27.

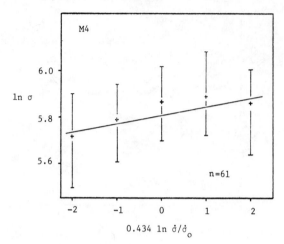

Figure 5 – Ln σ vs 0.434 ln $\dot{\sigma}/\dot{\sigma}_o$ for sample material M4 broken in
distilled water. Crosses represent median values,
error bars indicate the range of the data. The line is
from a linear regression using all of the data points
giving n equal to 61.

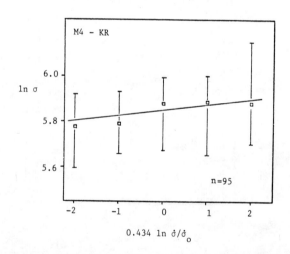

Figure 6 – Ln σ vs 0.434 ln $\dot{\sigma}/\dot{\sigma}_o$ for sample material M4 broken in
Krebs-Ringer solution. Boxes represent median values,
error bars indicate the range of the data. The line
is from a linear regression using all of the data
points giving n equal to 95.

TABLE II

Sample M8	300 MPa/sec	0.3 MPa/sec
DISTILLED WATER	399 MPa $\pm$ 36	392 MPa $\pm$ 27
KREBS-RINGER SOL'N	416 MPa $\pm$ 70	379 MPa $\pm$ 33
BOVINE SERUM	444 MPa $\pm$ 43	390 MPa $\pm$ 37

There does not seem to be a significant difference among the
strength values in different fluid environments.

6. CONCLUSION

If especially fine grained alumina is to be used in human
prosthetic devices, and if time-to-failure design diagrams are
to be constructed for proof testing purposes, then a more
precise understanding of the mechanisms which contribute to the
determination of n is required than is now available. In
particular this study indicates that in some cases more than one
failure mechanism or flaw-size distribution must be taken into
account in the data analysis. At this time it does not appear
that future tests need to be conducted in a simulated biological
environment.

REFERENCES

1. Griss, Peter, et al., J. Biomed. Mater. Res. Symposium, No.
    6, 177 (1975). 2. Townsand, Paul,
2. Townsand, Paul, Codman and Shurteff, Boston, Mass., Private
    Communication.
3. Wiederhorn, S. M., Tighe, N. J., and Evans, A. G., AGARD
    Report No. 651, presented at the 43rd Meeting of Structures
    and Materials, Panel of AGARD, Oct. 1976.
4. Greenspan, D. C., Palmer, R. A., Ritter, J. E., and Henck, L.
    L., Paper presented at the 3rd Annual Meeting of the Society
    for Biomaterials, New Orleans, 1977.
5. Wachtman, J. B., Jr., Capps, W., and Mandel, J., J. of Matls.
    7 (2) 188, 1972.
6. Kirstein, A. F. and Woolley, R. M., Journal of Research, 17C,
    1, 1967.
7. Evans, A. G. and Johnson, H., J. Matls. Sci., 10, 214, 1975.
8. Mandel, J., *The Statistical Analysis of Experimental Data*,
    Chapter 13, Interscience, Wiley, New York, 1964.

MODIFIED DOUBLE TORSION METHOD FOR MEASURING

CRACK VELOCITY IN NC-132 (Si_3N_4)

C. G. Annis, J. S. Cargill

Materials & Mechanics Technology
Pratt & Whitney Aircraft Group
Government Products Division
P. O. Box 2691
West Palm Beach, Florida 33402

The double torsion method for studying slow crack growth in brittle materials has been described in detail by Williams and Evans (1, 2), wherein it was shown that the specimen stress intensity factor is independent of crack length (see Appendix). The crack velocity, under constant torsional deflection, is inversely proportional to the square of the instantaneous load and directly proportional to the instantaneous rate of load relaxation with crack advance. It is therefore unnecessary to monitor crack length during the test.

The purpose of this paper is to suggest a simple modification to the specimen, and to present the results of elevated temperature crack propagation testing in NC-132, hot pressed Si_3N_4, using this modified technique.

The double torsion specimen, with its longitudinal crack-directing center slot, has become the specimen of choice in crack propagation testing of brittle materials because of its relative simplicity of fabrication and testing. The center slot however causes a certain intellectual discomfort because its effects, other than guiding the crack, are not well understood. This concern can be avoided altogether if the slot is eliminated. By introducing a crack into the specimen through a machined end notch with a cross section resembling the curved front assumed by a propagating crack, the crack will propagate along the specimen centerline if the test apparatus is accurately aligned. Misalignment, with its concomitant unequal torsions, and therefore uncertain crack front stress intensity, is readily manifested as a deviation from centerline propagation.

737

The specimens used to study NC-132 measured 3 x 1 x 0.1 inches,
with no channel. The crack starter notch detail is shown schematically
in Figure 1. Under constant cross-head displacement, a crack can start
in the thin section and proceed to full specimen thickness before fall-
ing below threshold stress intensity and arresting. The ceramic test
rig, with specimen in place, is shown in Figure 2. All testing was performed under constant crosshead displacement wherein the entire v, K
relationship is determined during one test. Induction heating using an
INCO X-750 susceptor was employed.

Figure 3 presents the results of testing at 2200°F and 1700°F.
Unlike the well behaved data at 2200°F, the propagation rates at 1700°F
exhibit considerable data scatter and poor test-to-test repeatability.
Acoustic emission studies of this material at room temperature indicate

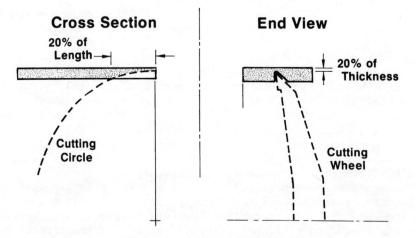

Figure 1. Notch detail.

Figure 2.   Four-point loading applies double-torsion.

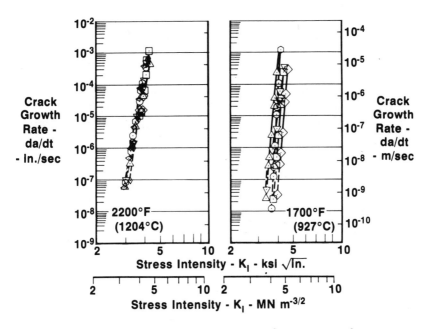

Figure 3.   Growth data at 2200°F and 1700°F.

continuous but sporadic propagation at the higher stress intensities, and discontinuous crack advance at lower K-levels. It is therefore assumed that at 2200°F the crack propagates continuously and at 1700°F and below, crack growth is discontinuous. For this material over the stress intensity range investigated crack velocity increases with increasing temperature.

Simple cyclic loading, using a mechanical eccentric to vary crosshead displacement, produced crack propagation as shown in Figure 4. Temperature effects on cyclic propagation are similar to those observed during monotonic loading.

Comparisons of measured crack propagation rates, in NC-132 with literature values (3) for NC-130, a similar hot pressed Si_3N_4, show NC-132 to have considerably higher crack velocities when compared at the same stress intensity at 2200°F. Perhaps this difference can be

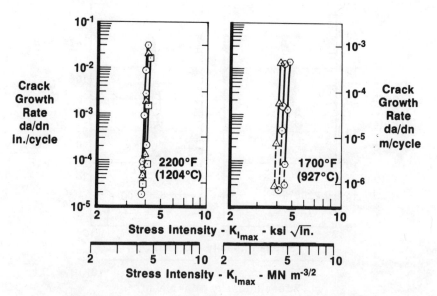

Figure 4.   Comparison of cyclic growth rate (da/dn) at 2200°F and 1700°F.

attributed to differences in impurity levels, e.g., MgO, CaO, which
form a viscous glassy phase at the grain boundaries, near 2200°F.
It will be noted, however, that the NC-130 data was obtained from a
slotted specimen, data for NC-132 from a specimen with no channel.
Because the v, K curves are quite steep, any small disagreement in
computed stress intensity would appear as a very large difference in
crack velocity.

Measured values for fracture toughness in NC-132 are compared
with literature values (3) for NC-130 in Figure 5.

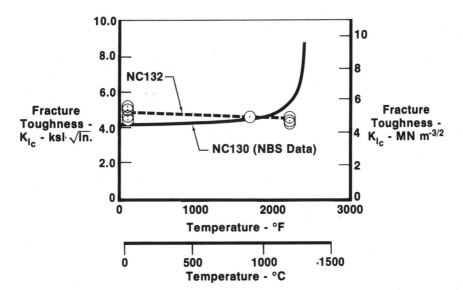

Figure 5.   Fracture toughness comparison of NC 130 and NC 132.

APPENDIX: Stress Intensity Determination for an Unslotted
            Double Torsion Specimen.

The following simple derivation of the expression relating
specimen geometry and applied load with stress intensity K, at
the crack front, parallels the derivation by Williams and Evans*
for the slotted specimen geometry.

Because of longitudinal symmetry about the crack, the double
torsion specimen can be considered as two elastic rectangular
panels, point loaded as in Figure 1. The rotation θ, of any shaft
under torque T, is proportional to its length L, and inversely proportional to the product of its polar moment of inertia J, and the
shear modulus G,

$$\theta \quad = \quad \frac{TL}{JG} \hspace{5cm} \text{equation 1}$$

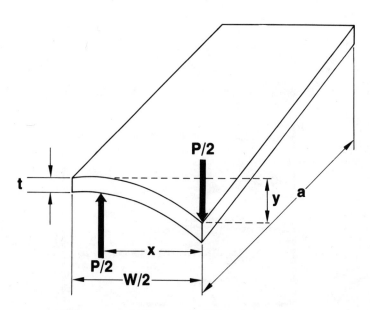

Appendix Figure 1.   Rectangular bar, point loaded in torsion (after
                     Williams and Evans, 1973).

* Williams, D. P. and Evans, A. G., "A Simple Method for Studying
  Slow Crack Growth," Journal of Testing and Evaluation, Vol. 1,
  No. 4, July 1973, pp. 264-270.

For a rectangular cross section,

$$J = (W/2)(t)^3/3 \qquad \text{equation 2}$$

Substituting into equation 1, and noting that for small angles, $\theta \doteq \tan\theta$

$$\frac{y}{x} = \frac{6TL}{Wt^3G} \qquad \text{equation 3}$$

where y is the vertical displacement under load $P/2$, and x is the torsional moment arm. Because the crack length a, is the effective length L, which experiences torsion, and since $T = (P/2)x$, equation 3 can be rearranged to give the elastic compliance, C:

$$C = \frac{y}{P} = \frac{3x^2 a}{Wt^3G} \qquad \text{equation 4}$$

The strain energy released through a unit increase in crack face area, $\mathcal{G} = \partial U/\partial A$, is related to compliance by equation 5,

$$\mathcal{G} = \frac{P^2}{2}\left(\frac{dC}{dA}\right) \qquad \text{equation 5}$$

The increment of area, dA, formed by an incremental crack extension, da is dA = t da, where t is the specimen thickness. Differentiating compliance C, (equation 4) with respect to crack length,

$$\frac{dC}{da} = \frac{3x^2}{Wt^3G} \qquad \text{equation 6}$$

Combining equations 5 and 6

$$\mathcal{G} = \frac{P^2}{2}\frac{3x^2}{Wt^4G} \qquad \text{equation 7}$$

Stress intensity K is related to $\mathcal{G}$ by

$$\mathcal{G} = K^2/E \qquad \text{equation 8}$$

and the elastic modulus E, is related to the shear modulus G, by

$$G = \frac{E}{2(1+\nu)} \qquad \text{equation 9}$$

where ν is Poisson's ratio.

744   C. G. ANNIS AND J. S. CARGILL

Finally, combining equations 7, 8, and 9 provides the desired result:

$$K = \frac{Px}{t^2} \left[\frac{3(1+\nu)}{W} \right]^{1/2}$$

where it is noted that for the double torsion specimen, stress intensity does not depend on crack length.

REFERENCES

1. Williams, D. P. and Evans, A. G., "A Simple Method for Studying Slow Crack Growth," Journal of Testing and Evaluation, Vol I, No. 4, July 1973, pp. 264-270.
2. Evans, A. G., "High Temperature Slow Crack Growth in Ceramic Materials," NBSIR 74-442, February, 1974.
3. Evans, A. G., "A Method for Evaluating the Time-Dependent Failure Characteristics of Brittle Materials and its Application to Polycrystalline Alumina," Journal of Materials Science 7 (1972) pp. 1137-1146, Materials Department, School of Engineering and Applied Science, University of California, Los Angeles, Cal.

SUBCRITICAL CRACK GROWTH IN GLASS-CERAMICS

B.J. Pletka and S.M. Wiederhorn

Institute for Materials Research
National Bureau of Standards
Washington, D. C.   20234

ABSTRACT

As part of a program to evaluate failure prediction theories
for ceramic materials, the crack propagation parameter, n, was
determined by fracture mechanics and strength techniques for a
lithium aluminosilicate and a magnesium aluminosilicate glass-
ceramic. In the magnesium aluminosilicate glass-ceramic, n was
influenced by the crack morphology, which probably resulted from
an interaction of the crack with particles present in the microstructure. For both glass-ceramics, values of n determined from
four-point bending and biaxial tension strength data were smaller
than values determined from crack velocity data. This result
raises doubts as to the direct applicability of fracture mechanics
data to failure prediction methods. It is recommended that for
any material, the equivalence of strength and fracture mechanics
data be proven before the latter are used for failure prediction
purposes.

1.   INTRODUCTION

According to theory, predictions of failure time of ceramic
materials should be independent of the experimental technique
used to characterize subcritical crack growth. Time-to-failure
predictions made from strength measurements should be identical
to those made from fracture mechanics measurements. This
hypothesis is based on the assumption that the propagation of
large cracks in fracture mechanics specimens is equivalent to the
propagation of small flaws in structural components. Since

some experimental data support this assumption [1-6] while others do not [5-7], additional experimental data is needed to determine whether fracture mechanics and strength techniques can be used interchangably for failure time predictions in structural ceramics.

To develop design diagrams that can be used to assess the mechanical reliability of structural ceramic materials[8-11], the crack propagation parameter, n, must be evaluated. The value of n can be determined by fracture mechanics techniques, in which the crack velocity, v, is measured as a function of the stress intensity factor, K_I. The relation between v and K_I is then described by the empirical relation:

$$v = v_o (K_I/K_o)^n \qquad (1)$$

where v_o and n are material constants for a given environment and K_o is an arbitrary constant used to normalize K_I.

The crack propagation parameter, n, can also be evaluated from dynamic fatigue experiments, in which the breaking strength is measured as a function of stressing rate. Data from these experiments can be fitted to the equation:

$$\sigma = B\dot{\sigma}^{1/(n+1)} \qquad (2)$$

where σ is the strength, $\dot{\sigma}$ is the stressing rate, and B is an experimental constant. B and n in equation (2) are determined by a least squares fit of the experimental data. Failure prediction diagrams are then determined by well-established theoretical procedures, either from equation (1) and K_{IC}, or from equation (2) and the strength of the solid in an inert environment [8-11].

As part of a program concerned with the evaluation of failure prediction theories for ceramic materials, glass-ceramic materials are being investigated. Glass-ceramics were selected because of their practical importance and because of their microstructure, which contains no porosity and a random orientation of small crystals. In this paper, equations (1) and (2) are used to evaluate and compare the crack propagation parameter, n, for two commercial glass-ceramics. Since accuracy in failure predictions depends on the accuracy of crack growth data [12,13], the influence of crack propagation morphology on n is also studied.

2. EXPERIMENTAL PROCEDURE

Two glass-ceramics were studied: a magnesium aluminosilicate nucleated with TiO_2 (C9606), and a lithium aluminosilicate nucleated with TiO_2 and ZrO_2 (Cer-Vit 126)*. X-ray diffraction analysis indicated cordierite as the primary crystalline phase in the magnesium aluminosilicate; some rutile and a third phase (possibly one of the magnesium aluminum titanates) was also present. In the lithium aluminosilicate, β-spodumene was the main crystalline phase, but unidentified minor phase(s) were also present. Both glass-ceramics were >95% crystalline.

Crack velocity data were obtained on precracked double torsion specimens (2x25x75 mm) that contained a side groove ($\sim$2 mm wide and 1 mm deep) to control the direction of crack propagation [14]. The specimens were placed on the test fixture so that the side groove was on the tensile surface of the specimen. Specimens were tested in distilled water using the stress relaxation technique [14], and crack lengths were measured between stress relaxation runs using dye penetrant techniques. Stress relaxation measurements were made with an electronic device that accumulated time-increment data as a function of load. These data were fed directly into a computer that analyzed them and reduced them to both tabular and graphical forms. To eliminate errors from extraneous relaxations, the fixture was first pre-strained with a dummy specimen, and held at load until no relaxation occurred. This procedure was then repeated with a precracked test specimen that was preloaded to within $\sim$10 percent of the load at which subcritical crack growth occurred. The stress intensity factor was calculated for plane strain conditions using a correction for the thickness of the specimen [15]. The parameter n was evaluated by a least squares fit of ln K_I upon ln v.

Two techniques were used to measure the strength of the glass-ceramics. Four-point bending measurements were made only on the magnesium aluminosilicate glass-ceramic. Specimens were bars (4x6x80 mm) that had been cut from the same billet as the

---

*The use of these glass-ceramics does not imply their endorsement for commercial applications by the National Bureau of Standards.

fracture mechanics specimens. A 400 mesh diamond grinding wheel
was used to produce specimens with a uniform surface finish.
Tests were conducted in distilled water at four stressing rates
($\sim$0.05 to 50 MN/m^{2}s) using a test fixture with a 50 mm major
span and a 10 mm minor span. Ten specimens were broken at each
stressing rate.

Strength measurements were made on both glass-ceramics
using a biaxial tension technique originally developed by
Wachtman et al. [16] (fig. 1). Specimens were discs (1-1.5 mm
thick and $\sim$32 mm in diameter) that were supported on 3 spheres
equi-spaced on a 12.7 mm radius. The specimens were center
loaded with a cylindrical ram until fracture occurred. These

Figure 1 - Biaxial tension fixture used in strength tests.

specimens were also cut from the same billet as the fracture
mechanics specimens and were ground with a 400 mesh diamond
wheel to control surface finish. Tests were conducted in distilled water using five stressing rates ($\sim$0.01 to 100 MN/m^{2}s).
A minimum of 9 specimens were broken at each stressing rate.

3.   RESULTS

3.1 Crack Velocity Measurements

Figure 2 illustrates typical crack velocity-stress intensity
factor curves for one of the lithium aluminosilicate double
torsion test specimens. The curves for successive relaxation

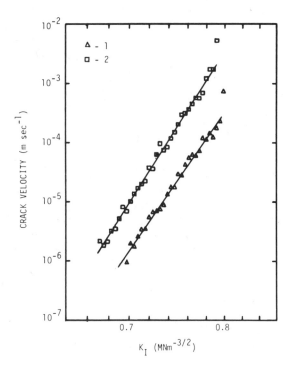

Figure 2 - Crack velocity, stress intensity factor data for two
          stress relaxation runs (identified in upper left of
          figure) on a lithium aluminosilicate double torsion
          specimen in distilled water at room temperature.

measurements shift to lower values of K_I as the crack length increases. The mean and median values of n are listed below in Table 1 for several specimens. A representative optical micrograph of crack morphology is shown in Fig. 3 for the surface containing the leading edge of a crack. The crack trace was relatively straight and there was no evidence of crack branching.

Figure 4 shows v-K_I data from five stress relaxations obtained on one double torsion specimen of magnesium aluminosilicate. It was noted that when the final crack length was very short, <20 mm, n values were invariably high ($\sim$120 for curve 1 of figure 4). However, as the final crack length increased, the value of n decreased (84 and 75 respectively for runs 2 and 3 of figure 4). In the first three stress relaxation measurements, a systematic shift of the curves to lower K_I was observed with increasing crack length. This shift in the v-K_I data has been observed by the authors on a variety of ceramic materials, suggesting that the double torsion specimen is not a constant K_I specimen. Transients, in which v increased as K_I decreased, were observed at the beginning of some of the stress relaxation runs for the magnesium aluminosilicate specimens (see runs 1 and 2 in figure 4), but not for the lithium aluminosilicate specimens. These transients suggest that crack propagation is more erratic in the magnesium aluminosilicate than in the lithium aluminosilicate ceramic.

Stress relaxation runs 4 and 5 lend further support to the erratic nature of the crack propagation in the magnesium aluminosilicate ceramic. In stress relaxation run number 4 there is a shift to higher K_I values compared to run number 3, and an increase in the slope of the v-K_I curve (n=167). A shift to even higher values of K_I occurs for run number 5. Furthermore, there is an acceleration in the crack velocity near the end of run number 5. This erratic behavior can be correlated with the morphology of the cracks in the magnesium aluminosilicate.

TABLE I

CRACK PROPAGATION PARAMETER n DETERMINED
FROM CRACK VELOCITY DATA

Glass-ceramic	Median	Mean and standard deviation
lithium aluminosilicate	43	46 ± 10
magnesium aluminosilicate	82	84 ± 9

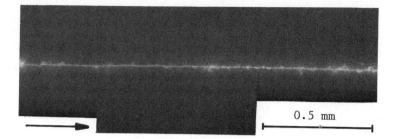

Figure 3 - Optical micrograph of crack morphology for the leading
crack edge in lithium aluminosilicate. Arrow indicates
crack propagation direction.

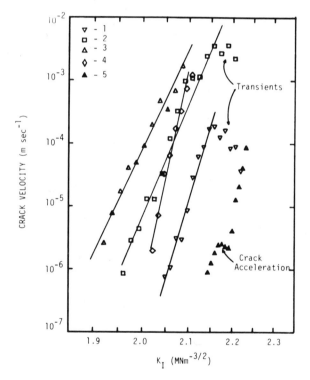

Figure 4 - Crack velocity, stress intensity factor data for five
stress relaxation runs (identified in upper left of
figure) on a magnesium aluminosilicate double torsion
specimen in distilled water at room temperature. See
text for a discussion of the various v-K_I curves.

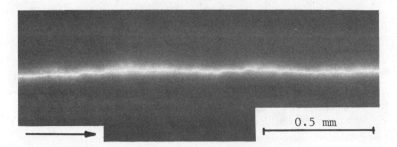

Figure 5 – Optical micrograph of leading edge of crack propagating
           in magnesium aluminosilicate, representative of stress
           relaxations 1–3. The crack appears to wander more
           than in lithium aluminosilicate (see fig. 3). Arrow
           indicates crack propagation direction.

     The crack morphology for the first three stress relaxation
runs is illustrated in figure 5. The meandering nature of the
surface trace of the crack indicates a strong interaction between
the crack and the microstructure of the ceramic. During the
first three runs this interaction was not sufficient to cause
scatter in the v–K_I data. In runs 4 and 5, however, crack
arrest and crack branching lead to high values of n, and unexpected shifts in the position of the v–K_I data. The crack
morphology corresponding to this type of crack propagation
behavior is shown in figure 6. The point of acceleration in run
5 probably corresponds to the arrest of the main crack, and
nucleation of a second crack a short distance behind the stationary
tip of the original crack (figure 6). This same morphology was
observed on other specimens. In one specimen, multiple crack
formation was present along the entire length of the crack,
figure 7. The mean value of n for this specimen, ∿125, was
higher than that obtained when multiple crack formation did not
occur (see Table I). These data suggest that the crack morphology
influences both the position of the v–K_I curve, and the values
of n obtained from these curves. No evidence of multiple crack
formation was observed at the trailing edge of cracks in the
double torsion specimens.

3.2 Strength Measurements

     The mean biaxial tension and four-point bend strengths for
magnesium aluminosilicate are plotted as a function of the
stressing rate in figure 8. The error bars indicate a confidence
limit of 95%. Using equation (2) a linear regression analysis
was performed to obtain n from both the mean and median value of

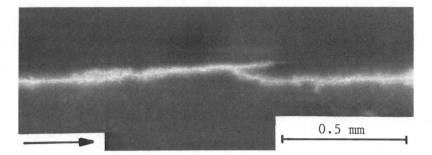

Figure 6 – During stress relaxation 4 in fig. 4, the leading edge
of the primary crack appears to have stopped propagating, and a new crack has nucleated adjacent to it
and begun propagating. Arrow indicates crack propagation direction, optical micrograph.

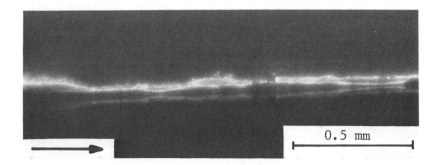

Figure 7 – Extensive multiple crack formation in the leading crack
edge is shown in this optical micrograph of a magnesium
aluminosilicate double torsion specimen. Faint line
towards bottom of micrograph is residual dye along wall
of side groove. Arrow indicates crack propagation
direction.

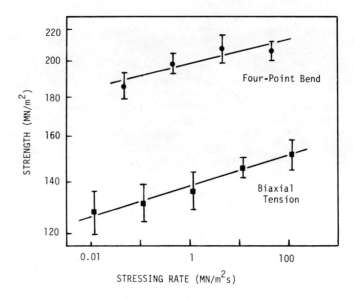

Figure 8 - Biaxial tension and four-point bend strengths
          as functions of stressing rate for magnesium
          aluminosilicate glass-ceramic.

the strength at each stressing rate. As can be seen from Table 2
it made little difference whether the mean or median value of the
strength was used to evaluate n. Systematic shifts in the value
of n were observed, depending on the experimental technique used
to determine n. These systematic shifts for the mean value of n
are summarized in figure 9. As can be seen, n values determined
from strength measurements are systematically smaller than those
determined from crack velocity measurements.

4. DISCUSSION

     In order to account for the influence of crack morphology
on n, the interaction of the crack with the microstructure has
to be considered. An optical micrograph of an unetched, polished,
thin-section of magnesium aluminosilicate shadowed with gold is
shown in figure 10. Particles, up to ∿8 μm in length and ∿2 μm
in width are seen to be present. Although the particles were
not identified, they are believed to be either rutile and/or one
of the magnesium aluminum titanates. It is suggested that the
dominant factor that determines the tortuous course of crack
propagation in the magnesium aluminosilicate ceramic is the
interaction of the crack with these particles. These particles
appear to block the crack motion in the magnesium aluminosilicate

TABLE II

MEASUREMENT OF CRACK PROPAGATION PARAMETER n
FROM DYNAMIC FATIGUE DATA

Glass-ceramic	Data obtained	Median	Mean
lithium aluminosilicate	biaxial tension	33 + 2	37 + 7
magnesium aluminosilicate	biaxial tension	47 + 6	51 + 5
magnesium aluminosilicate	4 pt. bend	70 + 15	63 + 21

ceramic resulting in local barriers that must be overcome for
crack motion to continue. Stress fields due to thermal expansion
anisotropy of the particles and thermal expansion mismatch
between the particles and the surrounding matrix [17] can be a

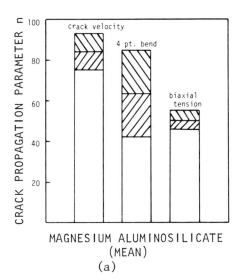

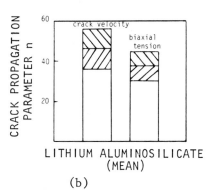

Figure 9 – Mean value of crack propagation parameter n in magnesium
          aluminosilicate (a) and lithium aluminosilicate (b)
          from crack velocity and strength measurements. Hatched
          regions indicate standard deviation of n for crack
          velocity data and standard deviation of the slope for
          strength measurements.

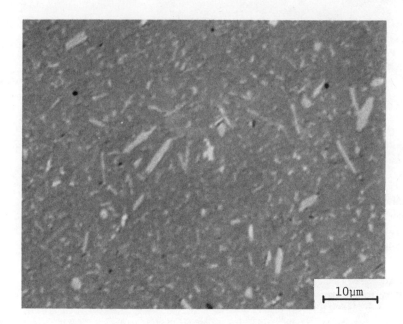

Figure 10 - Optical micrograph of unetched, gold shadowed thin-
section of magnesium aluminosilicate taken with
partially crossed polarizers. Note particles present
in the microstructure.

contributing factor to the interaction. This type of interaction
does not occur for the lithium aluminosilicate glass-ceramic
which has a relatively homogeneous microstructure. As a consequence, crack surfaces are smoother for the lithium aluminosilicate
ceramic, and a much more systematic crack propagation behavior
is observed.

    Transients in the crack propagation data shown in figure 4
for the magnesium aluminosilicate ceramic can be explained in
terms of particle, crack front interactions. A crack that
becomes momentarily arrested by particles will require higher
forces to break free from the particles in order to continue
propagating. When this occurs, the crack velocity is expected
to increase. This crack arrest process appears to be prevalent
for slowly moving cracks, since crack velocity transients are
often observed after the cracks have been arrested. As indicated
in figure 4, however, transients can also occur in the middle of
a stress relaxation run, indicating that crack tip-particle
interactions are also important at higher crack velocities.

    The interaction of cracks with particles can also be used
to rationalize variations of n from run to run. The interaction

of a particle with a crack tip can create a disturbance on the crack front that is not restricted to the immediate location of the initial interaction. If the disturbance causes the crack to deviate from its initial plane, crack propagation will become increasingly difficult as the disturbance spreads along the crack front. As a consequence, the crack velocity will decrease at a greater rate than if the crack-particle interaction had not occurred, and n will be observed to increase. If the interaction along the crack front is strong enough, secondary cracks will nucleate adjacent to the original crack, particularly if microcracks are present in this region. This crack nucleation would account for the multiple crack formation observed in the present work.

The macroscopic state of stress in a specimen also influences crack motion. In a uniaxial stress field, cracks can be held up by obstacles because of the strong orientation dependence of the stress intensity at the crack tip on crack-growth direction. The stresses required to force cracks to go around these obstacles result in erratic crack growth and a resultant increase in n as measured either by double torsion or uniaxial strength tests.*
By contrast, cracks in a biaxial stress field have a greater freedom of motion than those in a uniaxial stress field. Once a crack begins to grow in a biaxial field, the absence of a strong orientation dependence for crack growth allows these cracks to grow around obstacles. As a consequence, the importance of obstacles to crack growth in a biaxial tensile field is reduced, and the effect of obstacles on the value of n will be minimal. Thus, the value of n obtained in a biaxial strength experiment might be expected to be less than that obtained in a uniaxial strength experiment, or in a fracture mechanics experiment. This expectation is supported by experimental studies on the magnesium aluminosilicate glass-ceramic.

---

*A mechanism for the increase in n for the double-torsion test was given above. In a four-point bend test, the importance of obstacles to crack growth depends on the loading rate. For high rates of loading, the crack can become critical before serious obstacles are encountered. For slow rates of loading, more crack growth occurs before failure, increasing the chance of interaction with a serious obstacle. If the crack meets a serious obstacle, then more applied load will be necessary for critical crack growth. The parameter n will increase, since the slope of the dynamic fatigue curve will decrease.

The results of the present experiments raise doubts as to the general applicability of fracture mechanics data to failure prediction methods. Data obtained on the magnesium aluminosilicate glass-ceramic indicate a significant difference between the values of n obtained from crack propagation data and those obtained from biaxial tension measurements. Design diagrams based on these two sets of data would also show significant differences in predicted failure times of structural components. Predictions made from the biaxial tension specimens would indicate failure in a shorter time, for a given service stress, and consequently should be used for design purposes because the predictions are more conservative. By contrast, data from the lithium aluminosilicate indicate a smaller difference between the biaxial tension and the crack velocity data, suggesting that either set of data could be used for purposes of failure time prediction. In general, results from the present study suggest that the equivalence between strength and fracture mechanics data should be proven before the latter are used for failure prediction purposes.

ACKNOWLEDGMENT

Dr. B. Koepke of Honeywell, Inc., Corporation Research Center, Bloomington, Minn., kindly supplied the lithium aluminosilicate specimens.

REFERENCES

1. S. M. Wiederhorn, A. G. Evans, E. R. Fuller, and H. Johnson, J. Am. Ceram. Soc., 57, 319 (1974).
2. A. G. Evans and F. F. Lange, J. Mat. Sci., 10, 1659 (1975).
3. A. G. Evans and H. Johnson, J. Mat. Sci., 10, 214 (1975).
4. R. F. Caldwell and R. C. Bradt, J. Am. Ceram. Soc., 60, 168 (1977).
5. S. Mindess and J. S. Nadeau, Am. Ceram. Soc. Bull., 56, 429 (1977).
6. J. E. Ritter, Jr. and R. P. LaPorte, J. Am. Ceram. Soc., 58, 265 (1975).
7. J. E. Ritter, Jr. and K. Jakus, J. Am. Ceram. Soc., 60, 171 (1977).
8. A. G. Evans and S. M. Wiederhorn, Int. J. Fract. Mech., 10, 379 (1974).
9. S. M. Wiederhorn, Fracture Mechanics of Ceramics, (edited by R. C. Bradt, D. P. H. Hasselman and F. F. Lange) pp. 613-646, Plenum Press, New York, (1974).
10. S. M. Wiederhorn, Ceramics for High-Performance Applications, (edited by J. J. Burke, A. E. Gorum, and R. N. Katz) pp. 633-63, Brook Hill Publishing Co., Chestnut Hill, Mass. (1974).

11. S. M. Wiederhorn, N. J. Tighe, and A. G. Evans, <u>Mechanical</u>
     <u>Properties of Ceramics for High Temperature Applications</u>,
     AGARD Report No. 651, pp 41-55, December 1976.

12. S. M. Wiederhorn, E. R. Fuller, Jr., J. Mandel, and A. G.
     Evans, J. Am. Ceram. Soc., <u>59</u>, 403 (1976).

13. D. F. Jacobs and J. E. Ritter, Jr., J. Am. Ceram. Soc., <u>59</u>,
     481 (1976).

14. A. G. Evans, J. Mat. Sci., <u>7</u>, 1137 (1972).

15. A. G. Evans, M. Linzer and L. R. Russell, Mat. Sci. Eng.,
     <u>15</u>, 253 (1974).

16. J. B. Wachtman, Jr., W. Capps, and J. Mandel, J. Mater. <u>7</u>,
     188 (1972).

17. A. G. Evans, A. H. Heuer, and D. L. Porter, <u>Fracture 1977</u>,
     (edited by D.M.R. Taplin) pp 529-556, University of Waterloo
     Press, Waterloo, Ontario (1977).

DYNAMIC FATIGUE OF FOAMED GLASS

P. H. Conley[*], H. C. Chandan[+], and R. C. Bradt

Pennsylvania State University

University Park, PA 16802

ABSTRACT

The dynamic fatigue, or the stressing rate effect on the strength of foamed glass was studied in tension, in flexure, and in compression using bulk specimens. In addition to the expected differences in strength, the three methods also exhibited different stressing rate sensitivity, with the compressive strength showing the largest rate sensitivity and the flexural strength the least effect. Tensile strengths exhibited classic opening mode, K_I, rate sensitivity for a bulk soda-lime-silica glass, n $\approx$ 16. It is suggested that mixed mode (K_I, K_{II}, K_{III}) slow crack growth may be the mechanism of dynamic fatigue for non-tensile situations, and that the cellular microstructure of the foamed glass also assumes an important role in the effect. Questions are posed concerning the general dynamic fatigue effect for complex macrostress states.

INTRODUCTION

The dynamic fatigue or the stressing rate effect on the strength of brittle ceramics has been well documented for a variety of glasses[1], glass-ceramics[2,3], cement[4], oxides[5-7] and non-oxide materials[8]. The result is always that of higher strengths at higher stressing rates. Explanation has been the opening mode, (K_I), subcritical crack growth to the critical condition, (K_{Ic}). Evans[9] has integrated the classic (K_I,V) diagrams to yield the well recognized mathematical form of the

---

[*] Currently with Koppers Co., Pittsburgh, PA, 15219.
[+] Currently with Bell Labs., Norcross, GA., 30071.

effect of stressing rate on fracture strength:

$$\sigma_f = B \; \dot{\sigma}^{(1/1+n)} \tag{1}$$

where (σ_f) is the fracture stress, (B) is a constant, ($\dot{\sigma}$) is the stressing rate, and (n) is the slow crack growth rate parameter which also appears in the empirical macrocrack growth equation:

$$V = A \; K_I^n \; , \tag{2}$$

where (V) is the crack velocity, (A) is another constant, and (K_I) is the opening mode stress intensity. Charles[10] also derived an equation analogous to (1) for the stressing rate effect, but purely on a strength of materials/chemical reaction rate basis, so it is evident that subcritical crack growth restrictions, such as implied by equation (2) need not be invoked to yield the form of equation (1) which has been observed to describe the experimental results quite adequately.

It is a fair assessment that current concepts of dynamic fatigue generally accept that opening mode slow crack growth, (K_I,V), to the critical condition, (K_{Ic}), is a reasonable explanation for many ceramic failures under dynamic stressing conditions. However, there exist several experimental observations that do not conveniently ascribe to all aspects of this classically accepted failure process. These observations all center about the stressing rate effect on the strength in biaxial stress or in compression; that is, they suggest that the non-tensile rate effect may not be that of opening mode subcritical crack growth; for the slope, (1/1+n), does not always yield the same (n)-value as flexural ("tensile") strength or macroscopic crack growth measurements. These studies have been with concrete[11], glass-ceramics[3], a piezoelectric lead-zirconate-titanate[5], and for several alumina bodies[6,12]. However, for bulk soda-lime-silica glass, which does not contain a specific microstructure (ignoring small scale phase separation), the stressing rate effect appears constant from test to test, generally yielding an (n)-value of about 16.[13] Thus, the potential effects of microstructure, as well as macrostress state on dynamic fatigue of ceramics remains unresolved. The microstructural effects are naturally important; however, design with brittle ceramics for a multitude of macrostress states is also becoming increasingly common. Thus these effects appear to be even more crucial, for fatigue parameters are an integral part of lifetime failure predictions.[13] It is the objective of this study to examine some of these effects, on a limited scale, through the testing of bulk foamed glass specimens under different macrostress states at different stressing rates.

EXPERIMENTAL PROCEDURE

The foamed glass utilized in this study was a commercial product.[14] It was obtained in the form of five 4" x 18" x 24" slabs from a single manufacturing run. The average composition of this product on a weight percentage basis is: $70.7/SiO_2$, $7.5/MgO + CaO$, $12.4/Na_2O$, $3.8/BN$, $5.1/Al_2O_3 + FeO$, and about $0.5/C + Na_2SO_4$. The C and Na_2SO_4 are remanents from the foaming process, described in detail by Long.[15] Some physical properties are listed in Table I. This is close to a typical soda-lime-silica glass, akin to some of the commercial amber glasses, except for the foaming additions.[16] For all practical purposes, this foamed glass approximates many

TABLE I

Relevant Physical Properties of Foamglas HL/B*

Property	English Units	Si Units
Absorption of moisture (% by Volume)	0.2	0.2
Capillarity	None	None
Compressive strength	100 psi	6.89×10^5 N/m^2
Density	8.5 lb./ft.^{3}	136 kg/m^3
Flexural Strength	80 psi	5.51×10^5 N/m^2
Hygroscopity	No increase in weight at 90% relative humidity	
Modulus of Elasticity, (approximate)	1.5×10^5 psi	1.03×10^9 N/m^2
Shear Strength	50 psi	3.44×10^5 N/m^2
Water-Vapor Permeability	0.00 perm-in	0.00 perm-m

* Data from Pittsburgh Corning Corp. Port Allegany, PA, 16743.

previously, studied commercial soda-lime silica glasses in composition. However, its strength and density are much lower. The macrostructure of the particular foamed glass studied here is illustrated in Figure 1. The maximum cell size is approximately 1/8" diameter, but typically slightly smaller. There does not appear to be a consistent orientation of the cells throughout; however, there certainly exists some short range cell alignment for a number of different cells at various orientations throughout any given specimen.

Three types of strength tests were performed on this foamed glass. Using an Instron commercial testing machine, and also a high speed recorder at the faster stressing rates, specimens were broken by the Brazil or diametral compression test (a measure of the tensile strength), by the common flexural or bend test (often called the modulus of rupture), and by a simple compression test (compressive strength). The strengths and the stressing rates were determined from the equations in Table 2.

Actual specimens, as illustrated in Figure 2, were sawed from the original slabs simply by using a steel band saw and a steel core drill. The Brazil or diametral compression test specimens were core drilled 1 1/2" in diameter from the 4" thick dimensions

TABLE II

Equations for Strengths and Stressing Rates

Test	Strength	Stressing Rates
Brazil (tensile)	$\sigma_f = \dfrac{2\,P}{\pi\,D\,T}$	$\dot{\sigma} = \dfrac{E}{D}\left(\dfrac{dy}{dt}\right)$
Flexural	$\sigma_f = \dfrac{3\,PL}{2\,b\,d^2}$	$\dot{\sigma} = \dfrac{6\,E\,d}{L^2}\left(\dfrac{dy}{dt}\right)$
Compression	$\sigma_f = \dfrac{4\,P}{\pi\,D^2}$	$\dot{\sigma} = \dfrac{E}{H}\left(\dfrac{dy}{dt}\right)$

P is load at failure, D diameter, T thickness, (dy/dt) the crosshead speed, E the specimen elastic modulus, b the sample width, d its thickness, L the span length and H, the sample height. σ_f is the strength and $\dot{\sigma}$ is the stressing rate.

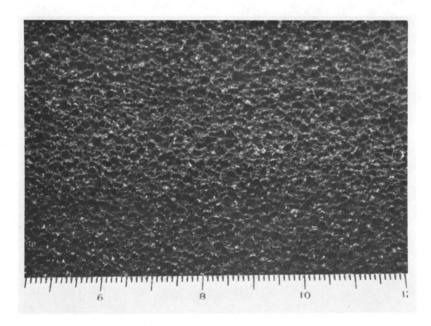

Figure 1. The typical "macrostructure" of the foamed
           glass utilized in this study.

of slabs. In order for the entire specimen to fit beneath the
test platens, 1" was sawed from each end, yielding a Brazil test
specimen 1 1/2" in diameter by 2" long. The compression specimens
were simply the 4" long 1 1/2" diameter cores. For the three
point flexural or bend test, rectangular bars were sawed 1" x 1" x 6"
and tested over a 4" span. In all cases the sample dimensions far
exceeded the foamed cell sizes. Prior to actual testing the loading points were first strengthened using hot tar and roofing felt
paper after ASTM specifications[17]. This is necessary to avoid
loading point crushing of this low strength material.

RESULTS

The tensile strengths, as measured via the Brazil or diametral
compression test are shown in Figure 3. A large number of specimens were tested, since no previous data existed for this test.
Figure 2 illustrates the typical specimen after this test, as well
as those from the other tests. The vertical crack in the Brazil
test specimen is typical of a tensile failure. It has been dis-

Figure 2.   Specimens utilized in this study, Flexural
top, Brazil lower left, and compression
lower right. Note the tar paper at the
loading points.

cussed by a number of other investigators. (18-20) It occurs rather
suddenly with an attendant abrupt drop in load, occasionally
followed by a reloading to even higher load levels, usually in a
non-linear fashion. When secondary loading occurred, it was not
utilized to calculate values shown in Figure 3. The strengths in
Figure 3 are only those from the initial maximum, occurring at the
end of the linearly elastic region. It is believed that this
failure is associated with the "intrinsic" flaws; whereas, the reloading maximum and attendant load drops may be connected with
flaws generated during the initial loading process. Attempts to
plot the secondary reloading maxima, by a number of different
methods, failed to produce any logical results as scatter was large
and no obvious rate trends apparent.

The tensile strengths of Figure 3 exhibit the low values expected for this porous cellular glass as well as a definitive trend
of increasing strength with increasing stressing rate. Both the
individual strengths for a specific stressing rate, and the rate

effect as evidenced by the slope exhibit some scatter, as expected
from any low strength material such as this. However, the crack
growth parameter (n) is about the expected tensile value, $15.9^{+6.8}_{-3.4}$,
frequently reported for opening mode (K_I) crack growth measurements of bulk soda-lime-silica glass.[13] It may be concluded that
failure of foamed glass in the Brazil test is indeed tensile as it
exhibits the expected opening mode strain rate effect, as well as
the expected low strength values.

The flexural strength results are shown in Figure 4. Less
specimens were utilized than in the Brazil test as there was not
enough material to permit as many individual stressing rates. Also
other results of flexural strengths of ceramics for different
stressing rates had confirmed log-log behavior. The general trend
is for strengths higher than the tensile strengths, as expected.
A definite stressing rate effect also exists as the strengths
group about a straight line of (n)-value $31.2^{+10.3}_{-5.9}$. Clearly this
(n) value is not the same as that for the tensile failure stressing rate effect as observed in the Brazil test. This was not expected. It was not expected since there exist at least four other
studies of the dynamic fatigue of soda-lime-silica glass in
flexure that all yield (n) values in the middle teens. In chrono-
logical order: Charles[10] observed 16 for solid rods (1958),
Ritter[21] reported a 13 for solid rods (1969), Evans and
Johnson[22] reported about 17 for microscope slides (1975), and
recently Chandan, Bradt, and Rindone[23] observed 17.2 for float
plate glass (1977). All of these bulk solid soda-lime-silica glass
studies yield values comparable to the opening mode, (K_I), (n)
value, as was earlier reported for tensile failures in the Brazil
test. However, the foamed glass reported in this study has a
flexural (n) value almost twice as large. Apparently the foamed
variety of soda-lime-silica glass is less sensitive to subcritical
crack growth in flexure than the solid glass. This difference in
(n) value must be attributed to the cellular microstructure of
the foamed glass, although the precise mechanics are not obvious.
It might be surmised that different sizes or shapes of the foamed
glass cells might yield even different (n) values. If this concept were extended to crystalline ceramics, it would predict
different (n) values for different microstructures in different
stress states, an extremely complex situation.

The compressive strengths are shown in Figure 5. These
strengths are generally the same as those in flexure and higher
than those in tension, but yet a third stressing rate sensitivity
is apparent. The (n)-value from these compression specimens is
$7.7^{+1.7}_{-1.2}$, about 1/4 that in flexure and only 1/2 that in tension.
The factor of two difference between compression and tension is
the same as that reported by Caldwell and Bradt[5] for a piezoelectric ceramic, although the specimen sizes and microstructures

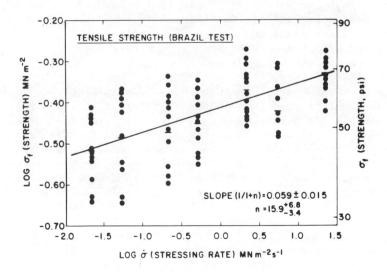

Figure 3.   The dynamic fatigue of foamed glass in
            tension as measured by the Brazil test.

bear little, if any resemblance. This factor of two may be
fortuitous, however, the (n)-value differences are substantial.
This disparity in (n)-values suggests that the exact mechanism of
the stressing rate effect in compression differs from that in
tension, as well as that in flexure.

The mechanism of the compressive strength/stressing rate
effect is not obvious from the lower (n)-values; however, other
compressive strength studies suggest that it may be mixed-mode
(I, II, or III) subcritical crack growth. For example, Sih[24,25]
discusses the occurrence of shear-mode crack propagation during
compression, clearly illustrating its existence. He concludes
that Mode I crack propagation does not exist in compression.
Other investigators have reported shear failures in compression.
Specifically, Adams and Sines[26] have observed some Mode II subcritical crack growth at flaws during compression testing, whereas
Petrovic and Mendiratta[27] have discussed mixed-mode fracture in
Si_3N_4. Hence, there is ample evidence of mixed-mode subcritical
crack growth as a possible mechanism leading to the low (n) value
obtained in compressive fracture.

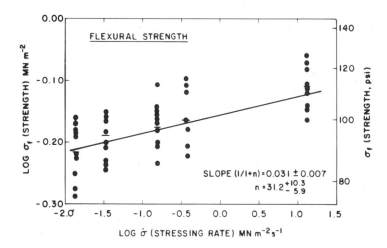

Figure 4.   The dynamic fatigue of foamed glass measured
            in the three point flexural test.

Figure 6 serves as a basis for comparison of the strengths
and the dynamic fatigue effects for this foamed glass in the three
different macrostress states. It is evident that the fatigue rate,
and probably the fatigue mechanisms differ. Only the Brazil tensile
test yields results which are in agreement with those generally
accepted as Mode I subcritical crack growth to K_{Ic} failure. This
leaves several possibilities for the dynamic fatigue effects in
flexure and compression. Accepting the premise that the effect is
one of subcritical crack growth, then it must be proceeding dif-
ferently in the other two tests than in the Brazil test. This
raises the additional question, "If the subcritical crack growth
is not of the opening mode variety (K_I), can the failure criteria
be other than (K_{Ic}), perhaps (K_{IIc}) or (K_{IIIc})?" The answer to
that is beyond the scope of this paper, but certainly merits
further consideration and investigation.

Figure 6 also permits some conclusions with regard to failure
prediction of ceramic structures via the integration of (K_I,V)
diagrams. Obviously, (K_I) (n)-values cannot be applied in a general
fashion to all possible stress states. For this particular foamed

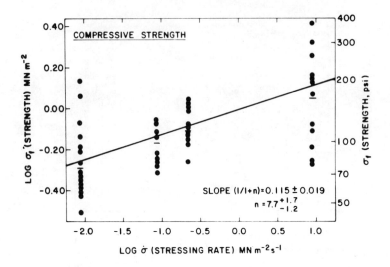

Figure 5. Dynamic fatigue of foamed glass in compression.

glass material and cellular microstructure such a practice would correctly predict a tensile condition, would be ultra-conservative for flexure, and could prove disastrous in compression. These preliminary results suggest that it may be necessary to conduct fatigue studies under service stress states, if accurate subcritical crack growth characteristics are to be determined for lifetime predictions.

SUMMARY AND CONCLUSIONS

The dynamic fatigue of foamed glass was studied for several different stressing conditions. It was observed that stressing rate effects varied with the macrostress state. They were compared through the fracture mechanics crack growth parameter (n). For this particular foamed glass $(n)_{flexure} > (n)_{tension} > (n)_{compression}$, in increasing order of subcritical crack growth. Thus when designing with this material and probably other ceramics as well, it is not possible to predict lieftimes for a general state of stress from a pure tension, or simple (K_I) opening mode experiment. The variation of the flexural (n)-value from that of

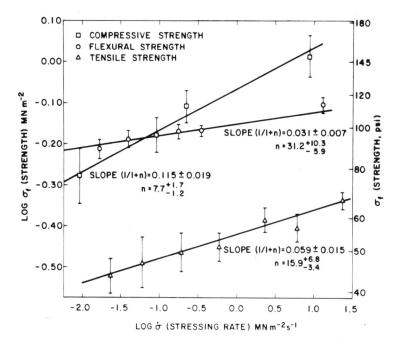

Figure 6.   A comparison of tensile, flexural, and compressive strengths and the dynamic fatigue
of foamed glass in these stress states.

solid specimens of soda-lime-silica glass suggests that the microstructure of the foamed glass assumes a crucial role in the subcritical crack growth behavior, other ceramic microstructures may
as well.

ACKNOWLEDGEMENT

The authors acknowledge the assistance and the supply of
specimens by Drs. H. B. Johnson and A. I. A. Abdel-Latif of the
Pittsburgh Corning Corp., Port Allegany, PA, 16743.

REFERENCES

1. J. E. Ritter, Jr., and C. L. Sherburne, J. Amer. Cer. Soc. <u>54</u>,
    (12), 601 (1971).
2. J. E. Ritter, Jr., and M. S. Cavanaugh, J. Amer. Cer. Soc. <u>59</u>,
    (1-2), 56 (1976).

3. B. J. Pletka and S. M. Wiederhorn, "Subcritical Crack Growth in Glass Ceramics", these proceedings.

4. S. Mindess and J. S. Nadeau, Bull. Amer. Cer. Soc., 56 (4), 429 (1977).

5. R. F. Caldwell and R. C. Bradt, J. Amer. Cer. Soc. 56 (3-4), 168 (1977).

6. E. M. Rockar, B. J. Pletka, and S. M. Wiederhorn, "Fracture of Alumina in a Simulated Biological Environment", these proceedings.

7. W. H. Rhodes, R. M. Connon, Jr., and T. Vasilos, Fract. Mech. Cer., Vol. 2, 709, Plenum (1974).

8. F. F. Lange, Ann. Rev. Mat. Sc., Vol. 4, 365, Ann. Rev. (1974).

9. A. G. Evans, Int. of Fract. 10, (2), 251 (1974).

10. R. J. Charles, J. A. P. 29, (12), 1657 (1958).

11. J. Takeda and H. Tachikawa, Mech. Beh. Mat., Vol. IV, 267, Soc. Mat. Sc. Japan (1972).

12. J. Lankford, J. Mat. Sci. 12, 791 (1977).

13. J. E. Ritter, Jr., "Engineering Design and Fatigue Failure of Brittle Materials" these proceedings.

14. Foamglas H/LB, Pittsburgh-Corning Corp., Port Allegany, PA.

15. B. Long, U. S. Patent 2,123,536 (1938).

16. W. D. Kingery, H. K. Bowen, and D. R. Uhlman, 109, Int. to Ceramics, Wiley (1975).

17. Section 3.34 ASTM Designation: C240-72, p. 191, Part 13 ASTM Annual Book of Standards, April (1973).

18. J. T. Fell and J. M. Newton, J. Pharm. Sci. 59, (5) 688 (1970).

19. A. Rudnick, A. R. Hunter, and F. C. Holden, Mat. Res. Stnd. 3, (4), 283 (1963).

20. O. Vardar and s. Finnie, Int. J. Fract. 11, (3), 495 (1975).

21. J. E. Ritter, Jr., J. A. P. 10, (1), 340 (1969).

22. A. G. Evans and H. Johnson, J. Mat. Sc. 10, 214 (1975).

23. H. C. Chandan, R. C. Bradt, and G. E. Rindone (to be published).

24. G. C. Sih, Eng. Fract. Mech. 5 (2) 365 (1973).

25. G. C. Sih, Int. J. Fract., 10 (3) 305 (1974).

26. M. Adams and G. Sines, UCLA-ENG-7537, 113 (1975).

27. J. J. Petrovic and M. G. Mendiratta, J. Amer. Cer. Soc. 59, (3-4) 163 (1966).

STATIC FATIGUE BEHAVIOR IN

CHEMICALLY STRENGTHENED GLASS

Carlton E. Olsen

IBM Corporation
5600 Cottle Road
San Jose, CA 95193

The static fatigue behavior of a glass which was chemically strengthened by ion exchanging K for Na is compared with an annealed glass. The calculated stress profile, as a function of both residual stress and load, is compared with the experimental stress level at which failures become time dependent.

INTRODUCTION

The successful use of glass in structural applications requires that the imposed stress level be carried for the service life of the part. The strength of many annealed glasses is time dependent due to the propogation of surface microcracks through a stress induced environmental interaction called static fatigue. An experimental study of the static fatigue of annealed glass by Mould etal 1-4 provided verification for many important parts of the theory of static fatigue proposed by Charles. 5-7

Chemical strengthening of the glass component reduces the chance of failure by increasing the strength. Such glasses are subjected to an ion exchange process which results in high residual compressive stresses on the surface.

Static fatigue of chemically strengthened glasses has received relatively little attention in the literature. A proof test of the long term strength reliability of ion exchanged glasses, similar to the work done by Wiederhorn and co-workers 8-12 on annealed glass would be valuable. Such tests

indicate the duration of time a glass component is likely to withstand an applied stress level after it has carried a stress overload for a specified time period. Components passing such a proof
test are considered to contain a flaw population of sufficiently
small length that they will not elongate to failure under the applied load in a time period of interest. A value for the stress
intensity factor, Kc, is needed for this type of fracture mechanics analysis technique.

Evans[21] reviews several methods for observing the crack
elongation as a function of load in annealed glass samples,
leading to a value of Kc. Unfortunately, these methods could not
be applied to chemically strengthened samples because the high
internal stresses cause a surface crack under load to instantly
propogate to failure.

Hagy [13] used a method suggested by Charles [6] to compare the
time dependent strength properties of annealed and chemically
strenthened glass. His investigation of the effect of loading
rate on the modulus of rupture (MOR) was interpreted by Ritter [14]
in terms of a factor called stress corrosion susceptibility.

Ritter [14] and Phillips [15] conclude that static fatigue should not
occur in chemically strenghtened glass until residual compressive
stresses are overcome by imposed tensile stresses which exceed
the static fatigue threshold stress.

No experimental verification was found in the literature for
the propogation of flaws in ion exchanged glass under static load.

It was the intent of this study to do the following:

1) Determine the load conditions which cause time dependent
    failure in chemically strengthened glass as a function
    of ion exchange depth.

2) Determine the combined stress profile of any sample due
    to ion exchange treatment and applied load.

3) Identify the critical depth into the surface where
    stresses change from compressive to tensile.

4) Apply the equation developed by Charles[6] which relates
    flaw length L to applied stress σ_{appl} to the ultimate
    strength σ_{theor}.

$$\sigma_{theor} = 2\, \sigma_{applied} \sqrt{L/\rho}$$

Where ρ is the crack tip radius.

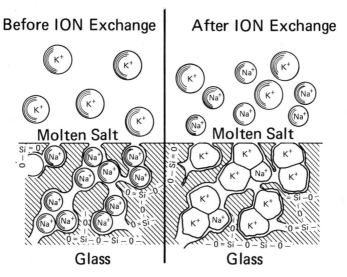

Figure 1. Compressive stress on glass surface due to ion exchange of K for Na (Nordberg[20])

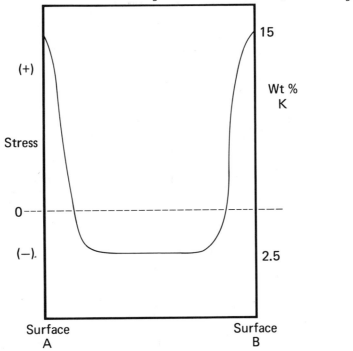

Figure 2. Section through ion exchanged glass show-ing composition and stress profiles.
+ = compressive stress

CHEMICALLY STRENGTHENED GLASS

The glass under consideration was a Corning sodium alumino-silicate composition (0317), which was immersed in a salt bath, allowing potassium to ion exchange with sodium on the surface. The larger size of the potassium ions caused crowding and distortion in the silicate network, which resulted in residual surface stresses in the glass. The change produced by ion exchange is illustrated in Figure 1.

The ion exchange process is diffusion controlled and the stress profile is directly proportional to the potassium concentration.[20] Figure 2 shows the combined stress and composition profile through the thickness of a section of glass. The residual compressive stress on the surface is offset by a small central tensile stress. The depths of zero stress are called the neutral axes. For the glass under consideration, the potassium gradient is controlled so that the neutral axis occurs 0.010 inches below each surface.

STRESS MODEL OF ION EXCHANGED GLASS COUPONS

When a stress is applied to a chemically strengthened glass, the neutral axes shift in response to the stress. A tensile load causes the neutral axis to move toward the surface and reduces the compressive pre-stress at the surface. Subsurface regions that were in a compressive field under no load conditions are exposed to tensile stresses when the load is applied. Clearly, the stress condition at a flaw tip is dependent on the original depth of exchange, length of the flaw and the magnitude of the load.

In order to relate time dependent failure to stress conditions, a program was written to calculate the stress distribution through the thickness of the test coupons, at the point of maximum deflection, for any applied load. Loads were applied in four point bending as shown in Figure 3.

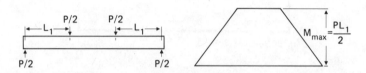

Figure 3.   Load and moment diagrams for 4 point bending.

Figure 5 illustrates the change in stress distribution for no-load,   100 and 200 lbs, respectively. The no-load condition shows a high compressive stress (85 ksi) at the surface with a central tensile stress of 5 ksi about 0.010 inches below each surface. As load is applied (eg., Figure 10 b), the left side represents the bottom surface which is placed in tension while the upper surface (right side) shows an increased compressive stress. At a load of 200 lbs, the outermost surface still has 50 ksi compression, but high central tensile forces (e.g., 40 ksi) approach the surface. This load usually causes fracture in an unabraded specimen.

The position of the neutral axis as a function of both ion exchange depth and imposed load was calculated on the basis of equal areas of compressive and tensile stress in a stress - thickness profile, as depicted in Figure 4.

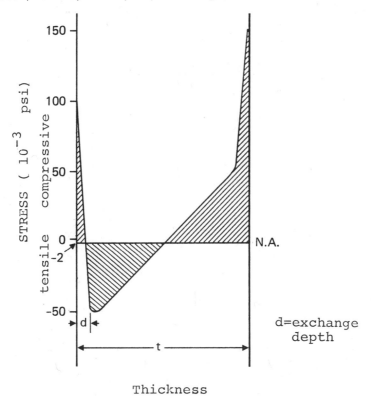

Figure 4   Cross section of sample coupon under a load showing
           stress profile and the depth of the neutral axis based
           on equal tensile and compressive stress area.

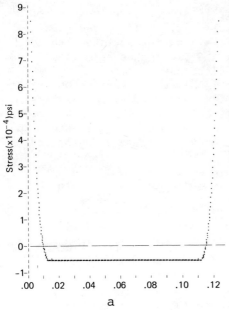

Figure 5. Stress profile through coupon thickness

a. No load on 10 mil exchange depth (measured photo-elastically)

b. Stress profile for 100 lb. load.

c. Stress profile for 200 lb. load (approaching fracture)

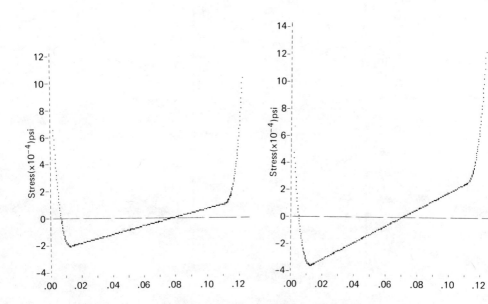

SAMPLE PREPARATION

The samples used in the experiment were prepared and individually numbered by Corning. The specimens were 1 x 3 x 0.125 inch coupons of 0317 glass in which K^+ was ion exchanged with Na^+. The original ion exchange depth was measured by Corning using photo elastic techniques. All samples had the same initial ion exchange treatment, in which the compressive stress layer extended to a depth of 0.010 inches below each surface.

Variations in ion exchange depth were produced by etching away the potassium rich outer layers with dilute HF acid. Measurement and verification of the profile resulted in samples having compressive layers 10, 8.34, 7.62, 6.49 and 3.25 mils deep.

After etching, all samples were uniformly abraded in a central area on the tensile side and stored in DI water prior to testing.

SAMPLE TESTING

A typical breaking stress was determined for each ion exchange depth by 4 point loading in an Instron tester at a constant cross head speed of .005 inches per minute (ipm).

A population of samples for each exchange depth was then tested as follows:

1) The sample was pre-loaded at .05 ipm to 75% of the estimated modulus of rupture (MOR) for the exchange depth.

2) Hold at static load for 2.5 minutes.

3) Increase the load in small, fixed increments by advancing the cross head at .005 ipm for a fixed time period.

4) Hold at new load for 2.5 minutes.

5) Automatically repeat steps 3 and 4 until the sample fails.

6) Record the sample number, load and hold time when when failure occurred. (Fortunately most (80%) samples failed at constant load rather than while the load was being increased.)

A population of samples for each exchange depth was loaded
to failure at liquid nitrogen temperatures using a 0.05 ipm load
rate to determine the MOR in the absence of conditions believed
to cause static fatigue.

RESULTS

TABLE 1

Computer Calculated Stress Profile

Exchange Depth mils	Applied Stress (psi)	Surface Stress (psi)	Depth of Neutral Axis (mils)	Depth of Max Tensile Stress (mils)	Max Applied Tensile Stress psi
10	0	130800	10.1	13.8	5260
	19500	111500	6.9	13.2	20500
	39000	92100	5.1	12.8	35800
	48800	82300	4.3	12.6	43600
	53600	75600	3.9	12.4	49000
	58500	72700	3.7	12.4	51300
8.34	0	91600	8.3	12.0	5200
	19500	72700	5.3	11.6	20500
	39000	53900	3.5	11.0	35800
	48800	44400	2.7	10.8	43600
	53600	39700	2.3	10.8	47500
	58500	38000	2.1	10.6	51300
7.62	0	77200	7.7	11.4	5260
	19500	58600	4.5	10.8	20500
	39000	39000	2.7	10.4	35800
	43900	35200	2.3	10.2	39700
	48700	30600	1.9	10.2	43600
6.49	0	57600	6.5	10.2	5200
	19500	39200	3.5	9.6	20500
	29300	30100	2.5	9.4	28100
	34100	25500	1.9	9.4	31900
	39000	20900	1.5	9.2	35800
3.25	0	18600	3.3	6.8	5200
	9750	9900	1.5	6.8	12800
	14600	5700	0.7	6.6	16600
	19500	1300	0.1	6.4	20500

TABLE 2

Measured Strength Values for Ion Exchanged Glass

Exchange Depth (mils)	AVG MOR (psi)	Std Deviation (psi)	No. of Spls	Avg* Fail- Time (sec)	MOR at Liq N_2	Std Dev	No. of Spls
10	52842	1367	20	39	55801	772	9
8.34	55495	3503	20	61	57584	4832	7
7.62	46083	2753	18	47	52204	1929	5
6.49	42422	3146	20	73	50170	3926	6
3.25	20111	3037	19	56	35802	5378	5

* Maximum possible failure time 150 sec

TABLE 3

Comparison of Measured Data With the Computer Model

Exchange Depth (mils)	MOR (psi)	Depth of Neutral Axis (mils)	Depth Where σ_{theor} 3x10^{6}psi (mils)
10.0	52840	3.9	3.95
8.34	55500	2.3	2.35
7.62	46100	2.1	2.20
6.49	42400	1.7	1.80
3.25	20100	1.	0.2

DISCUSSION OF RESULTS

In his previously cited experiment, Hagy[13] studied the effect of stress rate on the modulus of rupture of chemically strengthened glass coupons. He observed a decrease in the average fracture stress as the stress rate was decreased. He did not report failures occurring at constant load. Hagy observed about a tenfold increase (e.g., 4000 to 40000 psi) in the MOR at a given stress rate when annealed glass was chemically strengthened.

In this experiment, samples of several exchange depths were step loaded in regular increments, so that each sample carried each load for 2.5 minutes. If no failure occurred, the load was increased to a higher but constant value. It was noticed that samples failed at constant loads that they were able to support for some time period.

Such behavior in materials which do not undergo plastic deformation is indicative of a flaw growth mechanism. It is possible that in this case the mechanism is static fatigue.

It is recognized that the time period during which the sample was held was quite short when compared to product strength reliability lifetimes. It can be argued that the observed failure stresses might decrease if the hold time were extended.

For each sample of a given ion exchange depth, the load which caused failure was combined with the dimensions of that sample to determine that sample's MOR. The MOR value for all samples of a particular exchange depth were used to compute the mean MOR and standard deviation for that chemical strengthening treatment.

A program called STRESSCALC was written and used to calculate the stress level at any point throughout the thickness of a sample through a section of the length where maximum deflection occurred. The program superimposes the photoelastically measured stress profile resulting from the ion exchange depth with the stress condition caused by a specified load. Calculations were made at several load increments up to the load which caused failure for each of the five ion exchange depths.

The program listing gives the combined applied stress as a function of depth. It also calculates a theoretical stress value from an equation given by Charles[6] where:

$$\sigma_{theor} \cong 2\sigma_{appl.} \sqrt{L/\rho}$$

The expected theoretical strength is related to an applied stress, $\sigma_{applied}$, in a sample containing a crack of length L which has a radius ρ assumed to be 3 angstroms. Charles assumes that the maximum expected value for σ_{theor} should be $2\text{-}3\times10^6$psi.

STRESSCALC uses the combined applied stress value at each depth increment to calculate σ_{theor}, assuming that a crack of length L is present at that depth.

It was noticed that the theoretical stress is always compressive at the surface and becomes more tensile as the depth increases. When $\sigma_{theor} \gtrsim 3\times10^6$psi a failure condition exists at that depth if a flaw of that length is present. The failure criteria is satisfied at some depth for each combination of load and exchange depth even if failure doesn't occur. If the sample does fail, it is assumed that flaws were present at least to the depth where the value of $\sigma_{theor} \geq 3 \times 10^6$ psi. Since failure was observed at constant load, it is assumed that flaws were of sufficient length to reside in a tensile stress field so that they could elongate.

SUMMARY AND CONSLUSIONS

The modulus of rupture of chemically strengthened glass coupons was measured at static loads for several ion exchange depths. The stress field of each sample was calculated as a function of ion exchange depth and load.

The crack length required to have the crack tip reside in a tensile stress field was determined for loads up to the one which caused failure.

The time dependent failures observed at constant load suggest a flaw growth mechanism which could be static fatigue.

Studies of chemically strengthened glass which provide a means of determining the critical stress intensity factor Kc is still needed, in order to apply fracture mechanic's analysis techniques.

ACKNOWLEDGEMENTS

The author is pleased to acknowledge valuable discussions with Dr. Robert Atkin of IBM and Dr. J. McCartney of Corning Glass regarding this work. He is also grateful to Mr. Robert Wolmsely and Dr. George Henry for their assistance with certain aspects of the programming.

REFERENCES

1. R. E. Mould and R. D. Southwick, J. Am. Ceramic Society 42, 542 (1959).

2. ibid, 582 (1959).

3. R. E. Mould, J. Am. Ceramic Society, 43, 160 (1960).

4. R. E. Mould, J. Am. Ceramic Society, 44, 481 (1961).

5. R. J. Charles, Progress in Ceramic Science, Vol. 1, p. 1.

6. R. J. Charles, J. Appl. Physics, 20, 11, p. 1549 (1958).

7. R. J. Charles, J. Appl. Physics, 29, 12, p. 1657 (1958).

8. S. M. Wiederhorn and A. G. Evans (1973), NBSIR 73-147-Proof Testing of Ceramic Materials - An Analytical Basis for Failure Prediction.

9. S. M. Wiederhorn, NBS Report 8901, Effect of Environment on the Fracture of Glass, June 1965.

10. S. M. Wiederhorn, L. H. Bolz, J. Am. Ceramic Society, 53, 543 (1970).

11. S. M. Wiederhorn, J. Am. Ceramic Society, 50, 407 (1967).

12. S. M. Wiederhorn, A. Evans, D. Roberts, Fracture Mechanics Study of the Skylab Windows, Proceedings of Conf. on Fracture at PA State University (July 1973).

13. H. E. Hagy (1966), Central Glass Ceramic Res. Instit. Bulletin, 13 (1) 29.

14. J. E. Ritter, Physics and Chemistry of Glasses, Vol. II, No. 1, p. 16 (1970).

15. C. J. Phillips, Fracture, An Advanced Treatise, Vol. 7, Chapter 1, Academic Press (1972).

16. F. M. Ernsberger, Glass Industry, pp. 21-36 (June 1966).

17. R. J. Charles and W. B. Hillig, "High Strength Materials,
     V. Zackay editor, Wiley, p. 682-705 (1965).

18. R. J. Roark, Formulas for Stress and Strain, McGraw-Hill,
     p. 360-1 (1965)

19. A. J. Burggraaf, PhD Thesis, De Technische Wetenschappen
     Hogeschool Te Eindhoven September, 1965.

20. H. M. Garfunkel et al, J. Am. Ceramic Society, $\underline{47}$, 215
     (1964).

21. A. G. Evans, Fracture Mechanics of Ceramics, Vol I, pp 17,
     Plenum Press 1973.

PREDICTION OF THE SELF-FATIGUE OF

SURFACE COMPRESSION-STRENGTHENED GLASS PLATES

M. Bakioglu, F. Erdogan

Lehigh University, Bethlehem, Pa.

D. P. H. Hasselman

Virginia Polytechnic Institute and

State University, Blacksburg, Va.

ABSTRACT

A fracture-mechanical failure-prediction of the spontaneous fatigue of compression-strengthened glass plates in the absence of an applied load is presented. It is suggested that such a fatigue mechanism will occur for any surface crack with sufficient depth such that it penetrates into the tensile stress zone within the plate interior. The solutions for the stress intensity factor for plates with various internal stress distributions obtained previously are presented briefly. The conditions for no slow crack growth, crack arrest and catastrophic failure are established. The results obtained are illustrated on the basis of a surface-compression strengthened glass plate in a water environment at 25°C. The general results show that the problem of spontaneous fatigue can be reduced by increasing the plate thickness and by having moderate surface stress levels.

INTRODUCTION

The strength of brittle materials, particularly their impact and fatigue resistance, can be improved significantly by introducing residual stresses into the medium which are compressive at and near the surface. Such compressive stresses can be created by tempering, ion-exchange, or cladding with another material having a lower coefficient of thermal expansion. Particularly for glasses,

787

strengthening by either tempering or ion-exchange has found wide applications for a large variety of industrial and consumer products. However, glasses strengthened in this fashion appear to be susceptible to spontaneous fragmentation even during the complete absence of applied loads*. This behavior has been attributed to inclusions of NiS in the glass [1].

A qualitative consideration of other possible factors which could be responsible for this type of spontaneous fracture suggested that it also could be caused by the slow growth of surface cracks which may have been caused by an incidental impact at some time prior to the fragmentation. The driving force for this subcritical crack propagation is provided by the internal stresses. Slow crack growth would be possible if the initial surface crack is sufficiently deep so that the crack tip is in the tensile stress zone in the material interior.

Methods of failure prediction for brittle ceramics subjected to stress corrosive environments are well established and have been applied successfully to the prediction of static fatigue [2], strain-rate sensitivity [3], single-cycle thermal shock [4], as well as thermal fatigue [5,6]. It should be possible to carry out a similar failure analysis of the static fatigue of strengthened glasses subjected to internal stresses only, provided the proper fracture mechanics analysis is available. Such an analysis was reported by the present authors elsewhere [7] and illustrated with a numerical example for a glass plate with a single thickness. The purpose of this paper is to briefly report the analytical approach used in the analysis and for comparative purposes present additional numerical results for plates of different thicknesses.

FORMULATION AND SOLUTION OF THE CRACK PROBLEM

The two-dimensional problem under consideration is described in Figure 1. It is assumed that a homogeneous, isotropic, elastic plate of thickness h is under a given state of residual or internal stresses which is compressive near and at the surface and tensile in the interior. The plate is otherwise free from the external loads. Thus, the internal stresses are in self-equilibrium.

It is assumed that the plate contains an edge crack which is perpendicular to the surface and has a depth b, where $b_o < b < h$ as shown in Figure 1a. Because of the compressive stresses near

---

*Such a problem related to the spontaneous fracture of lenses in eyeglasses was encountered during a consulting case by one of the authors (DPHH).

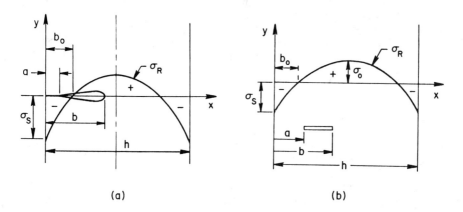

Fig. 1.   Surface crack configuration and residual stress distribution in surface-compression strengthened glass plate.

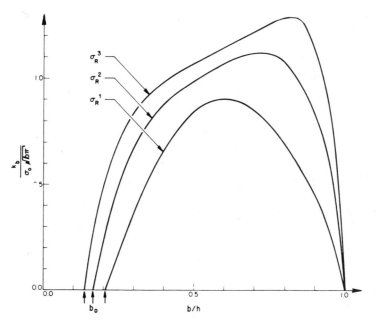

Fig. 2.   Stress intensity factor for surface crack in surface-compression strengthened glass plate for second order (σ_R^1), fourth order (σ_R^2) and sixth order (σ_R^3) internal stress distribution.

the boundary, the crack faces along $0 \leqq x \leqq a$ will be closed and will be under compression, the distance a being one of the unknown in the problem. The crack is open only between x=a and x=b, and because of the smooth contact of the crack surfaces, at x=a the "crack tip" has a cusp shape rather than being parabolic. This means that if the problem is treated as a contact-crack problem, at the end point x=a the stress intensity factor k_a must be zero. Thus, the condition:

$$k_a = o \tag{1}$$

provides the additional information in the solution of the problem to determine the distance a.

The problem shown in Figure la is solved for a given distribution of internal stress by considering an arbitrarily located crack along $a < x < b$ (a > o, b < h), as shown in Figure lb, and for a fixed b, determining the value at which $k_a = 0$, which also yields the corresponding value of K_b. Following the mathematical procedure reported in detail previously [7], stress intensity factors were generated for three symmetric internal stress fields expressed by

$$\sigma_R^1 (x) = \sigma_o [1-3(2x/h-1)^2], \tag{2}$$

$$\sigma_R^2 (x) = \sigma_o [1-5(2x/h-1)^4], \tag{3}$$

$$\sigma_R^3 (x) = \sigma_o [1-7(2x/h-1)^6] \tag{4}$$

where σ_o is the tensile stress at the midplane of the plate (Figure 1). A parabolic stress distribution (eq. 2) is typical for a tempered glass, whereas the 6th degree polynomial distribution (eq. 4) may be more representative of a glass strengthened by ion exchange. Figure 2 shows the numerical results for K_b for these three stress distributions.

For the calculation of fatigue-life the rate of slow crack growth as a function of stress-intensity factor or crack depth can be expressed generally by:

$$\dot{b} = G(K_b) = G[F(b)] = g(b) \tag{5}$$

For a glass with a fatigue limit, K_T, slow crack growth will occur only when

$$K_b > K_T \tag{6}$$

or

$$[F(b)]_{max} > K_T \text{ and } b_i > b_1 \tag{7}$$

where b_i is the initial crack length introduced into the plate at $t = o$ and b_1 is the minimum crack depth necessary for the initiation of slow crack growth obtained from (see insert in Figure 1)

$$F(b_1) = K_T, \ F'(b_1) > o \tag{8}$$

The slow crack growth would be eventually arrested if

$$K_T < [F(b)]_{max} < K_{IC}, \ b_i > b_1 \tag{9}$$

In this case, from eq. 5 the time to crack arrest (T_a) may be obtained from

$$T_a = \int_{b_i}^{b_a} \frac{db}{g(b)} \tag{10}$$

where b_a is the second root of eq. 8 i.e., (see Figure 2)

$$F(b_a) = K_T, \ F'(b_a) < 0. \tag{11}$$

On the other hand, if, in addition to the slow crack growth condition eq. 7, the condition

$$[F(b)]_{max} > K_{IC} \tag{12}$$

is satisfied, then the plate will fail catastrophically. The time to failure may be obtained from

$$T_f = \int_{b_i}^{b_2} \frac{db}{g(b)} \tag{13}$$

where the critical crack length b_2 indicated in Figure 1 is obtained from

$$F(b_2)=K_{IC}, \quad F'(b_2) > 0. \tag{14}$$

NUMERICAL EXAMPLE AND RESULTS

The analytical approach and results presented above were applied to soda-lime glass plates of thickness 2 and 4 mm with an internal stress distribution given by eqs. 2 or 4, exposed to a water environment at 25°C. As measured by Wiederhorn and Bolz [8], the rate of slow crack growth may be expressed as

$$(db/dt) = V_o \exp [(cK_b-E)/RT] \tag{15}$$

where V_o and c are constants, and the activation energy E = 1.088 x 10^5 Jmole^{-1}, the gas constant R=8.32 Jmole^{-1} °C^{-1} and temperature T = 298°K for the present study. As shown in Figure 3 the rate of slow crack growth for this glass in water can be described quite accurately by two bilinear regions between a fatigue limit,

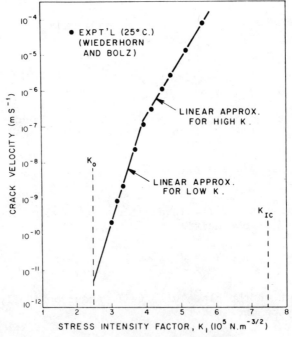

Fig. 3.   Slow crack growth in a soda-lime
-silica glass in water at 25°C.

$K_T \approx 2.49 \times 10^5$ N.m$^{-3/2}$ and the critical stress intensity factor, $K_{IC} = 7.49 \times 10^5$N.m$^{-3.2}$. For these two bi-linear regions V_o and C are [11]:

$$\ln V_o = -1.08, \quad C = 0.188 \text{ for } K_b < 3.62 \times 10^5 \text{N.m}^{-3.2}$$

$$\ln V_o = 10.3, \quad C = 0.110 \text{ for } K_b > 3.62 \times 10^5 \text{N.m}^{-3/2}$$

For an optical borosilicate glass exposed to moist air the crack-growth behavior is almost identical [9] to the bilinear crack-growth behavior chosen for the present example. Accordingly, the results obtained are not limited only to the specific glass chosen for the present example.

For the calculations of crack stability and times-to-failure, the initial crack-depth b was shown to be just sufficient to start the crack-growth i.e., at an initial stress intensity factor $K_i = K_T$ and corresponding initial rate of crack-growth db/dt $\approx 10^{-11.3}$ m.s.$^{-1}$. For the parabolic internal stress distribution and plate thicknesses of 2 and 4mm, the results are shown in Figures 4a and 4b, respectively. The corresponding results for the 6th order po-lymonial stress distribution are given in Figures 5a and 5b, respectively.

In addition to their quantitative significance the data shown in Figures 4 and 5 are quite illustrative of the self-fatigue and fracture phenomena which can be encountered in surface compression strengthened glass plates. The insert in Figure 4a is intended to assist in the interpretation. As defined earlier, the crack-depth b_1 for a given stress distribution is the minimum crack depth re-quired for slow crack growth to take place at a value of initial crack intensity factor $K = K_T$. It should be noted that the value of b_1 decreases with increasing level of surface stress. The stress levels σ_T and σ_c correspond to those stress levels at which during crack growth the maximum values of the stress intensity fac-tor K_b correspond to the fatigue limit K_T and the critical stress intensity factor K_{IC}, respectively.

As indicated in Figures 4 and 5, at low levels of surface stress $\sigma_s < \sigma_T$ no crack growth will occur, regardless of crack depth. At intermediate levels of surface stress, $\sigma_T < \sigma_s < \sigma_c$ such that $K_T < K_b < K_{IC}$, slow crack growth will take place. However, all growth will be in a stable manner until the crack is arrested at a value of b_a/h corresponding to $K_b = K_T$. The time for such crack arrest, T_a as expected, decreases with increasing level of surface stress.

At values of surface stress $\sigma_s > \sigma_c$ the crack will undergo

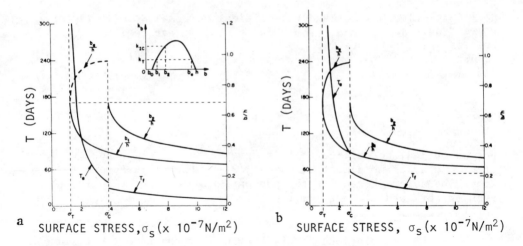

Fig. 4.   Time-to-crack arrest (T_a) and time-to-fracture (T_f) for surface compression strengthened glass plate with parabolic internal stress distribution in water at 25°C.   Plate thickness: a, 2mm and b, 4mm.

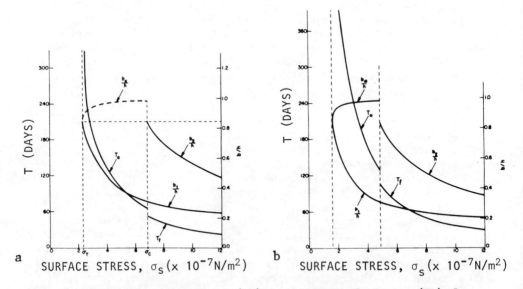

Fig. 5.   Time-to-crack arrest (T_a) and time-to-fracture (T_f) for surface compression strengthened glass plate with sixth order polynomial internal stress distribution in water at 25°C.   Plate thickness: 2, 2mm and b, 4mm.

slow crack growth in a stable manner until $K_b \geq K_{IC}$ at a value of
crack length b_2 and failure time T_f at which time catastrophic
fracture will occur (it should be noted that the time T_f decreases
with increasing level of surface stress). The discontinuity in the
time curve at $\sigma_s = \sigma_c$ is due to the fact that the subcritical crack
propagation for $\sigma_s < \sigma_c$ takes place for the increasing as well as
the decreasing branch of the K_b curve (see insert for 4a), whereas
for $\sigma_s > \sigma_c$, slow crack growth takes place only for the increasing
branch. The results shown in Figures 4 and 5, were calculated for
a crack depth, b_1 is just sufficient for slow crack growth to take
place. It is critical to note that for an initial crack depth b_i
$> b_1$, the stress levels required for slow crack growth and fracture
and corresponding times-to-failure are expected to be lower than
those indicated in Figures 4 and 5. Also, for lower values of the
fatigue limit K_T, smaller values of initial crack depth, b_1 than
those shown in figures 4 and 5 will be sufficient for slow crack
growth to occur.

     Comparison of the results for the two plate thicknesses shows
that on the basis of given values of b/h, the thicker glass plates
require a lower value of b_1/h, as well as lower values of surface
stress σ_T and σ_c for crack growth and catastrophic fracture to occur, accompanied by a slight corresponding increase in time-to-
crack arrest, T_a and time-to-catastrophic failure, T_f. However,
for a crack of given depth (i.e., non-equal values of b_1/h, the 4
mm thick plate requires a much higher value of surface stress for
the crack to become unstable than the 2 mm thick plate.

     Comparing the results for the two stress distributions a num-
ber of effects may be noted. First, for both plate thicknesses the
stress levels σ_T and σ_c are significantly higher for the 6th order
polynomial internal stress distribution than for the parabolic stress distributions. Secondly, for both plate thicknesses, the time-
to-crack arrect, T_a and time-to-failure, T_f are significantly higher
for the 6th degree polynomial internal stress distribution than for
the parabolic stress distribution. Thirdly, for low values of surface stress the 6th degree polynomial stress distribution requires
a higher value of crack depth for self-fatigue to occur than the
parabolic internal stress distribution. In contrast, at the higher stress levels, the 6th degree polynomial stress distribution
requires a lower value of initial crack depth for crack growth to
occur than the parabolic stress distribution. The above effects
are entirely consistent with the effects expected from the different stress distributions.

                              DISCUSSION

     From the analysis and numerical results presented in this
paper a number of general conclusions can be drawn. Most important-

ly, it is clearly demonstrated that spontaneous fracture resulting
from static self-fatigue does indeed, represent a possible mode of
failure of surface compression strengthened glasses. In fact, for
high values of internal stress level and initial crack depth the
failure times can be rather short.

Although the increased impact resistance of surface compression strengthened glasses must be considered a major advantage, the
possibility of spontaneous catastrophic fracture of such glasses
which may have received major surface damage, must be considered a
definite hazard. Even if the crack never becomes critical and propagated in a slow manner only, the remaining net ligament on crack
arrest could be just a small fraction of the plate thickness. This
could weaken the plate to such an extent that only a small applied
load would be required for complete failure, again giving rise to a
hazardous condition. For glass such conditions are particularly
hazardous since flaws of sufficient size to cause catastrophic failure are very difficult, if not impossible, to detect by current methods of non-destructive flaw detection techniques.

The incidence of the spontaneous failure due to self-fatigue
of surface compression strengthened glass plates, lenses, or other
objects can be reduced by developing glasses with high values of K_T
and K_{IC}, in combination with low rates of crack growth even in highly stress-corrosive environments. For a given glass, with flaws of
given thickness, increases in the thickness of the glass plate will
also reduce the tendency of self-fatigue. Furthermore, with the exception of very high stress levels, for maximum resistance to self-
fatigue internal stress distributions described by polynomials of
higher order than two are preferred to a parabolic stress distribution, for the number of reasons outlined earlier. Also, a moderate
value of surface stress level for the higher order polynomial is also preferred in order to reduce the probability for fatigue even for
deep cracks. An additional reason exists for moderate surface stress
levels. Fracture of brittle materials at higher levels generally
occur with the formation of numerous fragments with high kinetic
energy, which could result in personal injury. Fracture at low or
moderate stress levels generally results in fewer fragments with
low kinetic energy. As a trade-off, it appears advisable then to
have surface stress levels of sufficient magnitude to give an enhancement of impact resistance, but not sufficiently high to create
a hazardous condition if failure by self-fatigue should occur.

ACKNOWLEDGMENTS

This study was carried out as part of a research program spon-
sored by the Army Research Office (Triangle Park, N.C.) under Grant
DAAG29-76-0091 and by the National Science Foundation under Grant
ENG73-045053 A01. Manuscript preparation in part was also perform-

ed under the auspices of a Senior Scientist Award presented to one of the authors (DPHH) by the German Government.

REFERENCES

1. L. Merker, Glastechn. Ber. $\underline{47}$ 116 (1974).
2. A. G. Evans, J. Mat. Sc., $\underline{7}$ 1137-46 (1972).
3. A. G. Evans, Int. J. of Fracture, $\underline{10}$, 251-59, (1974).
4. R. Badaliance, D. A. Krohn, and D. P. H. Hasselman, J. Amer. Ceram. Soc., $\underline{57}$, 432-47 (1974).
5. D. P. H. Hasselman, R. Badaliance, C. H. Kim, and K. R. McKinney, J. Mat. Sc., $\underline{11}$, 458-64 (1976).
6. D. P. H. Hasselman, E. P. Chen, E. L. Ammann, J. E. Doherty, and C. G. Nessler, J. Amer. Ceram. Soc., $\underline{58}$ (11-12) 513-16, (1975).
7. M. Bakioglu, F. Erdogan, and D. P. H. Hasselman, J. Mat. Sc., $\underline{11}$ 1826-34 (1976).
8. S. M. Wiederhorn and L. H. Bolz, J. Amer. Ceram. Soc., $\underline{53}$, 543 (1970).
9. S. Wiederhorn in Fracture Mechanics of Ceramics, Ed. by R. C. Bradt, D. P. H. Hasselman and F. F. Lange (Plenum Press) 1975.

FRACTURE MECHANICS AND MICROSTRUCTURAL DESIGN

F. F. Lange

Structural Ceramics Group
Rockwell International/Science Center
Thousand Oaks, California 91360

1.0 INTRODUCTION

Two areas stand out among others where fracture mechanics concepts can be used in aiding the design of ceramic microstructures for optimum mechanical integrity. The first is the effect of microstructure size (viz. grain and/or inclusion size) on the formation of microcracks during fabrication. Sufficient evidence exists showing that microcracks developed by residual stresses arising during fabrication will only occur when the grain size/inclusion size exceeds a critical value. This evidence, combined with theoretical explanations is sufficient to suggest that both single- and poly-phase materials can be fabricated with high mechanical integrity despite the development of large residual stresses during fabrication. The second area is the effect of microstructure on fracture toughness. Review of this subject will include crack interaction with particulates, microcrack zones at crack fronts, fibrous fracture, and stress-induced phase transformations.

2.0 SIZE EFFECTS ON THE FORMATION OF MICROCRACKS

During cooling from the fabrication temperature, residual stresses can arise due to thermal expansion anisotropy, thermal expansion mismatch between different phases, and phase transformations. Past experience has generally indicated that residual stresses should be avoided since they usually lead to bodies that range in strength from generally not being strong, as is the case

for many tradition refractories and whitewares, to those which can
be crumbled as is the case for instabilized ZrO_2. Isolated and
more recent experience has shown that many ceramics known to
contain high residual stresses can be strong and fabricated without
developing microcracks. As illustrated below, the thread of
similarity between these isolated observations is concerned with
the size of the microstructural feature, viz., a critical size
exists which separates good and poor mechanical integrity.

2.1   Observations Concerning Single-Phase Materials

2.1.1   Thermal Expansion Anisotropy. Most noncubic ceramics
exhibit thermal expansion anisotropy. During cooling from the
fabrication temperature, the shape change of each grain desired by
the anisotropy is constrained by neighboring grains to produce
complex stress distributions. The explicit nature of the stress
distribution is unknown, but approximate stress levels have been
computed by Buessem [1] who modeled the polycrystalline body as
parallel arrangements of constrained grain pairs with arbitrary
orientations. This model results in a relation for the stress
perpendicular to the common boundary:

$$\sigma = \left[\frac{(\alpha_n + \alpha_n')}{2} - \bar{\alpha}\right]\frac{\Delta T\ E}{(1-\nu)} \quad , \tag{1}$$

where α_n and α_n' are the thermal expansion coefficients of the two
grains in a direction normal to the grain boundary, $\bar{\alpha}$ is the
average thermal expansion of the surrounding material, ΔT is the
temperature change and E, ν are the elastic constants of the
(assumed elastically isotropic) material. By their nature, such
residual stresses are highly localized to the volume of material
adjacent to the grain boundary.

It is well documented that the thermomechanical stressed due
to thermal expansion anisotropy cause microcracks to develop in
$Al_2O_3 \cdot TiO_2$,[2] $MgO \cdot 2TiO_2$,[3] $Li_2O \cdot Al_2O_3 \cdot 2SiO_2$,[4] Nb_2O_3,[5] Gd_2O_3,[6]
TiO_2,[7] and coarse-grained Al_2O_3.[8] Microcracks also develop in
BeO [9] which exhibits radiation induced expansion anisotropy.
Also well documented for most of these same materials, but little
appreciated, is the fact that microcracks need not develop when
the grain size is kept below a critical value.

The best documented evidence for this fact has been given by
Kuszyk and Bradt [10] for $MgO \cdot 2TiO_2$. These workers hot-pressed a
large billet of $MgO \cdot 2TiO_2$, annealed specimens for various periods
to promote grain growth, and then measured strength, elastic

modulus, sonic attenuation, thermal expansion and fracture toughness as a function of grain size. Their elastic modulus and attenuation results, shown in Fig. 1, concur with their other data and
direct observations to show that despite the high residual thermomechanical stress, microcracks do not develop in $MgO \cdot 2TiO_2$ when
the grain size is $\tilde{<}$ 1μm. More recently, similar grain-growth/
elastic modulus/internal friction/thermal expansion measurements

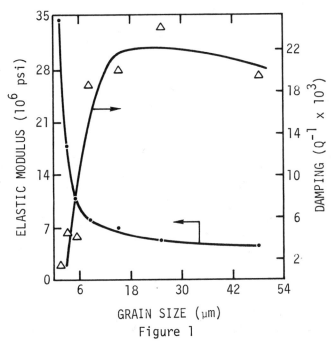

GRAIN SIZE (μm)

Figure 1

performed by Dole et al.[11] and have shown that microcracks do not
develop in HfO_2 for grain sizes $\tilde{<}$ 3μm. Similar, but more quantitative grain size/property data have been made for Nb_2O_3,[5]
TiO_2,[7] BeO_2,[9] and Al_2O_3.[8] Due to its lower thermal expansion
anisotropy relative to the other materials, Al_2O_3 requires considerable grain growth (> 300-500μm) before microcracks are
observed.[12]

    It would be nice to explain these observations by suggesting
that the magnitude of the residual stresses decrease with decreasing
grain size, but theoretical arguments do not support this hypothesis
and experimental evidence dismisses this argument as evident by the
results of Hickman and Walker[9] for BeO. These investigators
found that contrary to observation for coarse grained BeO (> 7.5μm),
fine grained material (<3μm) "failed to show any evidence of microcracking...." after considerable neutron irradiation, yet X-ray
line broadening experiments revealed the presence of large strains

until the material was purposely crushed to a fine powder. Thus, when the grain size is less than a critical value, microcracks do not form despite the presence of large residual stresses.

   2.1.2 <u>Phase Transformation</u>. Displacive phase transformations have plagued the fabrication of several important ceramics. The $\beta \rightarrow \alpha$ quartz transformation is one example.[12] Because of its tetragonal $\rightarrow$ monoclinic transformation, ZrO_2 could not be fabricated with any mechanical integrity until it was recognized that its cubic structure could be stabilized with the aid of one of several different additives. The tetragonal $\rightarrow$ monoclinic transformation involves a 7% volume increase accompanied by a shear strain of $\sim 18\%$.[14] Since the strain at fracture for high strength ceramics is $< 0.2\%$, it is not unreasonable to observe microcracking in unstabilized, polycrystalline ZrO_2.

   The constraint neighboring grains place on one another during such a volume and shape change must cause high stresses to arise which depend on the crystalographic nature of the transformation, and on the relative orientation of neighboring grains. Large stresses must exist to constrain the transformation. Explicit expressions have yet to be derived for either the stress distribution arising after the transformation or that required to constrain the transformation.

   Most recently, Gupta et al.[15] have shown that the ZrO_2 tetragonal $\rightarrow$ monoclinic transformation can be constrained to result in a polycrystalline, single-phase tetragonal material. Several fabrication parameters are necessary to achieve this result. First, small amounts of Y_2O_3 must be added to lower the transformation temperature, and second, the grain size must be $< 0.3\mu m$. When the grain size exceeds $0.3\mu m$, the amount of transformed monoclinic phase increases and microcracks develop as evident by a significant drop in strength and direct observation.

   Mutsuo and Sasaki[16] indicate that the crystolographic transformation responsible for the perzoelectric properties of $PbTiO_3$ also produces microcracks and that the transformation can occur without microcracking when the grain size is < 3.* Thus,

---

*Matsuo and Sasaki contend that the microcracks are due to thermal expansion anisotropy after the phase transformation. If they are correct, then their observations belong in the previous section. On the other hand, it is not unreasonable to correlate the microcracking to the transformation, i.e. depending on the magnitude of the volume change, the shear strain, and grain size, a displacesive transformation should be able to proceed without microcracking.

it can be concluded that the microcracking usually associated
with phase transformations can be prevented when the material is
fabricated with a grain size less than a critical value.

2.2 Observations Concerning Polyphase Materials

2.2.1 Thermal Expansion Mismatch. Differential thermal
contractions and the constraint of one phase on the other produces
thermomechanical stresses upon cooling from the fabrication
temperature. For this case, explicit expressions for the stress
distribution are available for either isolated spheres[17] or
ellipsoids[18] embedded with a matrix. The stress within the sphere,
the ellipsoid and most probably the irregular-shaped particle is
uniform, whereas the stress within the matrix rapidly falls off
with increasing distance from the partical/matrix interface. For
the case of the sphere with a radius R, the radial and tangatial
stresses within the matrix phase are*[17]

$$\sigma_r = \sigma_{max}(\frac{R}{r})^3$$
$$\text{for } r \geqslant R \qquad (2)$$
$$\sigma_\theta = -\sigma_{max}(\frac{R}{r})^3$$

where σ_{max} is the stress within the sphere given by

$$\sigma_{max} = \frac{(\alpha_p - \alpha_m)\Delta T}{k}$$

where

$$k = \frac{1 + \nu_m}{2E_m} + \frac{(1 - 2\nu_p)}{E_p}$$

and α_p, α_m are the thermal expansion coefficients , E and ν are the
elastic constants and p, m denote the particulate and matrix phase,
respectively. Since 95% of the total strain energy associated with
the embedded particle is contained within a spherical volume of
radius 2R,[19] adjacent particles can be considered isolated from
one another when their centers are separated by a distance of 4R.
If uniform spheres are located on close-packed face-center cubic
lattice points, the above criterion indicates that the spheres are

———————————————————

*Compressive stresses are indicated by a negative sign.

isolated from one another when their volume fraction is $\lesssim 0.10$.

It is widely recognized that when the thermal expansion mismatch is large, the resulting thermomechanical stress field results in microcracks. Sufficient evidence now exists to show that microcracking need not be a result of large thermal expansion mismatch as long as the particle size is maintained below a critical value.

Binns[20] published the first evidence and still the most comprehensive work showing that microcracking not only depends on the magnitude of the thermomechanical stress field, but also on the size of the particulate phase. He fabricated different series of composites containing either Al_2O_3 or $ZrSiO_4$ spheres of different sizes and volume fractions. The different glass matrices had a thermal expansion greater, equal and less than the particulate phase. Strength and elastic modulus measurements, and direct observations determined the presence of microcracks. For large particle sizes, the composites contained microcracks when $\alpha_p \neq \alpha_m$, but microcracks were not present in the same composites when the particle size was small. Davidge and Green [19] confirmed these important results.

Indirect evidence also exists linking particle size to microcracking in two phase systems. In the Si_2N_3-SiC system[21], strength and fracture toughness were measured as a function of the particle size and the volume fraction of the SiC phase. The resulting data indicated that when the particle size of the SiC phase was $\leqslant 5\mu m$, the SiC phase did not change the flaw-size characteristics of the Si_3N_4 matrix, but larger particles produced large microcracks. Other evidence for a particle size effect is even more indirect, e.g. the SiC-Al_2O_3 system[22] can be fabricated as strong bodies for volume fraction of Al_2O_3 between 0.02 to 0.15 despite a differential thermal expansion of $\sim 4x10^{-6}$

2.2.2 <u>Phase Transformations</u>. The effect of particle size on microcracking due to a phase transformation in a polyphase material has been recently given by Porter and Heuer[23] for the case of tetragonal ZrO_2 parcipitates in a cubic ZrO_2 matrix. When the size of the parcipitate exceed $\sim 0.2\mu m$, the matrix no longer constrains the transformation resulting in monoclinic parcipitates which are incoherently bonded to the surrounding cubic matrix. When the particle size is $\sim 0.2\mu m$, the matrix constrains the transformation, resulting in an array of tetragonal parcipitates within a cubic matrix.

2.3 Theoretical Relations

Experimental observations conclude that the microcracking phenomena associated with constrained volume/shape charges occuring during fabrication are not only dependent on the magnitude of the internal stress, but also dependent on the size of the microstructural feature which gives rise to the stress field. Two different, but complementary fracture mechanics analyses support this important conclusion.

2.3.1 <u>Energetic Approach</u>. The energetics of crack extension were examined and conditions for crack extension and arrest were determined by invoking Griffith's postulate which states that a crack will <u>not</u> extend while increasing the free energy of the system[24]. This analysis can be summarized as follows: Since the stress field is highly localized, the strain energy (U_{SE}^{o}) is finite and can be expressed in terms of the maximum stress (σ_{max}) and the volume associated with the microstructural features of diameter D

$$U_{SE}^{o} = k\ \sigma_{max}\ \frac{\pi}{6}\ D^{3}, \tag{3}$$

where k includes the appropriate elastic properties and dimensional constants. As a crack extends from the microstructural feature, the strain energy decreases as

$$U_{SE} = U_{SE}^{o}\ f(\mu), \tag{4}$$

where $0 \leqslant f(\mu) \leqslant 1$ and μ is the normalized crack size, c/D. Likewise, the surface energy (U_{S}) needed for crack extension can be set proportional to the area of the microstructural feature and the critical strain energy release rate, G_{c}, as

$$U_{S} = G_{c}\ D^{2}\ g(\mu), \tag{5}$$

where $g(\mu) \geqslant 0$ is the proportionality factor which depends on the normalized crack size, μ. When it is assumed that $f(\mu)$ exhibits a single inflection, a plot of the total energy change during crack extension, $U_{T} = U_{SE} + U_{S}$ also exhibits a single inflection as shown in Fig. 2 until either σ_{max} or D is increased to cause the development of a maximum and minimum. The maximum and minimum correspond to conditions of crack extension and arrest, respectively.

Since a maximum (or a minimum) in the U_{T} vs μ relation is required for crack extension, the condition where a maximum just begins to develop defines the first condition for crack extension. This condition is defined by $\partial U_{T} / \partial \mu = \partial U_{T}^{2} / \partial^{2} \mu = 0$, or by rearranging this result,

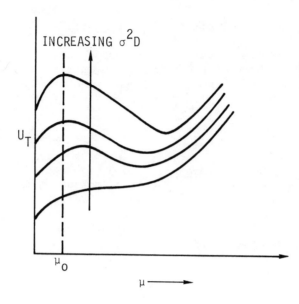

Figure 2

$$\sigma^2_{max} \, D = \text{constant.} \tag{6}$$

Thus, for a given material, a critical size exist below which crack extension can not occur, viz,

$$D_c = \frac{\text{constant}}{\sigma^2_{max}} \tag{7}$$

Once this first condition is satisfied, crack extension will occur when either σ_{max} or D is increased to shift the maximum in the U_T vs μ curve to coincide with the size of the pre-existing crack, μ_o as illustrated in Fig. 2. Thus, once the condition expressed in eq. 6 is satisfied, crack extension is statistical by nature of the distribution of pre-existing crack sizes associated with the microstructural feature.

2.3.2 <u>Stress Intensity Approach</u>. Expressions for the stress intensity factor can be obtained and crack extension is defined by the condition where $K \geqslant K_c$ [25]. Using the **principles** of super-position, the expression for the stress intensity factor can be obtained using [26]

$$K = \left(\frac{2}{\pi c}\right)^{1/2} \int_o^c \frac{\sigma(x)x^{1/2}}{(c-x)^{1/2}} \, dx \tag{8}$$

when the tensile stress, $\sigma(x)$, normal to the crack's trajectory
is known. This approach was taken by Evans[25] to examine the
criteria for radial crack extension from spherical inclusions within a stress field due to thermal expansion mismatch (eq. 2) when
$\alpha_m > \alpha_p$. The stress intensity factor for this condition is:

$$K = \frac{\sigma max}{2} D^{1/2} (\pi\mu)^{1/2} \beta ,$$ (9)

where $\beta = [1 - \frac{18}{4}\mu + 15\mu^2 - \frac{175}{4}\mu^3 + \ldots]$

When the stress intensity factor for the radial crack extending
from a spherical inclusion is plotted as a function of the normalized crack size, μ, a series of curves are obtained which exhibit
a maximum as shown in Fig. 3. As shown, the value of K increases
with increasing σ_{max} or D. Crack extension will occur when the
value of K for the given pre-existing crack size is equal to K_c
and crack arrest occurs at a μ where $K \leqslant K_c$ as shown schematically
in Fig. 3. Regardless of the pre-existing crack size, crack
extension can only occur when the K vs μ fucntion exceeds K_c which
first occurs when

$$K_c = \frac{\sigma_{max}}{2} D^{1/2} (\pi\mu)^{1/2} \beta_{max}$$

or, after rearranging

$$\sigma^2_{max} D \gtrsim 30K^2_c.$$ (10)

This expression again defines a critical sphere size below which
crack extension will not occur:

$$D_c \simeq 30 \frac{K^2_c}{\sigma^2_{max}}$$ (11)

2.3.3 <u>Improvements Required for Quantitative Predictions</u>.
Both the energetic and stress intensity approaches to crack
extension in highly localized stress fields agree with experimental
observation in qualitatively predicting a critical microstructure
size below which crack extension will not occur. As discussed
below, improvements are required for quantitative predictions.

The stress intensity approach is inherently limited by the
requirement that the forces, which give rise to the stress field,
are unaffected by the crack extension. This is certainly not the
case for crack extension in the highly localized stress fields
considered here and thus leads to an over estimate of K with

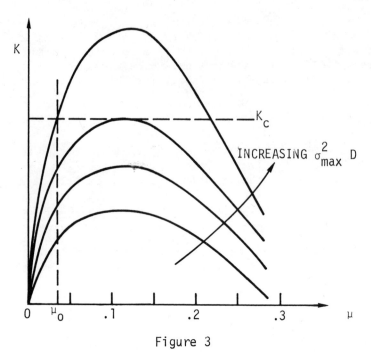

Figure 3

increasing crack extension. The energetics approach is currently
limited by the lack of knowledge of the function, f(μ). Both
approaches are limited to know stress distributions. Solutions to
crack extension in stress fields due, e.g., to thermal expansion
anisotropy, must await solutions to these more complex stress
fields.

    With the wide use and decreasing cost of computer modeling,
the energetics approach appears to offer the most convenient and
accurate way to solve current problems. In this approach, the
change in the strain energy associated with the microstructural
feature is computed as a function of the crack length to obtain
the strain energy release rate. As reviewed elsewhere[27], the
computer is used to model the stress system so that the strain
energy can be progressively calculated as a crack is increnenentally
extended along a given path. The strain energy release rate, G =
$\partial U_{SE}/\partial \mu$ is obtained by differentiating the resulting U_{SE} vs μ curve.
By equating this result to the derivative of eq. 4,

$$G = k\sigma_{max}^2 \frac{\pi D^3}{6} \; \partial f(\mu)/\partial \mu \qquad\qquad (12)$$

$\partial f(\mu)/\partial \mu$ can be determined and explicit expressions for crack
extension and arrest can be computed for the desired stress
distribution and crack geometry. Such expressions would then yield
quantitative information concerning the formation of cracks during

fabrication, e.g. the critical grain size to prevent microcracking in a given ceramic material.

Thus, two areas require extended work. First, the complex stress fields associated with thermal expansion anisotropy/mismatch and phase transformation must be obtained, and, second, explicit G expressions must be derived for pertinent crack geometries. Once solutions have been obtained in these two areas, materials engineers will be much better equipped to define the microstructural requirements necessary to fabricate strong single-phase and polyphase ceramics.

3.0 TOUGHENING MECHANISMS BASED ON CONTROLLED MICROSTRUCTURES

3.1 Crack Interaction with Second Phase Inclusions

Observations that crack fronts bow-out between second phase inclusions which act as pinning positions, suggested an energy absorbing mechanism based on the line energy of a crack front somewhat analogous to models suggested for dislocation-precipitate interactions.[28] The model lead to expression for the critical strain energy release rate, G_c, involving the interparticle spacing which was later modified to include the particle size[29]. The model also lead to determining G_c for a number of brittle matrix particle composites as a function of the volume fraction and the size of the particulate phase. One can conclude from these data that theoretical (see review, ref. 30) and experimental observations are only in agreement over a limited range of conditions, but sufficient agreement exists to suggest some validity of the line energy concept.

In general, G_c increases with both the volume fraction and the size of the particulate phase. Some systems, particularly those where the matrix phase is a polymer, exhibit a maximum G_c at a particular volume fraction. In those systems where strength and elastic modulus properties are also known, the increase in fracture toughness due to large particle sizes are generally offset by a larger increase in the apparent crack size, resulting in a net strength decrease. Although much more work is required to understand the crack front/particulate interactions, it presently appears that large particle sizes, which result in significant increases in G_c but promote lower strength must be avoided in favor of moderate particle sizes which lead to modest increases in fracture toughness and strength.

3.2 Microcracking at the Crack Front

Some investigators [31,32,33] have emphasized that toughening can be achieved by the growth of many microcracks which either pre-exist or are stress-induced adjacent to the primary crack front. This phenomenon has been best described by the experiments of Green et al., [31] who showed that the microcracked zone in relatively low strength, partially stabilized zirconia results in a non-linear elastic behavior prior to fracture and a stress-strain hysteresis (and thus, an energy dissipative process) in notched and unnotched test bars under cyclic loading. Using the analogy of a plastic zone developed in metals, they indicated that the radius of the microcracked zone (r), called the process zone, was related to the critical stress intensity factor and the stress required to induce microcracking or microcrack propagation, σ_{mc} by

$$r = \frac{1}{2} \frac{K_c^2}{\sigma_{mc}^2} \tag{13}$$

The formation and/or propagation of microcracks within the process zone can be thought of as an energy dissipation process due to the increased formation of fracture surface area. Claussen and co-workers [32,33] have championed the idea of stress-induced microcracking to explain the increased G_c observed when unstabilized ZrO_2 is added to Al_2O_3 [32]. Although their explanation may be valid, experimental evidence has yet to be obtained to invalidate the possibility of a stress-induced phase transformation which as indicated below, may not involve the formation of a process zone.

The effect of pre-existing microcracks on G_c and strength has been examined by Kuszyk and Bradt [10] for the case of $MgO \cdot 2TiO_2$. Microcracks were introduced into this material by increasing its grain size (see Sec. 2.1.1). The G_c and strength vs grain size results in Fig. 4 show that the absence of pre-existing microcracks leads to a low value of G_c and a high value of strength; G_c increases as the number of pre-existing microcracks increase, but at the same time, the strength significantly decreases. If a microcrack process zone was stress-induced in the fine-grained material, the zone size (see eq. 13) was apparently too small to have a significant effect on G_c, suggesting the need for existing microcracks to achieve high fracture toughness. Data for the fine-grained material tends to dismiss the suggestion that localized, residual stress fields, which are certainly present in this material, can increase fracture toughness [35].

In conclusion, pre-existing microcracks can be useful in increasing G_c and to add some stability to large cracks in materials where high strength is not a requirement, e.g. thermal insulators, etc. The author questions the present evidence that stress-induced microcrack has an effect on G_c[32,33,34]

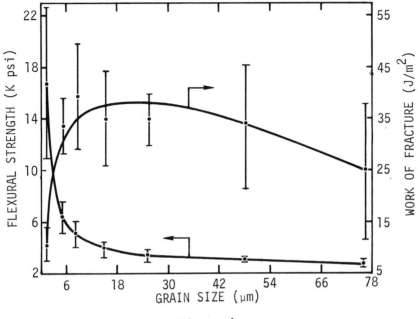

Figure 4

3.3 Fibrous Fracture

Fibrous modes of fracture, i.e. the pull-out of broken fibers from a matrix, has been recognized as responsible for the high fracture toughness of fiber composites[36]. The choice of so many non-cubic ceramic crystal structures, in which some may exhibit fibrous crystal growth, offers the possibility of in-situ fiber formation during fabrication. In-situ fiber formation is attractive because fiber strength is not degraded during fabrication and, at the same time, a large volume fraction of fibers can be produced within the dense body.

Silicon nitride is one of the few examples known to the author where increased fracture toughness appears to be derived from in-situ fiber formation. Dense Si_3N_4 can have either an equiaxial grain structure when fabricated with a β-phase powder or a fibrous microstructure when fabricated with an α-phase powder. Although the chemistry (content of second phases, etc.) is the same for both, the fibrous microstructure has four times the fracture toughness and twice the strength of the equiaxial microstructure[37]. Fracture surface observations show that fiber pull-out occurs. The apparent mechanism for fibrous grain growth in this system has been reported elsewhere[38].

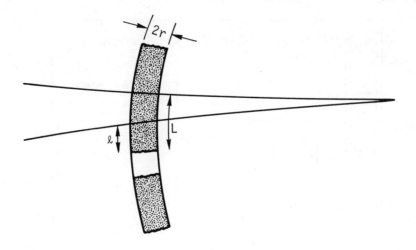

Figure 5

The microstructural features considered important in governing
the energy dissipated during fibrous fracture can be viewed by
examining the work performed during the pull-out process as initiated by Cottrell[36]. Consider a fiber shown in Fig. 5 of radius
r which fractures within the matrix and is pulled out by a force
(F) which is required to overcome the shear strength τ_i of the
fiber-matrix interface. If the shear stress τ_i is retained
during the pull-out process, then the force on the fiber is given
as $F = \tau_i 2\pi\ell$ and the work, W_N, required to pull out N fibers
which protrude by a length L is

$$W_N = N \int_0^L \tau_i 2\pi r\ell\, d\ell = N\tau_i \pi rL^2 \tag{14}$$

The number of fibers pulled out by a penny-shaped crack of diameter
2c is

$$N = \frac{V\pi c^2}{\pi r^2} \quad , \tag{15}$$

where V is the area (or volume) fraction of the fibers within the
matrix. Thus, the work required to pull out fibers as a function
of the crack size is

$$W_T = \pi V\tau_i rR^2 c^2, \tag{16}$$

where $R = L/r$, the average aspect ratio of the fibers protruding from the fracture surface. Sack[39] has shown that a stress σ_a applied to a body containing a penny-shaped crack lowers the free energy of the system by

$$W_L = - \frac{8(1-\nu^2)\sigma_a c^3}{3E_c} \qquad (17)$$

where ν and E_c are the elastic properties of the composite. The increase in energy by forming new fracture surface is*

$$W_S = \pi c^2 G_o, \qquad (18)$$

where G_o is the critical strain energy release rate of the matrix material. Adding these energy terms together and setting the first derivative of the summation equal to zero gives the condition for fracture. The resulting expression can be rearranged to give a fracture strength expression for the composite material

$$\sigma_c = \frac{\pi E(G_o + V\tau_i rR^2)}{4c}^{1/2} \qquad (19)$$

The critical strain energy release rate expression is

$$G_c = G_o + V\tau_i rR^2, \qquad (20)$$

illustrating the important microstructural features, viz. the volume fraction of the fibers oriented parallel to the applied load, the fiber radius and aspect ratio.

3.4 Stress Induced Phase Transformation

Recent work by several groups[15,23,40,41] have shown that certain ZrO_2 ceramics can be made to exhibit a high fracture toughness through controlled microstructural development. Sufficient evidence exists demonstrating that the toughness is due to the retension of tetragonal ZrO_2 below its transformation temperature and its stress-induced transformation to the stable monoclinic structure during crack extension.

Retension of the metastable phase depends on microstructural control. Two approaches have been used, each resulting in a different microstructure. In the approach initiated by Garvie et

---

*The increased surface area formed by fibrous fracture has been neglected to illustrate the effect of the pull-out work.

al.,[40] large grained ZrO_2 polycrystalline material partially stabilized with CaO is heat treated to precipitate ellipsoidal-shaped, sub-micron size tetragonal particles within much larger cubic grains. Both Garvie et al.[40] and Porter and Heuer[23] have shown that the size of the precipitate is critical for the retension of the metastable phase, as reviewed in Sec. 2.2.2. In the second approach, initiation by Gupta et al, sub-micron ZrO_2 powder is densified with additions of Y_2O_3 to produce polycrystal-line bodies which can comprise of a) all metastable tetragonal phase, b) a mixture of monoclinic and metastable tetragonal phases, or c) a mixture of metastable tetragonal and cubic phases, depend-ing on the amount of Y_2O_3 added (corresponding to the elevated temperature phase relations in the ZrO_2-Y_2O_3 system) and the sintering conditions[41]. The Y_2O_3 is known to lower the tetragonal to monoclinic transformation temperature, and it apepars to be, in part, responsible for the retension of the tetragonal phase[42,43]. Again, similar to observations in the first approach, microstructural size effects appear to be critical in retaining the tetragonal phase as reviewed in Sec. 2.1.2. The critical grain size appears to increase as the mole % of Y_2O_3 is increased[41].

As discussed in Sec. 2.3, retension of the metastable phase must be accomplished by the constraint of neighboring grains on one another, which in turn, can be accomplished if microcracks are not formed. Since the thermodynamics of microcrack formation depends on the size of the stress field (i.e. the stress gradient), microcracking will only occur when the microstructure size exceeds a critical value. Observations of extensive microcracking in the ZrO_2-Y_2O_3 system where conditions for the retension of the tetragonal phase were not present supports this explanation[41].

Evidence that the retained metastable tetragonal phase can be transformed by the stress field at a crack front has been obtained by both direct observation[23] and indirect x-ray diffraction exam-ination[15] of fracture surfaces. In addition, the surfaces of these materials can be transformed by either surface grinding or particle impact, and thus after proper calibration, these materials can be used to investigate the depth of damage introduced by various mechanical surface treatments[44].

The mechanics of the stress-induced phase transformation, its mode of energy absorbtion, and its effect on the critical strain energy release rate is not clearly understood. Workers who have studied the ZrO_2 transformation[45,46,47] have shown it to resemble[40] a martensitic transformation. This resemblance has been extended to make analogies with TRIP steels in which a stress induced marten-sitic transformation is believed responsible for the high fracture toughness of these alloys[48]. As reviewed by Christan[49], stress-induced martensitic transformations are a form of "...mechanical (non-elastic) deformation, to be compared with mechanical twin-

ning[49]". That is, the displacement of atoms from one structure
to another by the stress field of a crack would dissipate energy
during fracture suggesting one possible reason for the high
fracture toughness of metastable ZrO_2 ceramics.

Microcrack (or process) zones have also been suggested to
explain the fracture toughness of ZrO_2 ceramics[34]. In this line
of thought, the crack front produces microcracks to allow each
tetragonal grain or precipitate within the zone to transform. As
explained in Sec. 3.2, the formation and/or extension of the microcracks increase the surface area formed and thus amount of energy
dissipated. It was this author's first belief that microcracking
must necessarily accompany the stress induced transformation in
order to relieve the constraint, but further thinking has lead to
a new conclusion. Namely, the constraint on the volume increase
is simply released by the tensile stresses at the crack front.
As the crack passes the transformed material, stresses due to the
transformation can rebuild. Microcracking will not occur as long
as the size of the microstructure satisfies the requirements
detailed in Sec. 2. This thinking was inspired by the lack of an
observed microcrack zone. Porter et al[23] observe no microcrack
zone and preliminary observations by this author collaborates[41]
their observation.

Regardless of the mechanism invoked for energy absorbtion,
the microstructural features important to increasing G_c can be
examined by the following argument. Following an approach similar
to that introduced by Antolovich[50] for TRIP Steels,
the energy absorbed by the transformation as a function of the
radius c of a penny-shaped crack is

$$U_p = U_o V_p 2r\pi \ c^2 \tag{21}$$

where U_o = the energy absorbed per unit volume of stress-induced
transformed material, V_p is the volume fraction of the metastable
phase, and r is the radius of meterial from the crack front in
which the transformation occurs (see eq. 13, where σ_{mc} is the stress
required for transformation). Using the same approach detailed in
Sec. 3.3 for fibrous fracture, one can show that

$$G_c = G_o + 2U_o V_p r \tag{22}$$

indicating that the volume fraction of transformed material is the
important microstructural feature, as expected.

Fracture toughness data has been obtained using the indentation technique[51] for ZrO_2 + Y_2O_3 materials with phase compositions
that vary between pure tetragonal to pure cubic[41]. These data,
illustrated in Fig. 6 as K_c vs composition show that the highest

toughness occurs close to the pure tetragonal material and it
decreases with increasing amount of cubic phase supporting the
general conclusion stated in eq. 22.

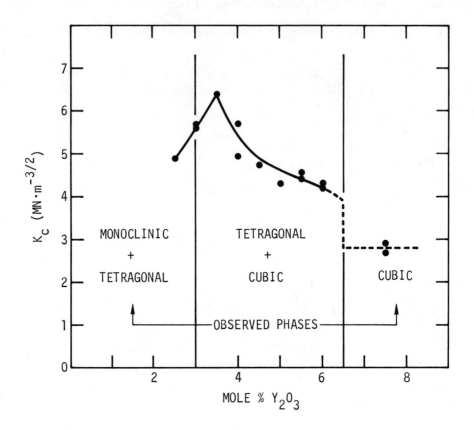

Figure 6

4.0   CONCLUDING REMARKS

    The author has attempted to outline several routes in which
fracture mechanics concepts can direct the design of ceramic
microstructures for optimum mechanical integrity. It is obvious
that the theories behind these directions are not sufficiently
explicit to define the optimum microstructure for an arbitrary
material. More obvious is the lack of knowledge concerning the
fabrication parameters which must be controlled to achieve many of the
desired microstructures. Further work should be directed towards
these deficiencies.

Despite the existing deficiencies, ceramic manufacturers do recognize the need for controlled microstructures to avert specific mechanical problems. For example, manufacturers of low expanding glass-ceramics (consisting of phases exhibiting strong thermal expansion anisotropy) occasionally produce low strength material containing microcracks. This problem, recognized as being caused by a large grain size, is remedied by bringing the heat treatment of subsequent batches back into control to produce the desired, small grain size, higher strength material. It might be expected that solutions to such problems are obtained through trial and error suggesting the need to generalize these problems with theory and implimenting the theory when manufacturing a new material.

ACKNOWLEDGMENT

The author wishes to acknowledge many useful discussions with A. G. Evans. Financial support was obtained from a Rockwell International Independent Research and Development program and the Office of Naval Research.

REFERENCES

1. W.R. Buessem, Mechanical Properties of Engineering Ceramics Ed. by W.W. Kriegel and H. Palmour III, pp. 127-48, Interscience, Inc. New York, 1961.

2. W.R. Buessen, N.R. Thielke, and R.V. Sarakauskas, Ceramic Age 60, 38 (1952).

3. E.A. Bush and F.A. Hummel, J. Amer. Ceram. Soc. 41, 189 (1958).

4. ibid, 42 388 (1959).

5. W.R. Manning, O. Hunter, Jr., F.W. Calderwood, and D.W. Stacy, J. Amer. Ceram. Soc. 55, 342 (1972).

6. S.L. Dole, O. Hunter, Jr, J. Nac. Mat. 59, 207 (1976).

7. H.P. Kirchner and R.M. Gruver, J. Amer. Ceram. Soc. 53, 232 (1970).

8. R.L. Coble, Ceramic Fabrication Processes Ed. by W.D. Kingery p. 219, John Wilen, New York, 1958.

9.   B.S. Hickman and D.G. Walker, J. Nuc. Mat. 10, 243 (1963).

10. J.A. Kuszyk and R.C. Bradt, J. Amer. Ceram. Soc. 56, 420 (1973).

11. S.L. Dole, O. Hunter, Jr. and C.J. Wooge, Private Communication.

12. F.F. Lange, unpublished results.

13. W.D. Kingery, Introduction to Ceramics, p. 392, John Wiley, New York, 1960.

14. J.E. Bailey, Proc. Roy. Soc. 279A, 395 (1964).

15. T.K. Gupta, F.F. Lange and J. H. Becktold, to be published.

16. Y. Matsuo and H. Sasaki, J. Amer. Ceram. Soc. 49, 229 (1966).

17. J. Selsing, ibid 44, 419 (1961).

18. J.D. Eshelby, Proc. Roy. Soc. 241A, 376 (1957).

19. R.W. Davidge and T.J. Green, J. Mat. Sci. 3, 629 (1968).

20. D.B. Binns Science of Ceramics, ed. by G.H. Steward, pp. 315-35, Academic Press, New York (1962).

21. F.F. Lange, J. Amer. Ceram. Soc. 65, 445 (1973).

22. F.F. Lange, "Strong, High-Temperature Ceramics" Ann. Rev. Mat. Sci. 4, 365 (1974).

23. D.L. Porter and A.H. Heuer, J. Amer. Ceram. Soc. 60, 183 (1977).

24. F.F. Lange, Fracture Mechanics of Ceramics, ed. by Bradt, Hasselman and Lange, pp. 599-613, Plenum Press, New York, 1974.

25. A.G. Evans, J. Mat. Sci. 9, 1145 (1974).

26. G.C. Sih, Handbook of Stress Intensity Factors, Lehigh Univ. Press, 1973.

27. V.B. Watwood, Jr.; Nuc. Engin. and Design II, 323 (1969).

28. F.F. Lange, Phil. Mag. 22, 983 (1970).

29. A.G. Evans, Phil. Mag. 26, 1327 (1972).

30. F.F. Lange, Composite Materials, Fracture and Fatigue, Vol. 5 ed. by L.J. Broutman, pp. 1-44, Academic Press, New York, 1974.

31. D.J. Green, P.S. Nicholson and J.D. Embury, J. Amer. Ceram. Soc. 56, 619 (1973).

32. N. Claussen, ibid 59, 49 (1976).

33. R. Rice, ibid 60, 280 (1977).

34. N. Claussen and J. Steeb, ibid, 59, 457 (1976).

35. A.G. Evans, A.H. Heuer and D.L. Porter, Fracture 1977, Vol. 1, ICF4, pp. 529-56, (1977).

36.   A.H. Cottrell, Proc. Roy. Soc. 282A, 2 (1964).

37. F.F. Lange, J. Amer. Ceram. Soc. 56, 518 (1973).

38. F.F. Lange, Nitrogen Ceramics, NATO Advanced Study Institute, Canterbury, England, Aug. 1976 (in press).

39. R.A. Sack, Proc. Phys. Soc. (London) 58, 729 (1946).

40. R.C. Garvie, R.H. Hannink and R.T. Pascoe, Nature 258, 703, Dec. 25, 1975.

41. F.F. Lange, Unpublished.

42. K.K. Srivastara, R.N. Patil, C.B. Choudhary, K.V.G.K. Gokhale and E.C. Subbarao, Trans. J. Brit. Ceram. Soc. 73, 85 (1974).

43. H.G. Scott, J. Mat. Sci. 10, 1527 (1975).

44. F.F. Lange and A.G. Evans, unpublished.

45. P. Duwez and F. Odell, J. Amer. Ceram. Soc. 33, 274 (1950).

46. G.K. Bansal and A.H. Heuer, Acta Met. 22, 409 (1974).

47. R.N. Patil and E.C. Subbarao, Acta Cryst. 26 535 (1970).

48. W.N. Gerberich et al. Fracture, ed. by P.L. Pratt, Chapman and Hall, London, p. 288, 1969.

49. J.W. Christian, Physical Metallurgy, Ed. by R.W. Cahn pp. 443-541, John Wiley, New York, 1965.

50. S.D. Antolovich, Trans. Met. Soc. AIME 242, 2371 (1968).

51. A.G. Evans and E.A. Charles, J. Amer. Ceram. Soc. 59, 371 (1976).

MICROCRACKING IN A PROCESS ZONE AND ITS RELATION

TO CONTINUUM FRACTURE MECHANICS

R. F. Pabst, J. Steeb, N. Claussen

Max-Planck-Institut für Metallforschung
Institut für Werkstoffwissenschaften
D-7000, Stuttgart, W. Germany

INTRODUCTION

Irradiation damage of BeO ceramic by fast neutrons causes an appreciable reduction in the strength and in the Young's modulus, whereas the density is relatively unaffected (1, 2,3). Moreover, the brittleness of unirradiated BeO transforms to a pronounced ductile behavior comparable with the nonlinear behavior of graphite or concrete. With a rigid testing device, stable crack propagation is always observed. The area below the load deflection curve (a measure of the fracture energy γ_w) increases with a decrease in the strain rate (Fig. 1), and the load-unload behavior exhibits considerable hysteresis. Heating in He at 1600 K results in an increase in the Young's modulus, in the strength and in the brittleness, compared to prior to irradiation. SEM pictures indicate that this behavior is due to the presence of a high density of micro- and macrocracks. Glucklich (4) refers to this mechanism of crack arrest or stabilization as "energy dissipation stability"; that is, the capacity of cracks and microcracks for dissipating energy and preventing catastrophic failure.

The possibility of energy dissipation stability exists for every real material, and the differences between metals, plastics and ceramics being only ones of degree.

Subcritical crack growth becomes more probable with an increase in the energy dissipation capability and an increase of the loading-system rigidity, or with a decrease in strain rate. The phenomenon of energy dissipation by microcrack formation is particularly apparent in multiphase

ceramics (5), and it seems reasonable to use it to enhance ductility and fracture toughness without a considerable reduction in strength, or even with an increase
in strength.

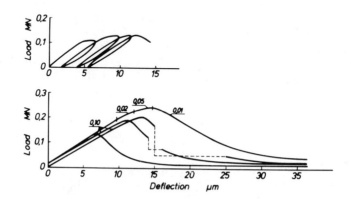

Fig. 1   Apparent ductility associated with "energy
          dissipation stability" during macro- and mi-
          crocrack formation. BeO ceramic irradiated with
          fast neutrons. Testing speeds:
          0.1;, 0.05; 0.02; 0.01 mm/min.

PROCESS ZONE CONSIDERATIONS

Analogous to plastic zone formation, the increase in crack
stability and fracture toughness of ceramic materials is
thought to result from the energy dissipation caused by
microcrack formation within a process zone. The idea of
a process zone has proved to be a concept which is useful for further application of linear continuum theory.

It should be emphasized that microcracking is accompanied by a limited amount of plastic deformation, as well
as by stress-induced phase transformations, friction effects, adhesion and bonding of the constituents, heat development and reduction of internal stresses which may
impede crack propogation.

A linear stress versus strain behavior involves infinitely high stresses. Once a crack is initiated (ideal Griffith material), there is no resistance to crack propagation. For real materials a nonlinear stress-strain function and a finite zone of stress release must be postulated to withstand the smallest load in a brittle material (6). The process zone may be described as follows:

a) a nonlinear discontinuos region where the stress distribution compatible with linear continuum mechanics is no longer valid. The stresses decrease at the boundary, and outside the region, the material is treated as a continuum;

b) a region where the macroscopic stress intensity factor K_I is defined only from the boundary to the continuum;

c) a region of crack nucleation and subcritical, discontinuous crack extension;

d) a region where the size and shape of the specimen correlate with the structure of the material.

PROPOSED SIZE AND SHAPE OF THE PROCESS ZONE

Size and shape at the point of instability.

If we define the process zone as the smallest region outside of which the material may be regarded as a continuum, no special shape need to be proposed. For the sake of convenience, the process zone shape is assumed to be cylindrical in ceramic and metal systems. To the authors knowledge, no direct means of measuring the shape exists at the moment.

Process zone sizes are computed at the point of instability and are related to the critical stress intensity factor for symmetric loading K_{IC}.

In thermoplastics, the craze zones may be regarded as process zones. They consist of elongated molecular chains and, in accordance with this structure, have a lower density and refractive index at onset of instability than the bulk material. It is thus possible in PMMA for instance, to measure the size and shape of the zone by an optical interference method. The size and shape are similar to those described by the Dugdale model (7) for a plastic zone:

$$S \approx \frac{\pi}{8} \left(\frac{K_{IC}^2}{\sigma_{ys}} \right) \quad \text{plane stress, where} \tag{1}$$

σ_{ys} = yield stress.

For metals, Krafft (8) derived a semi-empirical expression for the size d_T of a process zone:

$$d_T = \frac{1}{2} \left(\frac{K_{IC}^2}{E \cdot n} \right) \quad \text{where} \tag{2}$$

E = Young's modulus;
n = strain hardening exponent; and
d_T is considered to be a material constant which can be related to the spacing between the dispersed particles.

There is little information concerning the nature of the process zones of ceramic materials, and only rough computations can be made.

For real materials, the formulation of reference (9) giving

$$K_{IC} = \lim_{\rho \to 0} \frac{1}{2} K_t \sigma_N \sqrt{\pi \rho} , \tag{3}$$

is unrealistic (ρ = notch-root radius, K_t = stress concentration factor, σ_N = net section stress).

A finite radius ρ_c and consequently a finite structural parameter $\mathcal{E}_c$ must exist for the structure to be regarded as a continuum. We shall denote $\mathcal{E}_c$ as the process zone size. Consequently:

$$K_{IC} = \lim_{\rho \to \mathcal{E}_c/2} \frac{1}{2} K_t \sigma_N \sqrt{\pi \rho} , \quad \text{and} \tag{4}$$

$$\mathcal{E}_c = \frac{8}{\pi} \left(\frac{K_{IC}^2}{K_{t_c}^2 \sigma_N^2} \right) \tag{5}$$

(K_{t_c} = stress concentration factor for $\rho_c = \mathcal{E}_c/2$).

This implies, for a notch-tip radius $\rho < \mathcal{E}_c/2$ that K_{IC} is independent of ρ (ρ is imbedded in the process zone). $\rho c = \mathcal{E} c/2$ is the smallest possible effective radius for real materials (Fig. 2).

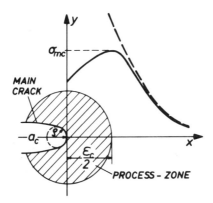

Fig. 2   Process zone $\mathcal{E}_c$ in front of a main crack (notch)

$\mathcal{E}_c$ may therefore be computed from K_{IC} measurements as a function of ρ, where K_{IC} is independent of ρ.

A rough computation of $\mathcal{E}_c$ can be made by surface crack evaluation. If we assume uniform damage of the surface to a depth a_c (no isolated cracks) the following may be stated:

$$a_c = \mathcal{E}_c/2 = \lim_{a_c \to \mathcal{E}_c/2} \left(\frac{K_{IC}^2}{\sigma_F}\right) \cdot Y, \text{ where} \qquad (6)$$

σ_F = strength, and
Y   = correction factor.

For a 4-point bend specimen, we obtain

$$a_c = \frac{1}{4}\left(\frac{K_{IC}}{\sigma_F}\right)^2, \qquad (7)$$

where σ_F is the bend strength of the unnotched specimen (Fig. 3.)

Computation of $\mathcal{E}_c$ by Eq. (7) assumes that $\mathcal{E}_c$ is "autonomous". The term autonomous means that $\mathcal{E}_c$ depends neither
on the load distribution nor on the crack length. K_{IC} and
$\mathcal{E}_c$ should be the same for a surface crack as for the end
region of a long crack (notch). This implies that data
(which might not be true for more ductile materials),
from Eq. (6) and (7) should be the same.

Comparison of the results using Eq. (6) and (7) gave the
same $\mathcal{E}_c$ values. ($\mathcal{E}_c \approx 90$ um for dense, 97% pure alumina
with an average grain size of 11 um and a maximum grain
size of ~ 35 um) (10). Measurements using notched specimens having a notch radius $\rho_c = \mathcal{E}_c/2$ revealed no effect on crack length down to $a/w = 0.04$, (w = specimen
width) regardless whether the material was porous or
dense (11).

Strangly enough, for sharp cracks this is not the case
(12). K_{IC} and therefore $\mathcal{E}_c$ increases with increasing
crack length. The sharp crack was produced by controlled crack propogation using a loading device. The crack
length was measured using a reflected light technique
and was controlled using a dye penetrant (12). Moreover,
K_{IC} data obtained for sharp cracks always proved higher
than those obtained for narrow notches. The data coincide for small lengths of crack and notches (Fig. 4).

The data obtained for sharp cracks imply an ever increasing $\mathcal{E}_c$ value with increase of crack length. This
disagrees with the common "notion" that the K_{IC}-fac-
tors estimated for notches should be higher than or
equal to those for sharp cracks. A similar behavior
was found by Strobel and Hübner (13), who explained
the phenomenon by postulating an enhanced secondary
crack formation at the crack tip during crack elongation (thereby increasing $\mathcal{E}_c$). No significant secondary
crack formation was observed in our experiments. It
therefore seems more rational to consider friction effects and adhesive forces at crack surfaces in order
to explain the process zone effectiveness. This unusual behavior may be due to the special method of
sharp crack propagation, i.e., controlled crack extension with a rigid device.

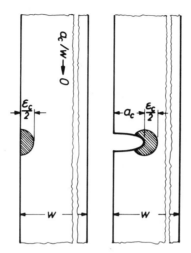

Fig. 3   Process zone ε_c in front of a main crack (notch) and at the specimen surface (a/w→0)

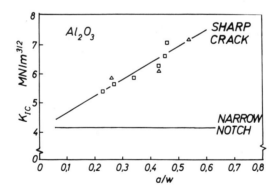

Fig. 4 K_{IC} as a function of a/w for a sharp crack and a narrow notch, □ 4-point bend test, △ 3-point bend test

Application of $\mathcal{E}_c$ to the Al_2O_3 + ZrO_2 and the Si_3N_4 + ZrO_2 Systems.

Most results for ceramics containing a dispersed second phase can be interpreted on the basis of microcracking in a process zone (14). Fig. 5 gives K_{IC}-data, the flexural strength and $\mathcal{E}_c$ for hot-pressed Al_2O_3-unstabilized ZrO_2 composites as a function of the volume fraction of ZrO_2 at the two extreme ZrO_2 particle sizes 1.15 /um and 6.4 /um.

Microcracks forme as a result of the high tensile stresses which develop in the Al_2O_3 matrix as a result of the transformation-induced expansion of the ZrO_2 particles. K_{IC} increases with increasing volume fraction up to a maximum. At volume fractions beyond the maximum the microcracks between the particles coalesce thus forming large cracks, and decreasing K_{IC}. Microcracks in composites with higher average particle sizes are operative at lower concentrations (14). Therefore, the maximum of the 6.4 /um composite is obtained at 3 vol.% whereas the maximum of the 1.25 /um composite is at 15 vol.%. With increasing volume fraction of ZrO_2 the finer particles agglomerate, forming large cracks which lead to a decrease of K_{IC} and flexural strength. As proposed by Eq. (7) $\mathcal{E}_c$ increases for increasing K_{IC} and decreasing flexural strength σ_F, the quality given by the ratio K_{IC}/σ_F. For the purpose of obtaining high fracture toughness combined with high strength, process zone size should be relatively small coupled with high effectiveness to dissipate energy.

At the best, the flexural strength should follow the same tendency as K_{IC}. This may be obtained with hot-pressed Si_3N_4-ZrO_2 composites (15).

The results of the measurements of K_{IC}, flexure strength and $\mathcal{E}_c$ are shown in Fig. 6. K_{IC} of the hot-pressed material increases with ZrO_2 volume fraction up to a maximum, also observed for Al_2O_3-ZrO_2 composites (Fig. 5) (14). As the strength follows the same tendency as K_{IC}, $\mathcal{E}_c$ is relatively small. Improved mixing conditions may have caused reduced ZrO_2 particle agglomeration and hence smaller microcracks have formed than those obtained with Al_2O_3-ZrO_2 mixtures.

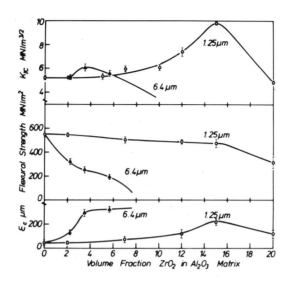

Fig. 5 K_{IC}, flexural strength σ_F, and process zone
ε_C for Al_2O_3-ZrO_2 composites as a function of
volume fraction ZrO_2

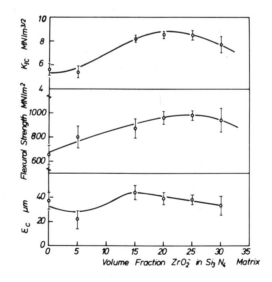

Fig. 6 K_{IC}, flexural strength σ_F and process zone ε_c
of hot-pressed Si_3N_4-ZrO_2 composites as
a function of volume fraction ZrO_2

The process zone of a running crack

It seems hopeless to advance any assumptions about
the size and shape of the process zone of a running
crack in ceramic materials, because of the uncertainty
in the K_{IC} variation with velocity. Brobergs (16) solution indicates a decrease in crack tip radius and a
decrease in K_{IC} with an increase of speed. Therefore $\mathcal{E}_c$
should decrease, reaching a limiting value determined
by the intrinsic K_{IC} value at a speed lower than the
Rayleigh wave velocity. It seems reasonable to assume
significant triaxial tension at the head of a rapidly
running crack which may be absent in the static case.
This can contribute to an enhanced brittleness.

The most remarkable measurements in this field were made
on thermoplastics (PMMA), quite recently (17). A large
increase in the crack speed occurs if K_{IC} is reached.
The cause of this abrupt increase is likely to be a
transition from isothermal to adiabatic conditions at a
critical crack speed, resulting in a rise in temperature in the crack-tip region. This leads to a reduction
of the materials resistance to fracture. The size of
the craze zone ahead of the tip and the heat generated
by the running crack are both markedly dependent on the
molecular weight of the PMMA. This may indicate that
the size and shape of the process zone in ceramic materials are a function of the structure i.e. microcrack
density and Young's modulus within the zone.

To the authors' knowledge, no direct measurement exists
for running cracks in ceramic materials. Based on the
measurements on thermoplastics earlier mentioned rough
assumptions can be made about the size and shape of a
process zone at different crack velocities (Fig. 7).

It should be pointed out that for dense ceramic materials it is more reasonable to assume that the K_{IC}-
factor is not significantly changed at high velocities.
Therefore $\mathcal{E}_c$ should be relatively unaffected. If the Material rate is sensitive as is the case for irradiated BeO
(high micro- and macrocrack density) or concrete, we may
assume a similar behavior as for PMMA. The questions of
heat development and thermal stresses still remains unresolved.

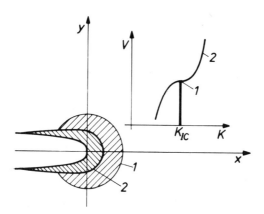

Fig. 7   Process zone size $\mathcal{E}_c$ at different major crack
velocities 1 and 2. K_I - V curve of ceramic
materials.

STRUCTURE AND CONTINUUM MECHANICS

The interactions between the stress fields surrounding
macrocracks and microcracks (as in a process zone) are
complex. It is reasonable to assume that, analogous to
plastic behavior, the tensile stress σ_y along the ligament is relieved and changed similar to that illustrated in Fig. 3, σ_{MC} being the peak stress within the finite zone $\mathcal{E}_c/2$ at the point of macrocrack instability.
Extension takes place when microcracks build at some
distance ahead of the main crack within the zone, link
up, and unite with the primary crack. $\mathcal{E}_c$ is a function
of size, shape and orientation of the single constituents
and the internal strength and stresses.

As pointed out the fracture toughness K_{IC} is valid only
from the boundary of the process zone to the continuum.
In order to characterize the interior of the process zone
the microscopic stress intensity factor k_{IC} and fracture
surface energies γ_{gb} and γ_c (γ_{gb} = grain boundary, γ_c =
cleavage) must be assumed (18) to account for microscopic
strains and stresses. But it should be difficult, if not
impossible, to approach analytically the macroscopic continuum fracture mechanics of K_{IC} and γ_I. In any case, the
structure must then be treated as a composite material

with microcracks running at an interface (grain boundary)
of dissimilar media or at cleavage planes of different
elastic constants. It is not true that a microcrack with
its plane at an angle of other than 90° to the maximum
tensile stress will grow in such a way that it becomes
perpendicular to the applied tensile stress if predominant failure modes (grain boundary, cleavage planes)
exist. So, an array of microcracks is normally not possible. With symmetric loading, a mixed-mode propagation
of the microcracks is implied by the composite behavior
with stress intensity factor K_{CR} other than K_{IC}. Together with $\mathcal{E}_C$, a field-energy concept instead of a maximum tensile-stress criterion should be used which pos-
tulates a criterion of crack instability (19).

CONCLUSIONS

Continuum mechanics is no longer valid at a distance
$\mathcal{E}_C/2$ from the major crack or notch tip. A volume of dimension $\mathcal{E}_C$ encompasses the microscopic flaws and hence
the stress singularity. The radius $\mathcal{E}_C/2$ is approximately of the same order of magnitude as the crack tip radius of a curvature ρ_C where K_{IC} is independent of ρ.
Although $\mathcal{E}_C$ cannot be determined precisely, it is
thought that $\mathcal{E}_C$ is more than an arbitrary empirical constant. $\mathcal{E}_C$ is characteristic for a given structure. Major
cracks propagate by nucleation and activation of subcritical flaws within $\mathcal{E}_C$. Subcritical flaw extension need
not be colinear with the main crack if predominant failure modes exist.

The fracture toughness of ceramics containing a dispersed
second phase has been analyzed using the concept of a
nonlinear microcrack zone $\mathcal{E}_C$ which precedes a major crack
tip. Fracture toughness increases with decreasing inclusion size and increasing volume fraction, which is equivalent to an increasing microcrack zone $\mathcal{E}_C$. The ideal
will be an increase in fracture toughness and strength
simultaneously this implies that the process zone $\mathcal{E}_C$ remains relatively small but highly effective.

REFERENCES

1. Pabst, R.F., Thesis, Univ. Stuttgart, 1972

2. Buresch, F.E., "Science of Ceramics", Vol. 7, 383, 1973

3. Buresch, F.E., Pabst, R.F., "Science of Ceramics", Vol. 6, XVI 13, 1973

4. Glucklich, J., Jet Propulsion Laboratory, TR 32-1438, Contract NAS 7-100, N 70-35917, 1970

5. Claussen, N., J. Am. Ceram. Soc. 59 (1-2), 49-51, 1976

6. Erismann, T.H., Material und Technik, 3, 120-125, 1973

7. Weidmann, G.W., Döll, W., Colloid und Polymer Sci., 254,205-214, 1976

8. Krafft, J.M., App.Mat.Res. (3), 33, 1964

9. Pabst, R.F., Fracture Mechanics of Ceramics, Vol. 2, 555, ed. Bradt, Hasselman, Lange; Planum Press, 1974

10. Pabst, R.F., Z. f. Werkstofftechnik / J. Mat. Tech. (1), 17, 1975

11. Pabst, R.F., (unpublished results)

12. Krohn, U., Elssner, G., Pabst, R.F., DGM-Tagung, München, June 1977

13. Strobel, H., Hübner, G., in Proc. "Festigkeit keramischer Werkstoffe", München, Dec. 1976, ed.: Deutsche Keramische Gesellschaft, (in print)

14. Claussen, N., Steeb, J., Pabst, R.F., Am.Ceram.Soc. Bull. (6), 559-62, 1977

15. Claussen, N., Jahn, J., submitted to J.Am.Ceram.Soc.

16. Broberg, K.B., Arkiv Fysik (18), 159, 1960

17. Döll, W., Int.J.of Fracture (12), 595, 1976

18. Buresch, F.E., DFG Arbeitsbericht We 183/78 Teil B, 1974

19. Wu, E.M., Composite Materials, Vol. 5, Fracture and Fatique, Strength and Fracture of Composites: 91, ed. L.J. Broutman, Academic Press New York, London, 1974

A STRUCTURE SENSITIVE K_{Ic}-VALUE AND ITS DEPENDENCE ON GRAIN SIZE

DISTRIBUTION, DENSITY AND MICROCRACK INTERACTION

Friedrich E. Buresch

Kernforschungsanlage Jülich, IRW

D-5170 Jülich, Germany

INTRODUCTION AND OBJECTIVE

Brittle materials, such as ceramics, acquire their strength by dissipative processes which take place in a non-linear elastic zone--called the process zone--ahead of the macrocrack tip. Various investigators have dealt with the theoretical as well as with the experimental aspects of this phenomenon (1-13). They provided evidence that the dissipative processes are predominantly based on microcracking, an event being linked to the formation and extension of a macrocrack.

These results have been obtained by statistical analyses, energy considerations, acoustic emission analyses and other experimental observations. The statistical approaches to brittle fracture mechanics have been fairly thoroughly evaluated by several researchers and are strictly based on linear elastic fracture mechanics and elastically non-interacting microcracks (1-3,9).

Following up an earlier report from our laboratory (4) this paper deals with the critical stored elastic energy density and the critical elastic energy release rate of the material inside the process zone as a consequence of a critical microcrack configuration in this region. Furthermore, we will elaborate on an analysis concerning the need of dissipative energy for the microcracking inside the process zone. The equations we have worked out provide evidence for the correlation between the stress intensity factor and the critical energy release rate with specific microstructural features (grain size, microcrack

density, elastic microcrack interaction, etc.) respectively; they
are in agreement with our experimental results. We expect the
model to be a useful tool in aiding the development of ceramics
with tailor-made microstructure.

THE TWIN-PARAMETER APPROACH TO BRITTLE FRACTURE PHENOMENA

The twin-parameter approach of notch fracture mechanics
opens the possibility for a better understanding of brittle
fracture phenomena by introducing a material specific stress
intensity factor depending on microstructural parameters
(Figure 1). Following Neuber (10), Weiss (11), Erismann (12)
and Buresch (4,13) it can be demonstrated that the fracture of
a certain ceramic is determined by critical values for the notch
fracture stress σ_{mc} and the size of the process zone ρ_c. It
means that the microcrack region in the immediate vicinity of a
crack tip is responsible for the non-linear behavior of ceramics.
This can be done in a homogeneous stress field by stress concentrations on pores as shown recently by Cooper (7). Alumina

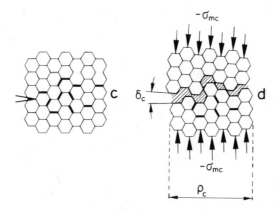

Figure 1.   A Model for Microcracking in the Process-Zone.

doped with unstabilized zirconia has a very high toughness as has
been shown by Claussen (8). This is caused by residual strains
as a consequence of thermal expansion mismatch, and phase transformation, which generates in a homogeneous stress field microcracks with the zirconia particles as origins. Therefore, the
stress intensity factor is given by

$$K_{Ic} = \frac{\sigma_{mc}}{2} \sqrt{\pi \rho_c}$$ [1]

It follows for the energy release rate in case of plane strain

$$G_{Ic} = \frac{\sigma_{mc}^2}{4E} \, \pi\rho_c(1-\nu^2)$$ [2]

This is equivalent to the Dugdale-Barenblatt model introducing the yield stress and the plastic zone size for σ_{mc} and ρ_c respectively. σ_{mc} is minus the cohesive stress. Below the values Equations [1] and [2] will be interpreted as purely material specific parameters independent from crack size and distribution of external loads.

The fracture criterion Equation [2] characterized by Young's modulus outside ρ_c can be described by critical structure sensitive parameters inside ρ_c. The dissipative processes inside ρ_c are a consequence of residual stresses and strains, caused by dislocation pile-up or anisotropic thermal expansion of the crystallites. In ceramics those stresses cannot be released by a homogeneous plastic deformation. Microcracking will occur if stress concentrations caused by external or internal stresses reach the microscopic stress intensity factor. Microcracking ahead of a crack tip is observed in brittle fracture materials like rocks, graphite, or steel at low temperatures (4).

Microcracking caused by external or internal stresses will occur when the normal stress component of a trans- or inter-crystalline cleavage plane reaches the specific cleavage stress. Favorably, microcracks will first be observed on cleavage planes normal to the external load. These microcracks are pinned on grain boundaries or other obstacles (Figure 1). With increasing K_I and super-imposed with residual stresses the microcrack density increases to the point where the notch stress and the process zone size reach material specific structure sensitive values σ_{mc} and ρ_c respectively, thereby joining favorable oriented microcracks with the main crack (Figure 1). In ceramics with low internal stresses and small grain size distribution the process zone is well defined and limited concerning the notch fracture stress σ_{mc} where the maximum normal stress component reaches the cleavage stress. In materials with high internal residual strain energy and a large grain size distribution like graphite especially with low cleavage energy, the process zone is spaced without sharp boundaries. This means that the load deflection diagram shows a severe deviation from linearity.

Microcracks lower Young's modulus E_ρ inside ρ_c depending on grain size, microcrack-length, micro-crack density and elastic microcrack interaction, respectively, by a factor f_ρ meaning, that $E_\rho = Ef_\rho(a_m,\beta,w)$ is constant in the region ρ_c. The critical stress intensity factor, e.g., for a 4-point bend specimen, with a through the thickness crack of length a_c and fracture stress σ_c is given by

$$K_{Ic} = \sigma_c \sqrt{a_c} \; y \left(\frac{a}{W}\right) \qquad [3]$$

Our model defines the toughness of ceramics on the basis of an energy criterion which means a critical value for the stored elastic energy density as well as for the energy release rate of the material inside the process zone. The stored elastic energy inside ρ_c characterize a crack resistance parameter which must be overcome by the moving crack. Therefore, cracks run in discrete paths as they are observed in many brittle materials like steel at low temperatures or graphite. The stored elastic energy at instability inside ρ_c is given by

$$W_{ec} = \frac{A(v)K_{Ic}^2 \; \rho_c}{16 \; E_\rho} \qquad [4]$$

where $A(v)$ is a function of Poisson's ratios for a given stress state. In the case of plane strain $A(v) \simeq 4$ for alumina. Following Rolfe et al. (15) for 4-point bend specimens with width W, critical crack length a_c and $x = (a_c/W)$ the polynomial $y(x)$ is given by

$$y(x) = 1,992 - 2,468x + 12,97x^2 - 23,17x^3 + 24,80x^4 \qquad [5]$$

From Equation [4] we derive the microscopic fracture criterion

$$F_c = -\frac{\partial W_{ec}}{\partial a_c} = \text{const.} = \frac{\sigma_c^2 \; \pi \; \rho_c A(v)}{16 \; E_\rho} \; y^*(x) \qquad [6]$$

introduced by Buresch (4). The polynomial is given by [4]

$$y^* = y(1.992 - 7,404x + 64,85x^2 - 162,15x^3 + 223,2x^4 \qquad [6a]$$

In (4) it is assumed that elastic energy release rates F_c and G_{Ic} and [2] respectively correspond with each other. That means with

$$\frac{\sigma_{mc}^2}{\sigma_c^2} = \frac{4}{\pi} \frac{a_c}{\rho_c} \; y^2 \qquad [7]$$

which is in an equivalent form on principle valid for the Dugdale Barenblatt model, that Young's modulus inside ρ_c is given by

$$\frac{E\rho}{E} = f_\rho = C \; \frac{\sigma_c^2}{\sigma_{mc}^2} \; y^*(x) \qquad\qquad [8]$$

or

$$f_\rho = \frac{C}{4} \frac{\rho_c}{a_c} \frac{y^*(x)}{y(x)} \qquad\qquad [8a]$$

where C is a proportional constant which includes the Poisson ratio. Following Equation [8] for a given material the micro-crack configuration inside ρ_c is constant. This is shown in Figure 2 for a/W ≳ 0.1 with measured values for a specific alumina (4).

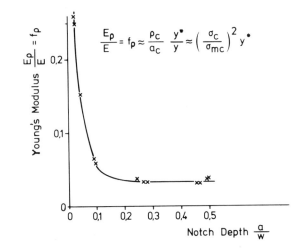

Figure 2.   Dependence of Effective Young's Modulus Inside ρ_c from Notch Depth.

The model is compatible with estimates referred to in Goodier and Field (1963). The diagram (Figure 2) suggests that the work done on the cohesive forces and assignable to new surfaces energy is the important part for very short cracks. For large macroscopic cracks it suggests that it is the work done in microcracking which dominates. With many problems involving solids with multiple cracks, it is reasonable to assume that the stress field in the vicinity of each microcrack is approximately the same as the stress field around a single crack in an infinite body. However, metallographically it has been found that the microcracks in front of the macrocrack tip of ceramics are separated by less than several microcrack lengths,

that means grain sizes. Therefore, they interact such as to
significantly alter the state of stress near a microcrack tip
as well as the strain energy associated with each microcrack.

Now we will improve the microcrack model and explore the
influence of the microcrack density, orientation and elastic
interaction respectively in detail.

INFLUENCE OF MICROCRACK INTERACTION PARAMETER ON TOUGHNESS OF CERAMICS

The nonlinear behavior in front of a crack or notch is
governed by the microstructure, especially the grain and pore
size distribution and microscopic fracture energy. We will
compare quantitatively the change in stored elastic energy
during microcracking inside ρ_c with the dissipative energy
needed for the generation of microcracks. We neglect other
terms of dissipative energy like dislocation motion, glide of
crack faces or chemical reactions. With the number of grains in
the unit of volume proportional to d^{-3} and the surface energy
for a penny-shaped crack proportional to $\gamma_s d^2$ the total dissipative surface energy inside ρ_c is given by

$$W_d = \frac{\pi \, \rho_c^2 \, \gamma_s \, \beta}{2wd} \qquad\qquad [9]$$

The symbols β and w denote the part of grains with microcrack
and orientation of the microcracks with respect to the main
normal stress respectively. β denotes alternatively the
concentration of dispersions which generate microcracks as a
consequence of thermal expansion mismatch or phase transformation.

In our model it is assumed that microcracking will occur if
the notch fracture stress reaches the critical value σ_{mc},
lowering the Young's modulus to E_ρ. Then the change in stored
elastic energy is given by

$$\Delta W_{ec} = \frac{\pi \, \rho_c^2 \, \sigma_{mc}^2 \, A(v)}{16 \, E} \left(\frac{1}{f_\rho} - 1\right) \qquad\qquad [10]$$

It follows with Equations [9] and [10] for critical values of
notch fracture stress, stress intensity factor and energy
release rate

$$\sigma_{mc} = 4 \sqrt{\frac{\gamma_s \, E}{dA(v)} \frac{\beta}{w} \frac{f_\rho}{1-f_\rho}} \tag{11}$$

$$K_{Ic} = 2 \sqrt{\frac{\pi \, \rho_c \, \gamma_s \, E}{dA(v)} \frac{\beta}{w} \frac{f_\rho}{1-f_\rho}} \tag{12}$$

and

$$G_{Ic} = 4 \frac{\pi \, \rho_c \, \gamma_s \, \beta \, f_\rho}{w(1-f_\rho)} \tag{13}$$

The term

$$I_\rho = \sqrt{\frac{\beta}{w} \frac{f_\rho}{1-f_\rho}} \tag{14}$$

is the elastic microcrack interaction parameter which denotes the change in stress intensity of one microcrack with respect to the elastic potential of the surrounding microcracks.

The elastic microcrack interaction is well known for an elastic solid by a rectangular array of cracks by Delameter et al. (14). The weakening of the solid can be expressed by

$$\frac{E_\rho}{E} = f_\rho = \frac{1}{1+\frac{2\pi a_m^2}{xy} B^*(\beta)} \tag{15}$$

where x and y denote the distance of microcracks in rows and stacks respectively. $B^*(\beta)$ is listed for several values of $y/2a_m$ and $x/2a_m$. B^* for a single row of cracks (that is when $y/2a \to \infty$) as a function of x/a_m is found by the expression (14)

$$B^* = 2 \left(\frac{x}{\pi a_m}\right)^2 \ln\left(\sec\frac{\pi a}{x}\right) \tag{16}$$

which is the well known equation for a row of collinear cracks.

Neglecting the difference in density of a three- and two-dimensional array of microcracks, the elastic microcrack interaction parameter equation [14] can approximately be given by

$$\sqrt{\frac{\beta}{w} \frac{f_\rho}{1-f_\rho}} = \sqrt{\frac{2}{\pi\, B^*(\beta)}} \qquad\qquad [17]$$

This is shown in Figure 3 for a rectangular array of micro-cracks of different densities. Only a parallel array of micro-cracks enhance the elastic microcrack interaction parameter.

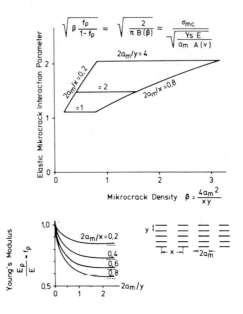

Figure 3.   Correlation of Theoretical Elastic Microcrack
             Interaction Parameter and Young's Modulus with
             Microcrack Density.   (Values of $B^*(\beta)$ are listed
             by Delameter et al. [1975]).

EXPERIMENTAL RESULTS

The influence of microstructural features like density, grain size distribution and elastic microcrack interaction shall be demonstrated by experimental investigation of two aluminas of different impurity level, which are 0.2% (denoted A) and 1% (denoted F) respectively. Plates of each material are sintered at temperatures of 1400°C, 1500°C 1600°C and 1740°C respectively. This results in densities and mean grain sizes in range of 2600 to 3900 kg/m^3 and 2 to 9 μm respectively. Stress intensity factors are determined with 4-point bend specimens. Only speci-mens with thickness B ⩾ 7 mm give real material specific values (4).   With measured values for K_{Ic}, E, ρ_c and d computed values for

$$\sqrt{\frac{\beta \, f_\rho}{w(1-f_\rho)}} = \frac{K_{Ic}}{2\sqrt{\dfrac{\pi \, \gamma_s \, E \, \rho_c}{d \, A(v)}}}$$ [18]

are shown in Figure 4 with $w = 1$ and $\gamma_s = 1$ J/m^2 for all materials.

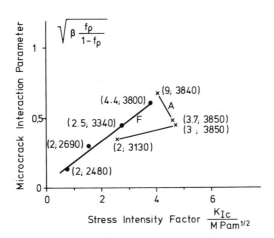

Figure 4.   Elastic Microcrack Interaction Parameter (Experimental) of Two Aluminas as a Function of Their Stress Intensity Factors. (Figures in parenthesis are mean grain sizes in µm and densities in kg/m^3).

DISCUSSION

Theoretically the toughness of a ceramic can be improved by a high density of parallel oriented microcracks as shown in Figure 3. This is experimentally demonstrated by Claussen et al. (8) with an alumina with dispersions of unstabilized zirconia and a duplex microstructure with a high density of parallel oriented microcracks. For its highest toughness values of about 10 MPam$^{1/2}$ our estimation for the value of the elastic micro-crack interaction parameter is about 2. For undoped ceramics the elastic microcrack interaction parameter is in comparison low. Our experimental results show the values in Figure 4 for aluminas of different grain size and densities. Generally, the interaction parameter is enhanced by grain size and density. It is, in comparison with Figure 5, concluded that intragranular pores can act as stress concentrators thereby enhancing the

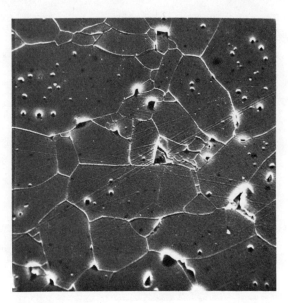

Figure 5.   Microstructure of an Alumina with a Mean Grain Size
            of 9 μm and a Density of 3840 1g/m^3. (Standard
            deviation of grain size distribution S = 100).

microcrack density. The microcracks are pinned on grain
boundaries. The number of intragranular pores increase with
increasing grain size following increasing sintering temperature.
As a consequence the length of microcracks increase resulting in
a diminished stress intensity factor (see Equation [12] and
Figure 5). It is assumed that the low value for the elastic
microcrack interaction parameter of undoped ceramics (Figure 5)
is a consequence of mostly collinear arranged microcracks inside
ρ_c.

The influence of parallel oriented microcracks on the toughness can also be estimated from our measurements concerning the
anisotropic fracture behavior of graphite. Here, specimens with
a macrocrack in the grain direction show a lower stress-intensity
factor in comparison to specimens with a macrocrack in the cross-
grain direction. According to metallographic observations we
assume that in the last case parallel oriented microcracks are
pinned on grain boundaries thereby enhancing the elastic microcrack interaction parameter (16).

The toughness of ceramics enlarges with increasing process
zone size. The process zone size depends on the grain size
distribution, especially the standard deviation of grain size
distribution as shown in Figure 6. In essence, the process

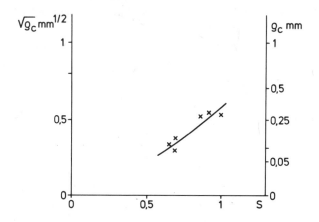

Figure 6.   Correlation of Process-Zone-Depth with Standard
            Deviation of Grain-Size Distribution of Alumina.

zone size is governed by the largest grains, which have the
largest microscopic stress intensity factor (4). Following
Equations [2] and [12] the process zone size is given by

$$\rho_c \simeq a_m \frac{G_{Ic}}{\gamma_s \, \pi \, (1-v^2)} \, I_\rho^{-2}$$ [19]

G_{Ic}/γ_s is a measure for roughness of the fracture surface by the
microcracks at instability.

     Normally high strength and high toughness exclude each
other. High strength is governed by a low mean grain size and
small grain size distribution. Alternatively, high toughness
can be a consequence of low mean grain size and large grain size
distribution.

     Concerning Figure 6 it is important to note that the
standard deviation of the grain size distribution which is
equivalent to the microcrack size distribution is proportional
to the standard deviation of the fracture stress distribution.
Therefore, the lower the standard deviation of the grain size
distribution, the higher is the Weibull modulus and the lower
the process zone size ρ_c (16). The process zone is an autonomous
region characterizing the toughness of ceramics independent of
external parameter like the crack size or the distribution of
forces. However, this is only valid if ρ_c is small in comparison
to characteristic geometric dimensions like the thickness B of a

4-point bend specimen. Material specific K_{Ic}-values only can
be measured if the inequality holds (4)

$$B \geqslant 50 \ \rho_c \qquad\qquad\qquad [20]$$

This shows Figure 7 very clearly for 4-point bend specimens with
thickness B equal to 3 and 7 mm respectively.

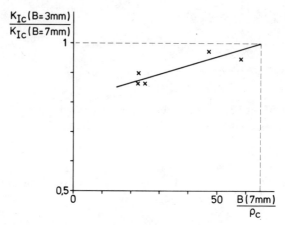

Figure 7.   Correlation of Relative Stress-Intensity Factor of
            Alumina with Relative Specimen Thickness.

ACKNOWLEDGMENTS

     The author wishes to thank Drs. H. Schuster, K.F.A., and
N. Claussen, M.P.I. Stuttgart, for valuable discussions.

LITERATURE

1. McClintock, F. A. and F. Zaverl, Jr. Submitted to Inter-
    national Journal of Fracture.

2. Evans, A. G., A. H. Heuler and D. L. Porter. I.C.F. 4,
    Waterloo/Canada, June, 1977 Proceedings in Press.

3. Hoagland, R. G., J. D. Embury and D. Y. Green. Scripta
    Metallurgica 9, 1975, 907.

4. Buresch, F. E. Science of Ceramics 7, 1973, 383 and 475.

5. Hoagland, R. G., G. T. Hahn and A. R. Rosenfield. Rock
    Mech. 5, 1973, 77.

6. Hübner, H. and W. Jillek. D.V.M. Bruchvorgäuge. Aachen,
    1975.

7.   Cooper, R. E.   I.C.F.4, Waterloo/Canada June, 1977,
     Proceedings in Press.

8.   Claussen, N. and J. Steeb.   Jour. Amer. Ceram. Soc. 58,
     1976.

9.   Leone, E. M. and D. M. Neal.   I.C.F.4, Waterloo/Canada,
     June, 1977, Vol. 3, 913.

10. Neuber, H.   Kerbspanungslehre Springer, Vlg. 1958.

11. Weiss, V. Fracture III. Academic Press, N.Y.   1971, 227.

12. Erismann, T. H.   Material und Technik, 1973, 120.

13. Buresch, F. E.   I.C.F. 4, Waterloo/Canada, June, 1977,
     Proceedings in Press.

14. Delameter, W. R., G. Hermann and D. M. Barnett.   Trans.
     ASME 1975, 74.

15. Rolfe, S. T. and S. R. Nowak, ASTM/STP, 463, 1970, 124.

16. Buresch, F. E.   To be published.

MICROSTRUCTURAL DEPENDENCE OF FRACTURE MECHANICS PARAMETERS IN CERAMICS

R. W. Rice, S. W. Freiman, R. C. Pohanka, J. J. Mecholsky, Jr., and C. Cm. Wu

Naval Research Laboratory

Washington, D. C. 20375

INTRODUCTION

Understanding the microstructural dependence of strength and related fracture mechanics parameters is essential to developing better ceramics. It is also essential to the proper application of fracture mechanics to predicting mechanical, e.g., stress-life, behavior of polycrystalline ceramics. Three basic aspects of microstructural behavior that must be considered will be discussed in this paper. The first aspect is the microstructural dependence of fracture energy as measured by the typical fracture mechanics test; i.e., for cracks large in comparison with the strength controlling microstructure. The second aspect is the modifications that must be made to fracture mechanics parameters when cracks are not large in comparison to the strength controlling microstructure, e.g., as in many components and test specimens. This issue was raised at the previous Fracture Mechanics of Ceramics Conference (1-3). The third aspect is the integration of the above aspects into an understanding of the overall strength-microstructure behavior; illustrated by a discussion of strength-grain size behavior.

MICROSTRUCTURAL DEPENDENCE OF FRACTURE ENERGY
Effects of Porosity

Several advances in the observation and/or theory

of the porosity and grain size dependence of fracture
energy have been made. Rice and Freiman (4), in analyzing the porosity dependence of fracture energy, recognized that: 1) it should be proportional to the
fraction of solid area fractured and 2) fracture will
generally follow the path of maximum pore and minimum
solid cross sectional area (and hence typical sterologi-
cal relations which give average areas are usually not
applicable). They then calculated fracture energy (γ)
as a function of volume fraction of porosity (P) for a
variety of idealized pore shapes and distributions assuming that γ was directly proportional to the minimum
solid area between pores. They showed that this almost
always gave γ as non-linear functions of P and that
over the porosity range of up to 20-50%, these can be
reasonably approximated by:

$$\gamma = \gamma_0 \, e^{-bP} \tag{1}$$

where $\gamma_0 = \gamma$ at P = 0 and b is a constant that depends
on the shape, orientation, and packing of the voids,
e.g., see Fig. 1. Intergranular voids, e.g., between
sintering spherical grains, fit this equation very well
in this range giving "b" values of 7-9 depending on the
stacking of the grains and hence of the voids. Although,
different "b" values can result from different mixes of
the various types of pores, it was shown that available
literature data for relatively pure bodies agreed reasonably well with the above equation and the "b" values
expected for the general character of pores in the
bodies (4, 5).

Major deviations to higher fracture energies than
predicted by Eq. 1 were observed at higher levels of
porosity, e.g., above ~ 25% porosity for larger grain,
less pure bodies, and 50% for finer grain, purer bodies.
Larson et al (6),for example, show Al_2O_3 refractories
can give work of fracture values up to ~ 10 times the
value for pure, dense Al_2O_3. Notched beam test results
were often roughly consistent with expected porosity
trends, but many were higher, e.g., up to ~ 50% greater ,
than pure, dense bodies. (This is one of several important differences between results from these test techniques.) The significantly higher than expected fracture
energies were attributed to microcracking as has been
clearly shown to extensively occur in porous rocks showing high fracture energies by Hoagland, et al (7).

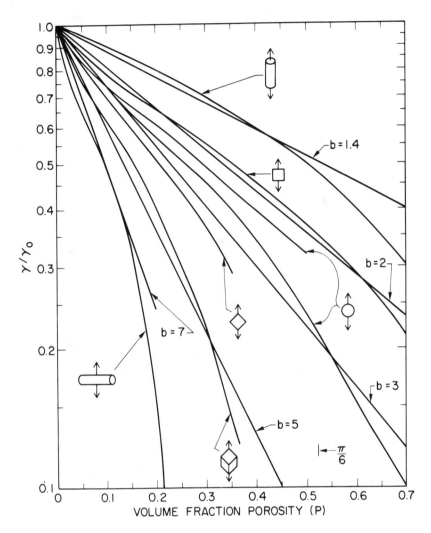

Fig. 1. γ/γ_0 vs volume fraction porosity for various
simple pore shapes. The arrows from each pore shape
indicate the direction of stressing relative to the pore
geometry. Note that the two curves shown for spherical
pores represent simple cubic stacking (lower curve) and
staggered stacking (upper curve).

Grain Size Effects

A recent survey by Rice (5) showed that dense polycrystalline bodies of cubic crystal structure showed no
significant dependence of fracture energy on grain size.
However, fracture energies of non-cubic materials typically rose over the limited grain size ranges studied,
except for $MgTi_2O_5$. This very anisotropic material also
showed fracture energies increasing with increasing grain
size, but then reaching a maximum, and decreasing (8).
More detailed analysis and further experimentation by
Rice and Freiman (9) showed the fracture energies of all
the non-cubic materials they studied, Al_2O_3, TiO_2, and
Nb_2O_5, first increased with increasing grain size, passed
through a maximum, then decreased in a fashion similar to
$MgTi_2O_5$ (Fig. 2).

Rice and Freiman (9) developed an analytical model
for the grain size dependence of the fracture energy of
non-cubic materials. The mathematical formulation of
this model is based on the adaptation by Rice and Pohanka (10) of Davidge and Green's model for spontaneous
fracture around second phase particles in glass (11) to
spontaneous fracture of polycrystalline materials of
anisotropic crystal structures. This model equates the
strain energy in and around a grain due to its strain
mismatch ($\Delta\epsilon$) with its neighbors, to the fracture energy
for cracking around the grain. In its simplest form,
this gives the grain size for spontaneous fracture (G_s):

$$G_s = \frac{12 \ E\gamma_B}{\sigma_i^2} \tag{2}$$

Where E = Young's modulus, γ_B = the grain boundary fracture energy, and σ_i = the internal stress ~ $\Delta\epsilon E$. The
fracture energy model is based on the addition of applied
stresses and internal stresses from the thermal expansion
anisotropy to result in microcracking. The concept is
that there are two opposite effects on the fracture
energy due to microcracking, namely that the number of
microcracks increases while the energy absorbed from
the applied stress per microcrack decreases, as grain
size increases. It is then assumed that the fracture
energy is the sum of energies due to microcracking and
fracture of unmicrocracked materials (γ_{pc}). The latter
is assumed to decrease with increasing grain size (G)
and associated microcracking according to ($1-G/G_s$); the
fracture energy should go to zero at G_s since the body
has, in principal, then totally microcracked due to the

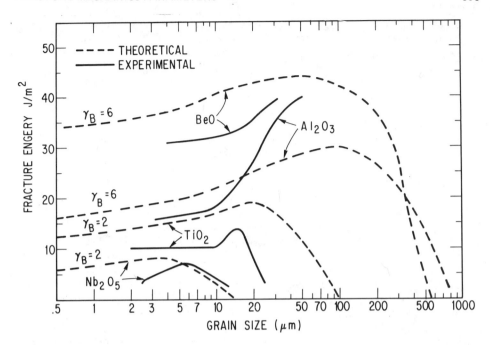

Fig. 2. Fracture energy as a function of grain size.
The range of experimental data is shown by solid lines
and the theoretical predictions are shown by dashed
lines. The grain boundary fracture energy (γ_B) values
used in the equation are shown on the left hand portion
of each theoretical curve. Current experimental work
indicates that the fracture energy of Al_2O_3 does indeed
decrease at larger grain sizes.

internal stresses. The resultant equation is:

$$\gamma = \gamma_{pc} (1 - G/G_s) + \Delta \epsilon \left[(12 \ E\gamma_B G)^{\frac{1}{2}} - \Delta \epsilon E \ G \right] \qquad (3)$$

Predictions based on this equation are in rather good
agreement with experimental data (Fig. 2). It should be
noted that plotting $(1/\gamma_B) [\gamma - \gamma_{pc} (1 - G/G_s)]$ vs G/G_s brings
all materials into a common band of behavior as predicted
by the theory. The above theoretical approach has been
modified for two phase systems and shows good agreement
with fracture energy data of WC-Co.

The above modeling is for materials of more limited
anisotropy where use of a single strain mismatch is reasonably appropriate. Preliminary analysis shows good

promise of modeling more anisotropic materials such as
MgTi₂O5 by considering microcracking from each of the
three quite different sets of expansion mismatches.

Experimental Evidence of Microcracking

The above model is clearly based on microcracking,
in contrast to a single simple crack that is typically
assumed in fracture mechanics analysis and tests. Except for limited (but sound) observations (12, 13) and
suggestions of such microcracking (1), there has until
recently been no study to verify whether or not this
fundamental assumption is valid, and very little effort
to study the detailed microstructural nature of crack
propagation. However, Wu et al (14) have shown that
microradiography and optical microscopy (in transparent
materials) can be used to study the nature of propagating
cracks in double cantilever beam specimens (Fig. 3).
Subramanian (15) and some of the present authors have
applied techniques for the direct SEM observation of
cracks (Fig. 4). The latter technique, while allowing
observation of the crack only on the surface of the
sample, gives higher resolution than the microradiographic technique and optical technique.

The extensive microradiographic studies complimented
by more limited optical microscopy clearly show that the
single simple crack assumed in fracture mechanics occurs
only in the minority of cases. Such single, simple
cracks having only some limited twist, occur in glasses,
probably in most single crystals and possibly in some
fine grain polycrystalline bodies. In addition to generally greater twisting of the crack, microradiography
shows all cubic polycrystalline materials with grain
sizes above the resolution limit of the technique (~
20 μm), exhibit some degree of crack branching with the
separation of the branches limited to at most a few
grains (Fig. 3).

Analysis of the results suggests that the degree
of this branching in cubic materials is related to their
degree of elastic anisotropy. The fact that the fracture
energy of cubic materials showed no significant dependence on grain size implies that such branching continues
to occur as grain size decreases below the limits of resolution of the microradiographic technique . Preliminary SEM observations corroborate this.

Microradiography also shows that a second phenomen-

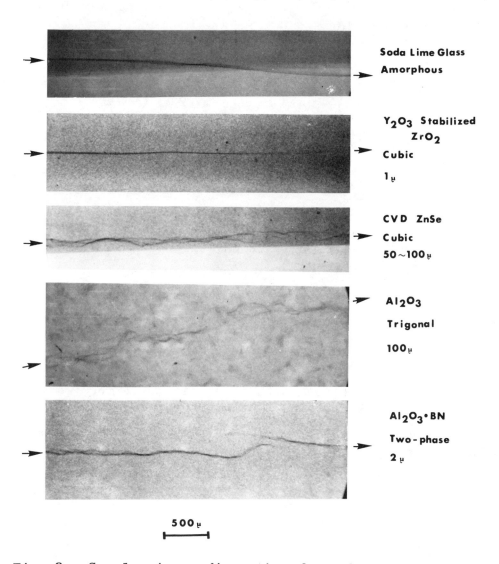

Fig. 3. Sample microradiographs of cracks in various materials, as indicated on the right hand side, which also lists their crystal or phase structure and the approximate grain size. Arrows indicate the direction of crack propagation.

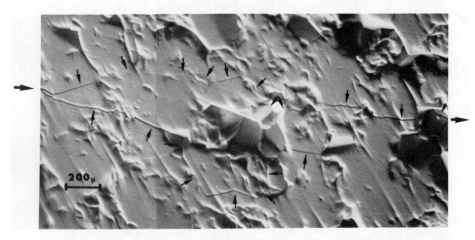

Fig. 4.   SEM of crack in dense Al_2O_3.   Note the small
arrows indicating portions of at least two branches of
the major crack which is propagating from left to
right (arrow).

on, crack branching on a multigrain, instead of few
grain scale, occurs at intermediate grain sizes in non-
cubic materials or bodies containing two phases with
substantial thermal expansion mismatch between them as
in some Si_3N_4 bodies and in recently developed composites, such as Al_2O_3 - BN.   These studies further suggest
that the ratio of the branch separation to the grain size
increases as the grain size increases, passes through a
maximum, then begins to decrease as the body approaches
sufficiently large grain sizes, to have pre-existing
cracks developing due to the internal stresses from the
thermal expansion mismatch.   This is consistent with
the above fracture energy trend and corroborates the
basis of the theoretical model explaining the grain size
dependence of fracture energy in non-cubic materials. It
has further been observed that Si_3N_4 - 15 w/o Y_2O_3[*] (γ
~ 80 J/m^2), shows more multigrain branching than the
typical Si_3N_4 - MgO bodies (γ: 25-50 J/m^2).   Such multigrain branching, especially in bodies having large mismatches in thermal expansion is consistent with such
bodies lying well above the linear γ-E trend as reported
by Mecholsky, et al (16) for most other ceramics.

    The above crack propagation observations, in addi-

_____

[*]Ceradyne Corp.

tion to supporting the fracture energy - grain size model for non-cubic materials, have two other implications. First, the added fracture area due to extra cracks, seen in all larger grain bodies is likely a major factor in fracture energies being substantially greater for polycrystalline than single crystal bodies. Second, the microradiographic and the E-γ results show MgO behaving like other cubic materials, indicating that microplastic flow has limited effect on fracture energy. Thus, there is no obvious reduction in crack branching or increase in fracture energy due to increasing plastic flow with increasing MgO grain size.

Effects of Plastic Deformation

Microplastic phenomena can, however, be significant as observed by Pohanka and Freiman (17) in $BaTiO_3$. There, substantial twinning, which occurred within single crystals in fracture energy tests below the Curie temperature, ocrrelated with higher fracture energies there than above the Curie temperature where no twinning occurred. Similarly, the significant increase in the fracture energy of $BaTiO_3$ polycrystals with increasing grain size below the Curie temperature, but not above (where it remains constant at the fine grain level which is the same above or below the Curie temperature) is consistent with increased twinning with increased grain size. However, such trends are also consistent with internal stresses from the phase transformation in the same fashion as due to thermal expansion anisotropy as discussed above. Further, the internal stresses, in addition to the applied stresses are a source of twinning, making separation of twinning and internal stress effects in polycrystals very difficult. Both are probably important, e.g., Rice and Pohanka (10) have reported that twinning significantly reduces the internal stresses due to phase transformation in $BaTiO_3$. However, the remaining stresses still lead to spontaneous fracture, but at a larger grain size (~ 800 μm) due to the reduced level of internal stress.

Effects of Phase Transformations

Phase transformation effects may also directly affect crack propagation. Both Garvey, et al (18) and Porter and Heuer (19) have suggested that one of the factors leading to toughening of partially stabilized ZrO_2 bodies is phase transformation of the non-cubic ZrO_2 particles in a zone around the crack tip. Porter and Heuer (19) have presented electron microscopy evidence to show that such a transformed zone does exist after passage of a crack. However, microcracking may also occur around such particles (20, 21). In fact, both microcracking and phase transformation are probably factors, with both being dependent on particle size. Similarly, Claussen, et al (22) have shown that both phase transformation and microcracking occur in Al_2O_3 with dispersed unstabilized ZrO_2 particles, with the relative roles of these apparently being particle size dependent.

Freiman and Pohanka (to be published) have observed significant increases in critical fracture energy of lead zirconate titanate bodies whose composition placed them close to the morphotropic boundary between the rhombohedral and tetragonal phases where a stress induced phase transformation near the crack tip would be expected. This is consistent with Burns' (22) earlier report, that high stresses around the crack tips in quartz caused the material immediately in the vicinity of a crack to be transformed to the high temperature form (then re-transformed back to the low temperature form as the high stresses around the crack tip passed with the propagation of the crack). Thus, phase transformations can result in two effects on fracture energy. If a phase transformation occurs in the vicinity of the crack, it can alter the fracture energy directly. If the transformation has already occurred, i.e., in the absence of a crack, it leaves internal stresses due to the mismatch strains between the grains and the body, that can alter the fracture energy due to their contribution to microcracking as discussed above. Again in either case, such effects should be grain size dependent

THE EFFECTS OF THE FLAW SIZE RELATIVE TO THE DIMENSIONS OF STRENGTH CONTROLLING MICROSTRUCTURE ON FRACTURE MECHANICS PARAMETERS

The application of fracture mechanics to polycrystalline ceramics has until recently been based on the assumption that fracture energies measured using cracks typically large in comparison with the microstructure are directly applicable to the prediction of strengths, life times, etc. of bodies containing normal flaws which are typically ~ 50 μm or less in depth. Only in the past few years has this basic assumption been questioned or investigated (1-4). This question of flaw effects has been found to be fundamental to understand-

ing fracture involving both extrinsic and intrinsic
heterogeneities. Extrinsic heterogeneities, i.e., variations in composition and isotropy are treated first.
Then intrinsic heterogeneities due to the pore and grain
structure are discussed.

Extrinsic Heterogeneities

Baumgartner and Richerson (23) were probably the
first to experimentally indicate effects of composi-
tional heterogeneities. They reported significantly
variable results of fracture energies calculated from
the size of inclusions in hot pressed Si_3N_4 and suggested
that this variability was possibly due to changes in the
local behavior of the material, primarily its fracture
energy. Subsequently, Freiman et al (24) have attributed
significantly higher measured fracture energy in fracture
mechanics tests to that indicated by actual flaw sizes
at fracture origins of specimens of $Si_3N_4-Y_2O_3$ to the
widely observed inhomogeneity in bodies of this composition. More recently Land and Mendiratta (25) have
considered the effect of locally varying character of
the material on fracture mechanics parameters. These
studies clearly show the need for broader attention to
the question of homogeneity to verify the applicability
of fracture mechanics to a particular system.

Consider next anisotropy frequently resulting from
some fabrication processes, e.g., CVD deposition, and
hot working such as extrusion and press forging. Lange
(26) was probably one of the first to observe an anisotropy in K_{IC} and strength in hot pressed Si_3N_4. Subsequently, anisotropy has been reported in hot pressed
Si_3N_4 and Al_2O_3 (27) and in β alumina (28). However,
one of the authors (Freiman) is finding that anisotropy
effects may depend significantly on fabrication and
possibly test parameters. In addition, Freiman et al (24)
have shown that an oriented, elongated grain structure is
not a sufficient condition for anisotropic fracture energy. They found no significant anisotropy in CVD (β SiC)
having a highly oriented, elongated grain structure. It
appears that a preference for grain boundary failure
(the CVD SiC failed transgranularly) and/or a fairly
anisotropic fracture energy behavior in the grains themselves is necessary for resultant anisotropy in textured
polycrystalline bodies.

Another heterogeneity is porosity. Rice (4, 29) has
suggested that when bodies contain a considerable amount

of porosity, with significant portion as large isolated
pores or pore clusters, e.g., as much of the porosity in
reaction sintered Si_3N_4 often is, failure commonly occurs from such large pores or pore clusters. Since the
spacing between the pores or clusters is typically substantially greater than their diameters, the critical
stages of failure occurs while the crack is propagating
out from the pore, unaffected by the surrounding porosity. Hence, the fracture energy measured directly on
the body which is lower due to the presence of porosity
has been suggested as being inappropriate for predicting
the mechanical failure of the body, i.e., fracture energy appropriate to the density of the material surrounding the large pores or pore clusters should be considered. While this appears to be a reasonable assumption,
experimental studies have not been conducted to verify
these effects.

Intrinsic Effects

Turning now to intrinsic effects, consider first
the heterogeneous internal stresses resulting from thermal expansion anisotropy and phase transformations. In
the previous section, it was shown that these stresses
are the fundamental cause of the variation of fracture
energy with grain size for cracks large in comparison to
the grain size. However, these stresses can directly
contribute to failure, and hence, alter the use of fracture mechanics in predicting failure. The first demon-
stration of this was by Pohanka et al (30), who showed
that the strength of $BaTiO_3$ with ~ 10 μm deep machining
flaws, was nearly 50% lower below the Curie temperature
where internal stresses were present than above the
Curie temperature where such stresses are absent. Further
study by Pohanka et al (31) revealed that this difference
in strength due to internal stresses decreased as the flaw
size increased, becoming zero for large flaws. This is
to be expected since on a long range scale, the internal
stresses must average to zero; however, due to the statistics of grain orientations there will be local multigrain size regions in which there will be more tensile
stresses than there are compressive stresses, while
other regions have higher compressive stresses to balance
these tensile stresses. However, since there is usually
a large distribution of flaws, those flaws in the vicinity
of regions having a higher net internal stress, $< \sigma_i >$,
adding to the applied stress, σ_a, will be the sources of
failure. Failure is then given by:

$$\sigma_a + <\sigma_i> = A \sqrt{\frac{E \gamma}{C}} \qquad (4)$$

where C = flaw size and A = geometrical factor. Pohanka
and Freiman (17) have shown that internal stresses in
bodies such as BaTiO₃ can be calculated without any
knowledge of the flaw character when one measures the
fracture energy with and without the internal stress,
i.e., below and above the transformation temperature
respectively. This can also be important in accounting
for other factors that may affect fracture energy and
strength such as twinning. Rice, et al (32), showed
internal stresses due to thermal expansion anisotropy
affect the mechanical behavior of all non-cubic materials.
These effects are ascertained by calculating the apparent fracture energy (γ_a), using A and C determined from
fractography in the normal Griffith equation:

$$\sigma_a = A \sqrt{\frac{E \gamma_a}{C}} \qquad (5)$$

This consistently gives γ_a's which start decreasing below the typical fracture energies measured by independent fracture mechanics, e.g., DCB, tests, at smaller
flaw sizes (Fig. 5). Use of Eqs. 4 and 5 simultaneously
allows $< \sigma_i >$ to be calculated as a function of C giving
an increasing contribution of internal stresses to failure as illustrated in Fig. 6. The $< \sigma_i >$ values projected
to C ~ 3G, where single crystal effects discussed below
are not yet significant, gives internal stresses that
approach those predicted from the known thermal expansion anisotropy of these non-cubic materials (i.e., σ_i ~
$E \Delta \epsilon$).

The above experiments have been performed on finer
grain materials since it is typically in these in which
definitive fracture origins can be identified for the
above analysis. Thus, an important area for further research is to determine whether increasing grain size
changes the rate at which the internal stresses contribute to failure stress as the flaw size decreases. Studies of large (e.g., ~ 100 μm) grain BaTiO₃ indicate that
the effects of internal stresses become insignificant
after the flaw size increases to a few times instead of
~ 100 times the grain size in fine (~ 5 μm) grain bodies.
However, whether this is due to the complications introduced by twinning or is an intrinsic effect of the distribution of internal stresses is not yet known.

The second intrinsic heterogeneity to be considered
is the grain structure itself. When the crack is no
longer large in comparison with the grain size, the

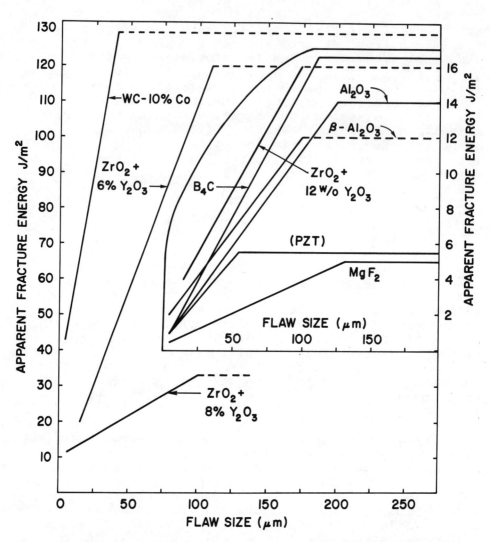

Fig. 5. Apparent fracture energy versus flaw size for various ceramic materials. The apparent fracture energy is calculated using the Griffith equation neglecting the possible contribution of internal stresses. Note the shift in scales of the curves in the insert. Some curves, mainly B_4C and ZrO_2 are preliminary analyses of more limited data.

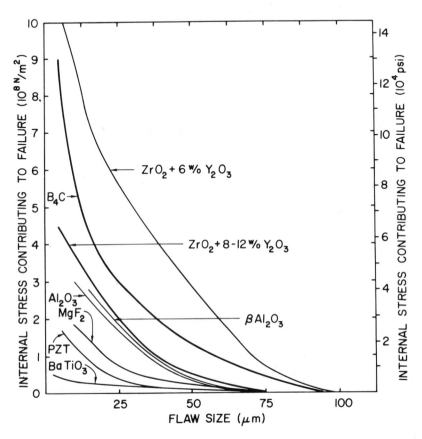

Fig. 6. The contribution of internal stresses to failure as a function of flaw size. The contribution of
the internal stress is calculated assuming that the
polycrystalline fracture energy remains constant, as
discussed in the text.

grain structure represents a significant heterogeneity
or perturbation of the continuum approximation that
fracture mechanics typically makes. There are basically
two extremes of continuum behavior that are involved.
The first is the fracture energy of single crystals or
bi-crystals for cracks respectively in single grains or
along grain boundaries, with these two fracture energies
being relatively close, e.g., within a factor of 2. At
the other extreme are the polycrystalline fracture energies. Clearly when the flaw is very small or very
large in comparison to the grain size, the two respective extremes apply. However, for intermediate crack
sizes, there must be a transition between these two
extremes. The possibility of using single crystal fracture energies to analyze the strength of large grain
polycrystalline bodies was apparently first considered
by Wiederhorn (33). Independently, Rice suggested that
such a transition could well be the cause of the transition in strength grain size behavior from a high dependence to a low dependence of strength on grain size, as
discussed later. At the last Fracture Mechanics of Ceramics Conference, Evans briefly considered such a transition in general terms (3) and Rice (1) and Freiman,
et al (2) considered it in terms of interpretation of
considerable amounts of data.

Since the last Fracture Mechanics of Ceramics Conference, direct evidence has shown failure from flaws
smaller than the larger grains of ZnSe (34) and BaTiO$_3$(30)
and that such flaws are consistent with single crystal,
but not polycrystalline fracture energies. Subsequently,
substantial measurement of this transition has been made
by both fracture analysis of specimens failed in strength
tests and by varying the number of grains across the web
of double cantilever beam specimens. Both tests showed
good correlation with one another and clearly demonstrated that there is a transition from single crystal*
to polycrystalline fracture energies and that this transition is typically completed when the flaw size is
approximately equal to, or a few fold, the grain size as
indicated in Fig. 7. The C/G ratio over which this
fracture energy transition appears to increase as the
ratio of polycrystalline to single crystal fracture en-

---

*Note that since grain boundary flaws cannot be determined
the transition from grain boundary to polycrystalline energies cannot be determined, but is believed to be similar to the single crystal-polycrystalline transition determined from intragranular flaws.

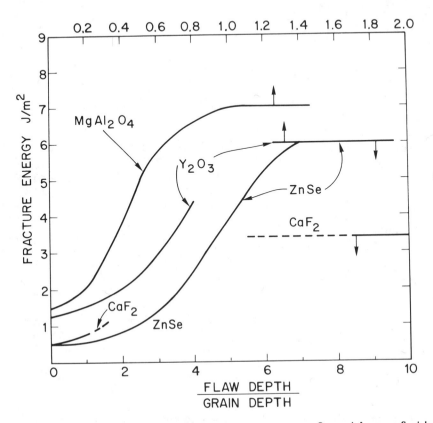

Fig. 7. Measured fracture energy as a function of the flaw size to grain size ratio. Cubic materials have been studied so that results are not complicated by effects of internal stress, as discussed. Note the two different flaw depth to grain depth scales and the arrow on each curve pointing up or down, indicating which scale is appropriate. The scatter in the data is too large to completely define the nature of the transition, e.g., it could be more abrupt than indicated by the smooth curve shown, however, the essential aspect of the data is that it clearly shows that a transition does occur.

ergies increases, indicating this ratio of fracture
energies which appears to be inversely proportional to
the number of low fracture energy crystal planes, is an
intrinsic factor. Evaluation indicates that texturing,
i.e., preferred orientation, can be a signficant extrinsic factor extending the C/G ratio over which the transition is completed.

This flaw size/grain size transition in fracture
energies provides a clear explanation why large grains
tend to dominate the mechanical behavior of bodies,
even if there is only one large grain present in the
material, i.e., such grains if favorably located provide
a lower energy path for crack growth. Also, the fact
that a single large grain can and often does determine
the mechanical behavior of a specimen is another important heterogeneity that must be considered in applying
fracture mechanics. Thus, it is necessary to be assured
that no single grain sufficiently large to allow failure
at a fracture energy significantly below the polycrystalline value is present in order to apply fracture energies
measured on a body itself for predicting its strength behavior or for prediction of life times. Thus, for example, Freiman et al (34) have shown that single crystal
fracture controls crack growth and can mean 20-fold
shorter lives in ZnSe (Fig. 8). McKinney, et al (35)
have shown that single crystal fracture energies are
consistent with delayed failure behavior of ZnSe.

There is the possibility that some flaws within
large grains may propagate to the grain boundary, arrest, and restart only when the stress intensity reaches
a value for polycrystalline growth. However, this
scenario becomes impossible for flaws smaller than some
fraction of the grain size. Thus, if the initial C/G
ratio is > the polycrystalline/single crystal fracture
energy ratio, the increase in crack length as these
flaws grow to the grain boundary produces K_I's higher
than the polycrystalline K_{IC}. However, frequently the
fracture energy does not rise to the polycrystalline
value, until at least the second set of grains are
reached. Further, the kinetic energy associated with
the growth of a critical flaw also reduces the chances
of crack arrest. Thus, arrest of intragranular cracks
at the first grain boundary is unlikely for C/G less
than $\sim \frac{1}{2}$.

Composites

A very similar transition to that of single crystal

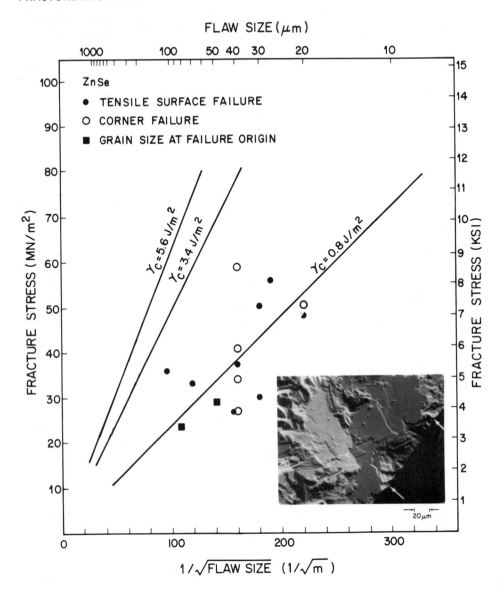

Fig. 8.   Fracture stress vs 1/√flaw size for CVD ZnSe.
Slope of the curve passing through the origin is directly
proportional to the fracture energy of the material.
Note that data scatters about a line of γ_C equal to
0.8 J/m^2 compared to a value measured on single crystal
ZnSe of 0.6 J/m^2.   Inset shows flaw within large grain
of ZnSe, corroborating the use of single crystal values
of γ_C.

to polycrystalline fracture energy occurs in composite
materials where one has a transition from a matrix fracture energy to the composite fracture energy. A recent
survey by Rice (4) showed that almost all ceramic composites, whether containing dispersed ceramic particles
or fibers for toughening purposes, while often having
measurably increased fracture toughness, almost always
had significantly decreased strengths with increasing
fiber or particle content. Analysis showed that when
strengths decreased the flaw size in the matrix was consistently smaller than the spacing between the particles
or fibers in the composites, and when strengths increased,
the spacing between the particles and fibers in the composites was less than the flaw size in the matrix. Hasselman and Fulrath (36) were probably the first to indicate this in their pioneering ceramic composite
studies. They reported that there was no strength increase until the mean free path between the particles
in their composites was equal to or less than the flaw
size in the matrix. This effect can also be seen by
considering Lange's line tension theory of fracture energy for composites (37) and also by considering stress
intensity increases with crack growth as in the above
case of the effects of flaw size to grain size ratios
on single crystal to polycrystalline fracture energy
transitions. However, the above considerations have
typically not been used by investigators evaluating the
results of their composite materials. Fracture energies
have been determined and typical fracture mechanics procedures used to calculate flaw sizes in the composite
bodies using strengths measured on the bodies, despite
the fact that strengths were often significantly decreased
while fracture energies were significantly increased compared to those of the matrix.

Internal stresses due to phase mismatch in composites and related two phase, e.g. "glass-ceramic" and
cermet bodies should contribute to failure in the same
way they do in single phase bodies of anisotropic crystal
structure. Analysis of Suzuki and Hayashi's (38) WC-Co
data shows this (Figs. 5, 6) giving a $< \sigma_i >$ value at C ~
3G that is consistent with stress levels estimated from
strain mismatches between WC and Co. Similarly, Freiman
and Rice (39) have shown that such stresses may be important in some "glass-ceramic" bodies. Thus, internal
stresses should be a limiting factor in the strength of
composite and related two phase bodies. For typical
flaw dimensions (10-50 μm), this limit is especially
important for particle sizes of a few microns or more.

Many of the model composites and some "glass-ceramic" systems have had particle or grain sizes sufficiently large to seriously question the typical analysis of their behavior neglecting internal stress effects.

STRENGTH-GRAIN SIZE RELATIONSHIPS

Rice (1,5,40) has proposed a two-branch strength-grain size model, e.g., Fig. 9, as have Rhodes and Cannon (41). Bradt and colleagues (42-44) have used such a model with a somewhat different interpretation for their machining-strength results. Some basic aspects to this model appear to have been proposed in an earlier review by Emrich (45).

The transition from grain boundary or single crystal to polycrystalline fracture energies, discussed earlier, supports the concept suggested by Rice, as well as Rhodes and Cannon, that the transition between the two regions of strength-grain size behavior occurs when $C \sim G$. Extrinsic factors, e.g., the degree of texturing, and intrinsic factors, e. g., the ratio of single crystal to polycrystalline fracture energies, can vary this transition some. Thus, it may not occur in some cases until the flaw sizes are a few times the grain size.

Three other effects on the strength-grain size relationships should be considered. The first is the often observed continued increase of strength with decreasing grain size due to the stresses aiding failure. Machining flaws are observed not to change significantly as a function of the grain size for a given material and machining operation. Thus, bodies of increasing grain size will have smaller flaw size to grain size ratios and hence, an increased contribution of internal stresses to failure leading to some strength decrease. Second, the expected strength increases from increased fracture energies at intermediate grain sizes in non-cubic materials is typically not realized for two reasons: 1) while microcracking can occur around a crack large in comparison with a microstructure such as is typically used in fracture mechanics tests, the strength is likely to be dominated by a local collection of microcracks or internal stresses that can act as a most severe flaw, i.e., giving the lowest effective fracture energy rather than the highest in a fashion similar to that suggested previously by Rice (1, 4), and 2) increasing direct contribution of internal stresses to failure. Third, in the large grain regime, strengths will not continue to de-

crease indefinitely with increasing grain size, but will
reach a limiting value. An upper approximation of this
limit is the typical single crystal strength, but lower
grain boundary than single crystal fracture energies and
contributions of internal stresses (as previously suggested by Rice and McDonough (46), or resultant cracking
in addition to porosity and impurities can lead to
strengths lower than single crystal values.

It is useful to consider three more detailed cases,
supporting the above model, reported in the past few
years. The first is Al_2O_3 data of Tressler, et al (42)
which Rice (47) has pointed out agrees with the present
model; i.e., use of single crystal fracture energies
gave flaw and grain sizes about equal to the intersections of the two sets of branches. More recently, Vircar
and Gordon (28) reported two-branch strength-grain size
behavior with two different finer grain branches due to
anisotropy from preferred orientation in their β-Al_2O_3.
Their plot showed the two sets of branches intersected
at grain sizes of ~ 120 μm in contrast to their calculated flaw sizes of 60-90 μm. This is actually fairly
good agreement with the model considering the effects
that flaw shape and orientation, and factors affecting
the single crystal to polycrystalline fracture transition can have on the strength-grain size transition.
However, their grain size measurements were significantly
increased by the tabular shape of grains in their larger
grain bodies, which failed by transgranular (mostly
basal cleavage) fracture along the long axis of the
grains. Although the flaws are probably not as large
as their large grains (50-100 x 200-400 μm), even if they
were, the smaller cross-sectional dimension of the grain
(e.g., along the basal cleavage plane) would be the upper
limit of the critical flaw size, since it is always the
smaller flaw dimension that dominates strength behavior
(the larger dimension only influences the flaw geometry
factor) (1,48). Recognizing this reduces the grain size
for intersection, e.g. by ~ 2, giving C ~ G, and indicating a fracture energy in their large grain bodies of
~ 3 J/m^2, which would appear reasonable for single crystal cleavage.

The second case to consider is that of hot pressed
SiC. Cranmer, et al (43, 44) plotted their data for
different machining of various grain sizes with Prochaska
and Charles' (49) data. However, Prochaska and Charles'
plot is also based on the length of their tabular grains
rather than the width or half-width (depending upon grain

orientation), as previously pointed out by Rice (1).
Taking this into consideration significantly shifts the
curve of Prochaska and Charles, e.g., as suggested in
Fig. 9 so that the intersection of the two sets of curves
would occur with closer agreement between the flaw and
grain sizes. Although, there may be some question of
combining data of Cranmer, et al and Prochaska and
Charles, because of different modes of surface finish,
the former also had tabular grains in their largest
grain bodies, so that data would be shifted to the right
in a similar fashion as Prochaska and Charles. In any
event, the changed slope in this re-analysis of Prochaska
and Charles' data also changes the indicated fracture
energy $\sim$ 3-7 J/m^2 (assuming a penny-shaped flaw) which
is close to the 3-5 J/m^2 estimated by Rice from flaws at
fracture origins of smaller size single crystals.

SUMMARY AND CONCLUSIONS

Extensive study has revealed a great deal about the
microstructural dependence of fracture mechanics parameters in ceramics. The porosity dependence of fracture
energy can be often approximated by $\gamma_0 e^{-bP}$, where b
depends on the character of porosity. With higher porosity, larger grain sizes, or more second phase, microcracking can become important, significantly increasing
fracture energies above those expected for that level of
porosity, and even above those for a fully dense body of
the same material. Fracture energy is found to have no
significant grain size dependence for cubic materials,
but to first increase, pass through a maximum, then decrease as grain size increases for non-cubic materials.
An analytical model was developed based on microcracking
due to the combined effects of internal and applied
stresses from the thermal expansion anisotropy, that
generally agreed with this grain size dependence for
non-cubic materials. Direct examination of cracks supports the microcracking concept and shows that crack
propagation is often much more complex than commonly
assumed.

Analysis of the fracture energies applicable to
the strength behavior of ceramics has been made by determining the character of actual fracture origins and
comparing this with fracture mechanic predictions. From
this it is unequivocally clear that fracture mechanics
often cannot be used as a simple black box technique to
predict the mechanical behavior of ceramics. Such simple

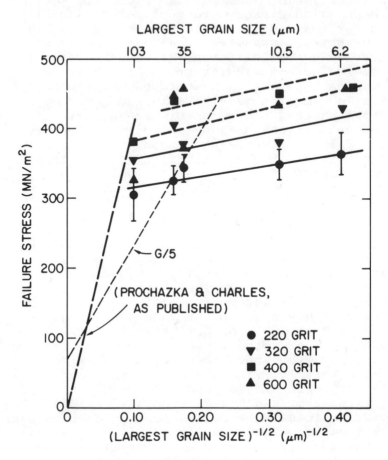

Fig. 9.   Strength versus grain size.   The solid lines
(from work of Cranmer, et al, on SiC) illustrate the
nature of the model proposed by Rice and others, namely
that there can be two branches to strength-grain size
behavior from flaw effects alone.   The intersection of
the two sets of curves will occur when the grain size
and the flaw size are approximately equal.   Note the
revision of Prochaska and Charles' SiC data, as discussed in the text.

application does indeed work in a variety of cases, e.g. glasses, single crystals, and many polycrystalline bodies. However, there can be many important deviations in poly-crystalline bodies due to extrinsic factors such as chemical and microstructure heterogeneities which can be an important, but often neglected issues. Intrinsic effects can also be quite important. Thus, for example, if a body either consists of grains large enough so that a single flaw can be contained within a grain, or even contains only one such grain, then a fracture energy less than that of the polycrystalline body, approaching in the extreme that for single crystal cleavage or easy grain boundary fracture, must be used. Similar effects occur for flaws in composite ceramic bodies. Internal stresses, another intrinsic effect, have rather complex effects on failure. On the one hand, they are respon-sible for the variation of fracture energy with grain size in typical fracture energy tests and yet at the same time, they can increasingly contribute directly to the failure stress as the flaw size becomes smaller in comparison with the grain size. The latter effect tends to dominate, presumably with the worst combination of cracks or internal stresses leading to failure and thus, either counterbalancing or precluding the increased fracture energy, due to microcracking from these stresses, having an effect on strength.

Finally, the strength-grain size model based on the concept that two different branches of the curve on a Petch-type plot meet when the grain size is approxi-mately equal to the flaw size was further substantiated. The transition from single crystal or grain boundaries to polycrystalline fracture energies is a dominant fac-tor in the value of the ratio of the flaw size to grain size at which two curves meet. The orientation of the flaw is another factor. Internal stress effects are cited as probable reasons for some continued increase in strength with decreasing grain size at finer grain sizes and for lower than single crystal strengths at large grain sizes.

REFERENCES

1. R. W. Rice, p. 323 in Fracture Mechanics of Ceramics, Ed. by R. C. Bract, D. P. H. Hasselman, and F. F. Lange, Plenum Press, N.Y., 1974.

2. S. W. Freiman, K. R. McKinney, and H. L. Smith, Ibid, p. 639.

3. A. G. Evans, Ibid, p. 17.

4. R. W. Rice and S. W. Freiman, presented at the Sixth
    International Materials Symposium: Ceramic Microstruc-
    tures, 1976. Proceedings to be published by John
    Wiley and Sons (in press).

5. R. W. Rice, in Properties and Microstructure, Ed. by
    R. K. MacCrone, to be published by Academic Press,
    New York (in press).

6. D. R. Larson, J. A. Coppola, and D. P. H. Hasselman,
    J. Am. Ceram. Soc. 57 (10), 417, 1974.

7. R. G. Hoagland, G. T. Hahn, and A. R. Rosenfield,
    in Rock Mechanics 5, 77, 1973.

8. J. A. Kuszyk and R. C. Bradt, J. Am. Ceram. Soc. 56
    (8), 1973.

9. R. W. Rice and S. W. Freiman, to be published.

10. R. W. Rice and R. C. Pohanka, submitted for publi-
    cation to J. Am. Ceram. Soc.

11. R. W. Davidge and J. T. Green, J. Mat. Sci. 3, 629,
    1968.

12. M. J. Noone and R. L. Mehan, p. 201 in Fracture Mech-
    anics of Ceramics, Ed. by R. C. Bradt, D. P. H.
    Hasselman and F. F. Lange, Plenum Press, N.Y., 1974.

13. D. J. Green, P. S. Nicholson, and J. D. Embury,
    Ibid, p. 541.

14. C. Cm. Wu, S. W. Freiman, R. W. Rice, and J. J.
    Mecholsky, to be published.

15. K. N. Subramanian, to be published.

16. J. J. Mecholsky, S. W. Freiman, and R. W. Rice,
    J. Mat. Sci. 11, 1310, 1976.

17. R. C. Pohanka, S. W. Freiman, and B. A. Bender,
    accepted for publication by the J. Am. Ceram. Soc.

18. R. C. Garvie, R. H. Hannink, and R. T. Pascoe, in
    Nature 258, 703, 1975.

19. D. L. Porter and A. H. Heuer, J. Am. Ceram. Soc. 60
    (3-4), 183, 1977.

20. R. W. Rice, J. Am. Ceram. Soc. 60 (5-6) 280, 1977.

21. D. L. Porter and A. H. Heuer, Ibid, p. 78.

22. R. L. Burns (deceased), unpublished manuscript based
    on thesis studies in the Dept. of Mineral Eng.,
    Stanford Univ.

23. H. R. Baumgartner and D. W. Richerson, p. 367 in
     Fracture Mechanics of Ceramics, Ed. by R. C. Bradt,
     D. P. H. Hasselman, and F. F. Lange, Plenum Press,
     N.Y., 1974.

24. S. W. Freiman, A. Williams, J. J. Mecholsky, and
     R. W. Rice, presented at the Sixth International
     Materials Symposium: Ceramic Microstructures, 1976,
     Proceedings to be published by John Wiley and Sons
     (in press).

25. P. L. Land and M. G. Mendiratta, J. Mat. Sci. 12,
     (7), 1421, 1977.

26. F. F. Lange, J. Am. Ceram. Soc. 56 (1), 518, 1973.

27. G. K. Bansal, presented at the Sixth International
     Materials Symposium: Ceramic Microstructures, 1976,
     Proceedings to be published by John Wiley and Sons
     (in press).

28. A. V. Virkar and R. S. Gordon, J. Am. Ceram. Soc.
     60 (1-2) 58, 1977.

29. R. W. Rice, to be published.

30. R. C. Pohanka, R. W. Rice, and B. E. Walker, Jr.,
     J. Am. Ceram. Soc. 59 (1-2), 71, 1976.

31. R. C. Pohanka, R. W. Rice, B. E. Walker, and P. L.
     Smith, Ferroelectrics 10, pp. 231-235, 1976.

32. R. W. Rice, R. C. Pohanka, and W. J. McDonough, to
     be published.

33. S. M. Wiederhorn, p.317, in Ultrafine-Grain Cer-
     amics, Ed. by J. J. Burke, N. L. Reed, and V. Weiss,
     Syracuse Univ. Press, Syracuse, 1970.

34. S. W. Freiman, J. J. Mecholsky, R. W. Rice, and
     J. C. Wurst, J. Am. Ceram. Soc. 58 (9-10), 406,
     1975.

35. K. R. McKinney, J. J. Mecholsky, and S. W. Freiman,
     submitted for publication.

36. D. P. H. Hasselman and R. M. Fulrath, J. Am. Ceram.
     Soc. 50 (8), 399, 1967.

37. F. F. Lange, Phil. Mag. 22, 983, 1970.

38. H. Suzuki and K. Hayashi, Planseeberichte für
     Pulvermetallurgie, 53, 24, 1975.

39. S. W. Freiman and R. W. Rice, accepted for publi-
     cation, J. Mat. Sci.

40. R. W. Rice, p. 287 in Ceramics for High Performance Applications, Ed. by J. J. Burke, A. E. Gorum, and R. N. Katz, Brook Hill Pub. Co., Chestnut Hill, Mass., 1974.

41. W. H. Rhodes and R. M. Cannon, AVCO Corp. Summary Report, prepared under Naval Air Systems Command Contract, N00019-73-C-0376, 1974.

42. R. E. Tressler, R. A. Langensiepen and R. C. Bradt, J. Am. Ceram. Soc. 57 (5), 226, 1974.

43. R. C. Bradt and R. E. Tressler, presented at the Sixth International Materials Symposium: Ceramic Microstructures, 1976, Proceedings to be published by John Wiley and Sons (in press).

44. D. C. Cranmer, R. E. Tressler, and R. C. Bradt, J. Am. Ceram. Soc. 60 (5-6), 230, 1977.

45. B. R. Emrich, Air Force Materials Laboratory Report ML-TDR-64-203, Sept. 1974.

46. R. W. Rice and W. J. McDonough, p. 394 in Mechanical Behavior of Materials, Vol IV, Soc. of Mat. Sci., Japan, 1972.

47. R. W. Rice, J. Am. Ceram. Soc. 58 (3-4), 154, 1975.

48. G. K. Bansal, J. Am. Ceram. Soc. 59 (1-2), 87, 1976.

49. S. Prochaska and R. J. Charles, p. 579, in Fracture Mechanics of Ceramics, Ed. by R. C. Bradt, D. P. H. Hasselman and F. F. Lange, Plenum Press, N. Y., 1974

ROLE OF STRESS-INDUCED PHASE TRANSFORMATION IN ENHANCING STRENGTH AND TOUGHNESS OF ZIRCONIA CERAMICS

T. K. Gupta

Westinghouse Research and Development Center

Pittsburgh, Pennsylvania 15235 U.S.A.

ABSTRACT

In analogy with metal systems, the stress-induced phase transformation is shown to be an important mechanism for arresting crack
propagation in a ceramic system. This is demonstrated in zirconia
ceramics which can be made with varying amounts of metastable tetragonal phase. The metastable phase can undergo a stress-induced
tetragonal-to-monoclinic phase transformation, arrest crack propagation by absorbing energy at the front of a crack, and can markedly
increase the strength and fracture toughness of zirconia ceramics.

INTRODUCTION

It is well known that ceramics fail in a brittle manner when
subjected to a mechanical or thermal stress. For these materials,
the propagation (extension) of a crack is nearly instantaneous once
a crack is initiated by an applied load. As a result, failure in
ceramics, unlike that in metals, is often catastrophic. This is the
major limitation of most single-phase homogeneous ceramics.

In an attempt to make a ceramic which is both strong and tough,
two schools of thought have emerged. In one approach, ceramics of
increasing strength have been fabricated in an effort to increase
the toughness. The disadvantage of this approach is that the vast
kinetic energy that is available in a stronger ceramic during crack
propagation produces a fracture which is, nevertheless, catastrophic
and even more spectacular. Failure occurs without a warning. In
the other approach, the ceramics are designed to promote a stage of
stable crack propagation following the stage of crack initiation.

This approach requires that a crack arrester be present in a ceramic matrix (1). Increasing attention has been devoted in recent years to enhancing the strength and fracture toughness of ceramics by providing crack arresters in the structure. This has resulted in the study of several composite systems wherein a second phase has been "deliberately" added to a ceramic matrix by hot pressing, e.g., SiC particles in Si_3N_4 matrix (2), ZrO_2 particles in Al_2O_3 matrix (3), etc. This approach has yielded ceramics with a higher toughness but not necessarily a higher strength. The reason for this behavior is that the strength of a brittle material is determined by the weakest point in the structure, and the weakest point in a hot-pressed composite is the interface region between the dispersed phase and the matrix (1).

It is shown in this paper that a still different approach is available for increasing the strength and toughness of ceramics. This approach is based on the phenomena of stress-induced phase transformation, wherein a phase transformation in front of a crack is capable of absorbing energy that is otherwise available for crack propagation. The zirconia alloy system appears to be the most promising ceramic to study this phenomena.

STRESS-INDUCED PHASE TRANSFORMATION IN ZrO_2 ALLOY SYSTEM

Stress-induced phase transformations are known to arrest crack propagation and increase the toughness of materials. In a metal system, TRIP steels belong to this class of materials which contain metastable austenite at room temperature. On fracture, the metastable austenite transforms to stable martensite by absorbing energy at the crack front. Crack propagation is thus arrested and consequently, the TRIP steels show high fracture toughness.

In a ceramic system, the analogous material to TRIP steels would be the alloy systems based on zirconia. The ZrO_2-Y_2O_3 phase diagram, Figure 1, is remarkably similar to the type of phase diagram often encountered in steel metallurgy. The higher Y_2O_3 content phase of cubic ZrO_2 is reminiscent of austenitic steel, wherein nickel or manganese is added to stabilize the face centered cubic (or austenitic) iron, and provides the class of strong, tough, and usually oxidation-resistant austenitic steels. The lower Y_2O_3 content alloys, wherein tetragonal-to-monoclinic reaction occurs, is reminiscent of martensitic transformation in steel in which the properties of steel are determined entirely by the heat treatment procedure. The tetragonal-to-monoclinic transformation in pure ZrO_2 is also known to be martensitic. The reason for the cracking of pure ZrO_2 on thermal cycling is the large volume change associated with this martensitic reaction. High carbon steels also crack on cooling. However, high carbon steels can be

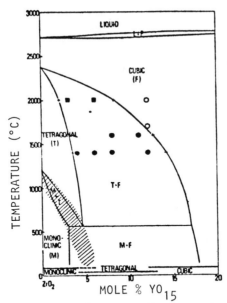

Figure 1. Phase diagram of ZrO_2-Y_2O_3 System
[After H. G. Scott (13)].

cooled to room temperature without cracking if the phase transformation is by controlled nucleation and growth kinetics. From the phase
diagram analogy it has been suggested that such a controlled reaction
may also be possible in the zirconia alloy system. Granting that
possibility, it may be further suggested that the metastable tetragonal zirconia, in analogy with metastable austenitic steel, should
also undergo a stress-induced phase transformation and absorb energy
during crack propagation, thus imparting toughness to the ceramic.
The ZrO_2 alloy system thus offers a unique opportunity for studying
one of the most important reactions so basic to the structure and
property control in metallic systems. It is this rationale that has
led to the use of ZrO_2 alloy as the "model" material for the present
study. This same material has usefulness in practical systems.

RETENTION OF METASTABLE-TETRAGONAL PHASE IN POLYCRYSTALLINE ZrO_2

The key to demonstrating the phenomena of stress-induced phase
transformation in zirconia alloy systems lies in demonstrating the
existence of a metastable tetragonal phase at a temperature well
below the equilibrium transformation temperature of 1000-1200°C.
By carefully choosing the particle size of the initial powder and
the concentration of alloying oxide, and by varying the sintering
temperature and time, it is possible to fabricate dense, sintered
bodies of zirconia containing varying amounts of tetragonal phase (4).

The X-ray line traces of a nearly 100% tetragonal ZrO_2 having a bulk density of $\sim$5.6 g/cm^3 are shown in Figure 2. The line traces are in agreement with the traces reported in the literature for both the powder (5) and the solid (6) ZrO_2 containing tetragonal phase. A standard Norelco X-ray diffractometer, with Ni-filtered CuK_α radiation, was used to determine the phases present. Line traces were obtained over a wide range of 2 θ values, but the emphasis was placed on that part of the profile where 2 θ = 25°-35°, as this contained the two strongest lines of the monoclinic phase at 28.3° ($11\bar{1}$) and 31.7° (111) and the strongest line of metastable tetragonal phase at 30.5° (111). The percentage of phases was calculated from the relative intensities of the two monoclinic peaks ($11\bar{1}$ and 111) and the tetragonal peak (111) after correcting for the background counts. A similar approach was taken by Mitsuhashi et al (5), for powder ZrO_2 containing tetragonal phase. It should be noted that there is no monoclinic phase in this particular specimen. Furthermore, should a cubic phase be present in the specimen, peaks would appear in places marked by the arrows. The inspection of the peaks indicate that the amount of cubic phase, if at all present, would be very small. For all practical purposes, the tetragonal phase can be considered to be the major crystalline phase.

This conclusion is further substantiated by examining the bulk structure of the sintered body. For this, the tetragonal zirconia specimens were carefully surface polished* and X-rayed as a function of specimen depth. A typical result is shown in Figure 3, where a specimen ($\sim$1.5 mm thick) which contained   90% tetragonal phase at

---

*Tetragonal-to-monoclinic transformations were held to minimum by surface polishing.

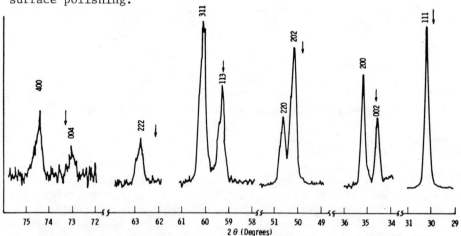

Figure 2.   X-ray diffraction pattern of nearly 100% tetragonal zirconia

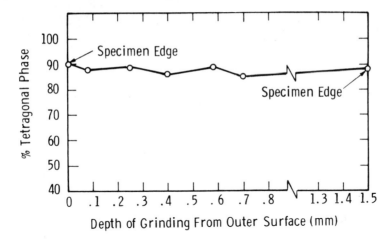

Figure 3. Tetragonal phase as a function of specimen depth

the outer surface is seen to maintain approximately the same composition through half the thickness of the specimen. This suggests that the tetragonal structure is the major phase in the sintered body.

An alternate approach to stabilize the tetragonal phase has been taken by other workers (6,7) in this field. Here, the tetragonal zirconia has been dispersed as a minor phase in a stabilized cubic ZrO_2 matrix. The tetragonal zirconia precipitates, ~1000 Å, were formed by heating, quenching and isothermal annealing of CaO or MgO stabilized zirconia, exactly in the same fashion as the heat-treated steels. The ultimate objective appears to be the same in both cases--to use the tetragonal phase as a crack arrester.

DEMONSTRATION OF STRESS-INDUCED PHASE TRANSFORMATION

Analogous to TRIP steels, the dense polycrystalline ZrO_2 ceramic containing a high concentration of tetragonal phase showed an irreversible tetragonal-to-monoclinic phase transformation on mechanical grinding as illustrated in Figure 4. The as-sintered specimen (Fig. 4a) showed the strongest tetragonal 111 peak and a very weak 11$\bar{1}$ monoclinic peak. When the same specimen was reduced to powder by grinding (Fig. 4b), the 111 monoclinic peak appeared and 11$\bar{1}$ peak became relatively strong. At the same time, the relative concentration of the tetragonal phase was decreased. It is obvious that the martensitic phase transformation occurred by subjecting the specimen to mechanical stress. Such transformation

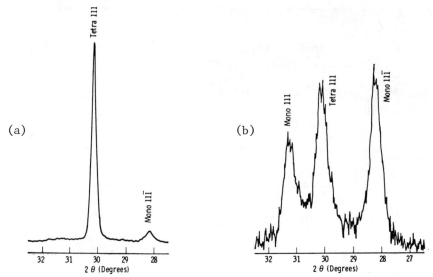

Figure 4. Effect of mechanical grinding on tetragonal-to-
monoclinic phase transformation in zirconia ceramic:
(a) before grinding, and (b) after grinding of the same
specimen (Mono = monoclinic phase, Tetra = tetragonal phase)

was also observed when the specimens were broken. The freshly
fractured surface of ZrO_2 contained greater amounts of monoclinic
phase than what was present at the outer as-sintered surface.
Table I illustrates a typical case where a specimen containing high
tetragonal phase (∿97%), as determined by the X-ray analysis of the
as-sintered surface, was first broken and then ground to powder.

TABLE I

Phase Changes of Tetragonal Zirconia
Ceramic Upon Fracture and Grinding

Characteristics of Specimen Subjected to X-ray Analysis	% Tetragonal	% Monoclinic
As-sintered surface	97	3
Broken surface of the same specimen	56	44
Ground powder of the same specimen	48	52

The X-ray line trace analysis gave ~56% tetragonal phase for the broken surface and 48% for the ground powder. The decrease in tetragonal content suggests that the stress at the crack front caused phase transformation.

FRACTURE TOUGHNESS OF ZIRCONIA CERAMICS CONTAINING TETRAGONAL PHASE

It was proposed earlier that if a phase transformation could be induced in front of a crack, a considerable increase in toughness could be achieved in ceramics. The increase arises from the ability of the tetragonal phase to absorb energy from a propagating crack and thereby arrest its growth. A stage of stable crack propagation is thus assured due to phase transformation in front of a crack. Consequently, the energy demand curve for fracture in ceramics containing a metastable phase deviates considerably (1) from that without a crack-arresting metastable phase. The critical stress intensity factor can then be written as:

$$K_c = \left| (2\gamma + U)E \right|^{1/2}$$

where U is the energy absorbed due to phase transformation, γ is the energy absorbed due to the creation of a new surface, and E is the elastic modulus.

Fracture toughness was determined on polished specimens by the indentation technique with the Vickers indenter. From a measurement of hardness, H, indentation crack length C and the indentation impression radius a, the value of K_c was determined by using the calibration curve developed by Evans and Charles (8). Fracture toughness data for several specimens are shown in Table II. The table also contains the amount of tetragonal phase present on the surface of the polished specimens.

It is seen that the values of K_c range between 6 and 9 $MN \cdot m^{-3/2}$. Evans and Charles (8) claim that the indentation technique enables fracture toughness data to be obtained within an accuracy of 10-30%. To resolve this uncertainty, fracture toughness was also obtained on two pre-cracked, double-beam cantilever specimens (0.1 x 1.5 x 3.0 cm) containing >80% tetragonal phase. The average K_c value obtained by this technique was 6.4 $MN \cdot m^{-3/2}$. Thus the two measurements agreed within experimental error.

These values of K_c are higher than those of most other structural ceramics and substantially higher than that of partially stabilized zirconia containing no metastable phase, Table III. This latter value was roughly estimated from the reported values of fracture surface energy γ_i and the Young's modulus E for a partially stabilized

TABLE II

Estimate of Fracture Toughness by Indentation*

Specimen No.	$\bar{a}$ (m)	$(\frac{\bar{c}}{a})$	H ($GN \cdot m^{-2}$)	$\frac{\phi(\frac{H}{\phi E})^{0.4}}{H\sqrt{a}}$ ($GN^{-1} \cdot m^{3/2}$)	$\frac{K_c \phi(\frac{H}{\phi E})^{0.4}}{H\sqrt{a}}$ (dimensionless)	K_c ($MN \cdot m^{-3/2}$)	% Tetragonal on Polished Surface
68	1.38×10^{-4}	1.31	12.1	5.04	0.0457	9.07	88
66	1.41×10^{-4}	1.99	11.6	5.11	0.0363	7.11	>79
52	1.40×10^{-4}	1.66	11.7	5.10	0.0400	7.84	>88
59	1.43×10^{-4}	2.39	11.2	5.18	0.0314	6.07	>92
11	1.37×10^{-4}	1.74	12.3	5.00	0.0386	7.71	>85
21	1.42×10^{-4}	1.29	11.4	5.18	0.0457	8.82	>86
60	1.40×10^{-4}	1.31	11.7	5.10	0.0457	8.96	>87

*Note: Load = 50 kg, $E \approx 145$ $GN \cdot m^{-2}$, $\phi \approx 3$

TABLE III

Fracture Toughness of Various Materials

Materials	K_c (MN·m$^{-3/2}$)	References
ZrO_2 (metastable phase)	6-9	Present study
ZrO_2 (PSZ-stable phase)	1.1	9
Si_3N_4	4.8 – 5.8	10
SiC	3.4	11
B_4C	6.0	8
Al_2O_3	4.5	Unpublished data
Sapphire (single crystal)	2.1	8
Spinel (single crystal)	1.3	8

zirconia containing a stable monoclinic phase (9). The K_c value for the partially stabilized zirconia containing the monoclinic phase from the literature value is estimated to be $\sim$1.1 MN·m$^{-3/2}$, whereas the observed K_c values for ZrO_2 containing the metastable tetragonal phase are between 6 and 9 MN·m$^{-3/2}$. Since evidence that the tetragonal-to-monoclinic transformation occurs in the vicinity of the crack during fracture has been obtained on these same materials, it is reasonable to suggest that the stress-induced transformation is responsible for the material's high fracture toughness as outlined in the previous arguments.

STRENGTH OF ZIRCONIA CERAMICS CONTAINING TETRAGONAL PHASE

Like fracture toughness, an improvement in strength is also expected as a result of a phase transformation in front of a crack. The critical fracture stress, σ_c, required to cause crack propagation can be expressed as

$$\sigma_c = \left| \frac{(2\gamma + U)E}{2\pi c} \right|^{1/2}$$

where c is the critical crack length and the quantity $UE/2\pi c$ is the contribution to strength due to stress-induced phase transfromation.

Strength data were obtained using disc specimens (1.9 cm diameter and ∿0.15 cm thick) which were placed in symmetrical biaxial flexure (12). The specimens had similar densities. The biaxial tensile strength as a function of metastable tetragonal phase content is illustrated in Figure 5. Several important observations can be made from the data. First, when the amount of tetragonal phase is low (<10%), the observed strength is very poor (<100 MPa). Second, as the metastable tetragonal content is increased to ∿30%, there is a rapid increase in strength. This is believed to be due to stress-induced phase transformation as discussed below. Finally, a constant high strength (600-700 MPa) is maintained when the apparent tetragonal content is between 30 and 100%.

The contribution of stress-induced phase transformation to high strength can be inferred from the strength vs grain size data presented in Figure 6. At low strength (< 200 MPa), the primary phase is monoclinic (>90%), so there is little contribution to strength from phase transformation, and therefore, the strength shows an increase with decreasing grain size from ∿0.42 to ∿0.34 μm, as is common with conventional ceramics. At high strength (600 MPa), where the primary phase is tetragonal, there is again a slight

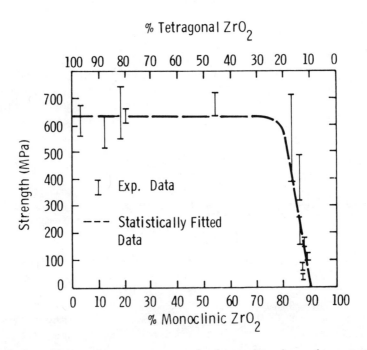

Figure 5.   Strength vs tetragonal phase in zirconia ceramics

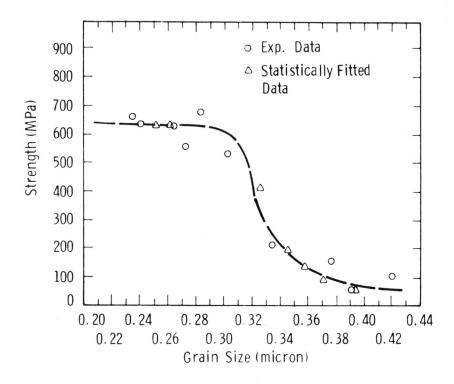

Figure 6.   Strength vs grain size in zirconia ceramics

increase in strength with decreasing grain size from ∿0.30 to ∿0.24
μm, in accordance with the behavior of conventional ceramics.   The
most significant observation in the present study is the rapid
increase in strength, from 200 to 600 MPa, for a modest decrease
in grain size from ∿0.34 to ∿0.30 μm.   In view of the strength vs
grain size behavior at high and low strengths, this three-fold
increase in strength in this narrow grain size range cannot be
attributed to the small grain size alone.   It is concluded that
the rapid discontinuous increase in strength from 200 to 600 MPa
is due to the rapidly increasing contribution of stress-induced
phase transformation to strength as the tetragonal phase increases
from 10 to 30%.   Further increase in tetragonal phase does not seem
to increase the strength significantly.   This behavior is not clearly
understood at present.

SUMMARY

It is shown that the stress-induced phase transformation can occur in ceramic systems. The zirconia alloy system appears to be a "model" candidate material for studying the above phenomena. In this system, it is possible to obtain a metastable tetragonal phase at a temperature well below the equilibrium transformation temperature. The metastable phase can arrest crack propagation by absorbing energy at the front of a crack, and thereby cause a tetragonal-to-monoclinic phase transformation. Fracture toughness and strength are markedly improved as a result of this stress-induced phase transformation during crack propagation. The investigation described in this paper suggests a new approach in the development of strong and tough ceramics.

ACKNOWLEDGMENT

The author wishes to thank W. D. Straub for preparing the specimens, R. C. Kuznicki for X-ray analysis, F. F. Lange for fracture toughness measurements and R. B. Grekila, B. R. Rossing and J. H. Bechtold for many helpful discussions.

REFERENCES

1. T. K. Gupta, "A Qualitative Model for the Development of Tough Ceramics," J. Mat. Sci. 9, 1585 (1974).

2. F. F. Lange, "Effect of Microstructure on Strength of Si_3N_4-SiC Composite System," J. Amer. Ceram. Soc. 54, 614 (1971).

3. N. Claussen, J. Steeb and R. F. Pabst, "Effect of Induced Microcracking on the Fracture Toughness of Ceramics," Cer. Soc. Bull., 56 (6) 559 (1977).

4. T. K. Gupta, "Sintering of Tetragonal Zirconia and its Characteristics," to be presented at the Fourth Round Table Conference on Sintering in Dubrovnik, Yugoslovai, Sept. 5-10, 1977.

5. T. Mitsuhashi, M. Ichihara and U. Tatsuke, "Characterization and Stabilization of Metastable Tetragonal ZrO_2" J. Am. Cer. Soc., 57, (2) 97-107 (1974).

6. R. C. Garvie, R. H. Hennick and R. T. Pascoe, "Ceramic Steel?" Nature, 258, 703 (1975).

7. G. K. Bansal and A. H. Heuer, "Precipitation in Partially Stabilized Zirconia," J. Amer. Ceram. Soc., 58 (5-6) 235 (1975).

8. A. G. Evans and E. A. Charles, "Fracture Toughness Determination
    by Indentation," J. Am. Cer. Soc., $\underline{59}$ (7-8) 371 (1976).

9. D. J. Green, P. S. Nicholson and J. D. Embury, "Fracture Tough-
    ness of Partially Stabilized ZrO_2 in the System $CaO-ZrO_2$" J.
    Am. Cer. Soc. $\underline{56}$ (12) 619 (1973).

10. F. F. Lange, "Relation Between Strength, Fracture Energy and
    Microstructure of Hot-Pressed Si_3N_4," J. Am. Cer. Soc., $\underline{56}$, 518,
    (1973).

11. A. G. Evans and F. F. Lange, "Crack Propagation and Fracture in
    SiC," J. Mat. Sci., $\underline{10}$, 1959-64 (1975).

12. T. R. Wilshaw, "Measurement of Tensile Strength of Ceramics,"
    J. Amer. Ceram. Soc., $\underline{51}$, (2) 111 (1968).

13. H. G. Scott, "Phase Relationship in the Zirconia-Yttria System,"
    J. Mat. Sci., $\underline{10}$, 1527-1535 (1975).

TOUGHNESS AND FRACTOGRAPHY OF TiC AND WC

J.L. Chermant, A. Deschanvres and F. Osterstock

Laboratorie de Cristallographie et Chimie de Solide
L.A. 251. Equipe Matériaux-Microstructures
Université de Caen
14032 Caen Cedex, France

INTRODUCTION

The brittleness of ceramic materials, in particular transition metal carbides, renders them very sensitive to small defects. Consequently the results obtained in conventional mechanical testing exhibit a large scatter and are not always representative of the material itself.

The methods of fracture mechanics have been improved considerably in recent years and can be applied in a study of brittle materials (1)(2)(3). These methods give results which can be compared with those obtained in conventional tests and hence enable the change in brittleness to be observed for a series of materials.

In this work we have studied a series of materials of generally decreasing brittleness. We have determined the characteristic parameters for bending, compression and toughness. The mechanical tests were supplemented by an optical and scanning electron optical observation of the fracture surface, in order to determine the fracture mechanism.

EXPERIMENTAL DETAILS

The materials used were transition metal carbides with 0.5 % vol. cobalt or without bonding metal supplied by METALLWERK PLANSEE, TiC, and UGINE CARBONE, WC. The mean grain sizes were 20 μm and 6 μm for TiC, and 6 μm for both bonded and unbonded WC.

For K_{IC} tests the specimens were 4 x 8 x 40 mm^3. They were tested in three point bending, the distance between the lower knives being 32 mm. The crack was introduced by spark erosion (4). The fracture stress in bending (σ_{rf}) and Young modulus (E), were measured

in three point bending on specimens 2 x 5 x 20 mm^3. In this case the
distance between supporting knives was 16 mm. The deflection is measured by an electroinductive transducer fixed to the support for
the lower knife edges (5). In compression we used rectangular specimens 4 x 4 mm^2 and 8 mm tall. The occurence of any plastic deformation was detected by a comparator with an accuracy of 1/1000 mm.
This gave a detection limit of less than 0.01 %.

All the mechanical tests were carried out on a 100 kN TINIUS
OLSEN, Type LOCAP, testing machine.

The values for the critical stress intensity factor, K_{IC}, and
elastic energy release rate, G_{IC}, have been obtained by the analytical method (6). In this case for three point bending :

$$K_{IC} = \frac{3 \; Fr \; L \; Y \; \sqrt{a}}{2 \; Le^2} = \sigma_r \; Y \; \sqrt{a}$$

$$G_{IC} = \frac{K_{IC}^2}{E} \; (1 - \nu^2)$$

where   Fr   : is the rupture load
        L, l, e : specimen length, width and thickness respectively
        a   : crack depth
        E   : Young modulus
        ν : Poisson number
        Y $= 1.93 - 3.07 \frac{e}{a} + 14.53 \left(\frac{e}{a}\right)^2 - 25.11 \left(\frac{e}{a}\right)^3 + 25.80 \left(\frac{e}{a}\right)^4$

The scanning electron microscope used was a Jeol JSM P15.

RESULTS

Mechanical Features

The main results of the mechanical tests are presented in Table
I. All fractures occured in the elastic region. Summarizing the
following trends were observed :

Bending : The materials based on WC fracture at much higher
stresses than those based on TiC. The addition of 0.5 % cobalt increases markedly the fracture stress. This is without doubt due to
the fact that the cobalt phase blunts some of the stress concentrations which would appear between the grains. This trend is confirmed by Fig. 1a, where the change in some of the mechanical characteristics is showed up to 5 % vol. cobalt. The fracture stress of TiC
is strongly influenced by the mean diameter of the carbide crystals.

Compression : In this case also the WC based materials fracture at higher stress then those based on TiC. However the addition
of 0.5 % cobalt has virtually no influence on the fracture stress.
At higher cobalt compositions there is however an effect (Fig. 1b).
Due to lack of various grain sized materials, it was not possible to
observe the effect of the grain size on the rupture stress of TiC.

Toughness : The values of K_{IC} obtained are 3.8 MPa $\sqrt{m}$ for TiC ;
7.5 MPa $\sqrt{m}$ for pure WC and 8.9 MPa $\sqrt{m}$ for WC-0.5 % Co. This corresponds to G_{IC} values of 45 J/m^2, 80 J/m^2 and 90 J/m^2 respectively.
The reduction in the brittleness of various materials is thus capable of being assessed by these measurements. For WC materials the
increase of K_{IC} with addition of cobalt is very marked (Fig. 1c).
It is interesting to note that the value of K_{IC} tends towards the
same limit when the cobalt composition tends to zero, irrespective
of the grain size (0.7 - 1.1 and 2.2 µm) (5). It should be pointed
out that in the case of alumina it is reported in the literature
that K_{IC} may decrease (7) or increase (8) with an increase in grain
size. However, only a decrease of K_{IC} with decreasing size is plausible. It was not possible in this work,due to a lack of homogeneous material of different grain size to study the change in K_{IC}
with grain size of TiC.

The critical size for a semi-elliptical crack, a_c, calculated
by the expression (9) :

$$a_c = \left[\frac{K_{IC}}{\sigma_{rf}}\right]^2 \frac{1}{1.21\pi}$$

is 20 µm for pure WC, 17 µm for 0.5 % Co and 16 µm for TiC. For TiC
the critical defect size is that of a whole carbide crystal. For WC
the value a_c corresponds either to many times the average grain, or
to that of a very large grain (Fig. 4).

Fractography

A fractographic study has been made in which either the fracture
line was analysed on a polished surface by interference microscopy or
the fracture surface was observed directly by scanning electron microscopy. We have made a systematic study of the three types of
fracture studied i.e. K_{IC}, bending and compression tests for the
different materials. Likewise we have taken as reference the fracture surface of a toughness specimen.

Toughness : For the two concentrations of WC, the crack propagation is predominantly intergranular. However some of the large
crystals may be observed in which transgranular fracture has occured
(Fig. 2). This observation is in agreement with the extrapolation
of the results of a previous investigation (4) where we observed the
change in proportion of intergranular fracture with the proportion
of cobalt. The fracture surface of TiC materials is different :
there is a mixture of equal proportions of inter and transgranular
fracture (Fig. 3). These two types of fracture are arranged in regions. A quantitative discrimination based on size of grain fractured is difficult. It appears, however, that the regions of transgranular fracture are due to an accumulation of large crystals. The
majority of intergranular debonding occurs around small crystals.

		WC 0.5 % Co	WC	TiC	TiC
$\bar{D}$	μm	6	6	20	6
E	MN m^{-2}	680 000	700 000	400 000	400 000
σ_{rc}	MN m^{-2}	3 300	3 240	2 070	
σ_{rf}	MN m^{-2}	970	770	415	590
G_{IC}	J m^{-2}	90	80	45	
K_{IC}	MPa $\sqrt{m}$	8.9	7.5	3.8	
a_c	μm	17	20	16	

Table I. Values of Young modulus, E, compression rupture stress, σ_{rc}, bending rupture stress, σ_{rf}, critical strain energy release rate, G_{IC}, critical stress intensity factor, K_{IC}, and critical defect size, a_c, for the materials studied.

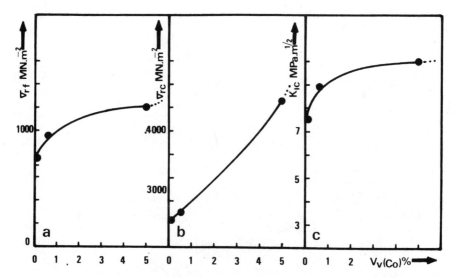

Fig. 1 : Variation of mechanical properties of WC based materials with volume fraction of cobalt:
a) rupture stress in bending, b) rupture stress in compression, c) critical stress intensity factor.

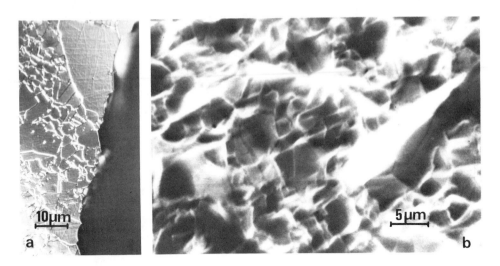

Fig. 2 : Fracture path (a) and fracture surface (b) of
        a pure WC toughness specimen.

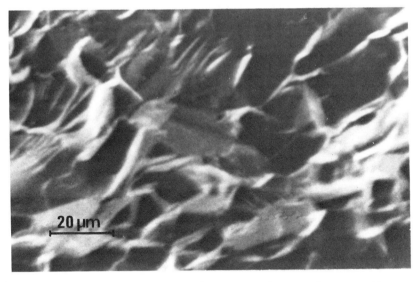

Fig. 3 : Transgranular and intergranular regions on the
        fracture surface of a TiC toughness specimen.

Bending: In the case of alloys based on WC, fracture is initiated by rupture of the large crystals (Fig.4). This has been observed on the fracture line and on the fracture surface. The fracture surface is very irregular and there are many crack networks. This is because fracture is initiated at several points simultaneously. At the center of the specimen, the fracture resembles that observed in toughness tests and corresponds to the propagation stage.

The fracture may also be initiated at a single site (Fig.5). Then it is due to an accumulation of large crystals and impurities. There is fracture of the large crystals as well as debonding, facilitated by the presence of impurities. The propagation zone resembles that in toughness testing with transgranular fracture of large crystals and debonding of the small crystals.

We have observed for TiC specimens fractured in bending regions of intergranular or transgranular fracture for both the grain sizes studied. The transgranular fracture however predominates. Also debonding of large crystals can be observed. Transgranular fracture is of two types:

-fracture in which cleavage produces a river pattern. This arises from fracture occuring on a preferred cleavage plane (100) and (110) (Fig.6a) (11) (12).

-ondulated transgranular fracture. This pattern seems to be Wallner lines (Fig.6b).

Compression : In general during a compression test, initiation of fracture occurs at relatively low stresses compared with the fracture stress(2400 MPa for WC and WC-0.5%Co, and 1700 MPa for TiC). This initiation stress is detected by the appearence of slight crackings. These crackings appear suddenly bearing in mind initiation and stable propagation. At the end of this process there exists in the specimen a network of cracks and at the critical load the specimen bursts.

For WC and WC-0.5% Co, the fracture features are basically intergranular debonding and transgranular fracture . It was possible to see also, on the surface of polished specimens prior to fracture, fractured and debonded crystals. Slip lines could not be detected (Fig.7).

In TiC fracture occurs essentially by debonding during intergranular sliding. The steps joining the different initiation planes are similarly predominantly intergranular. The fractures only very rarely have a transgranular character (Fig.8).

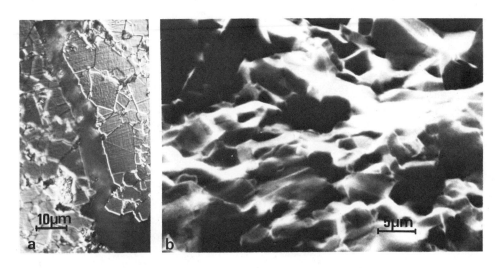

Fig. 4 : Transgranular rupture of large crystals as fracture initiation of a pure WC bending specimen:
a) optical micrograph.   b) SEM micrograph.

Fig. 5 : Fracture initiation at an accumulation of large
crystals and impurities (WC 0.5 % Co ; bending).

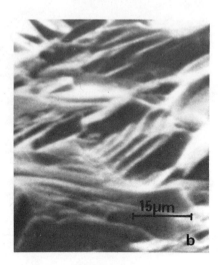

Fig. 6 : Transgranular rupture of TiC crystals
        a) river pattern. b) ondulated pattern(Wallner lines).

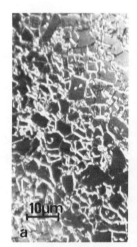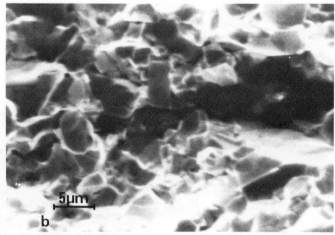

Fig. 7 : Polished surface (a) and fracture surface (b)
         of WC compression specimens stressed at
         3000 MPa (a), and 3200 MPa (b).

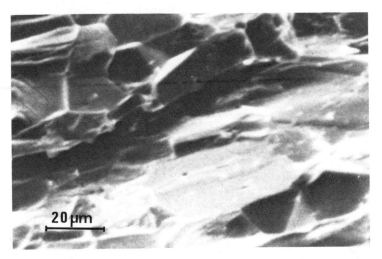

Fig.8 : Fracture surface of a TiC compression specimen .

DISCUSSION

The values of toughness for these carbides (7.5 MPa$\sqrt{m}$
for WC ; 3.8 MPa$\sqrt{m}$ for TiC) are very low, confirming that these
hard materials are indeed brittle. The values are comparable with
other carbides for example SiC (12). TiC is the most brittle material we have studied. Furthermore its fractographic features are
very sensitive to the applied stress state. This shows that the
boundaries between carbide-carbide and the TiC crystals themselves
have essentially the same strength. In the case of bending, fracture
appears to be initiated by transgranular fracture of the crystals.
This hypothesis is based on the very strong dependance of the
rupture stress on the mean diameter of the crystals and by the fact
that the critical defect size is of the same order as the crystal
size. Lack of homogeneous materials of different grain sizes prevented a verification of the $D_{TiC}^{-\frac{1}{2}}$ dependence of σ_{rf}.

For composites based on WC the predominance of intergranular
fracture in toughness and bend tests shows that the grain boundaries
are much weaker than the crystals. The very strong change in fracture stress with small addition of cobalt shows that the cobalt penetrates the grain boundaries and undoubtedly destroys the continuity
of the WC skeleton. This change is observed in compression only
at higher cobalt concentrations. This could be explained by the
fact that in compression the crystals are forced against each other
and the plasticising influence of the cobalt is thus suppressed.
This is confirmed by the fact that a constant value of σ_y 0.05%/C_{WC}
(C_{WC} is the contiguity of the carbide phase) is obtained up to 20%

vol. Co (5)(13). Furthermore, within this composition range, a $D_{WC}^{-1/2}$ dependance of σ_y 0.05%/C_{WC} is observed. This could also explain the occurence of various transgranular cracking as a rupture initiation in the whole volume of the specimen.

Based on fractographic evidence it can be stated that for WC fracture is initiated by fracture of the large crystals. The fracture then propagates trangranularly in large crystals and intergranularly between the others. It should thus be possible to establish a critical diameter for transgranular fracture.

The mechanism of fracture in TiC is more complex due to the fact that it is always both inter and transgranular irrespective of the grain size of the carbide crystals at least for the range studied here. This may be due to the similar strength of crystal boundary and of the crystal itself.

Our results for pure carbides are in agreement with those obtained for composites based on these carbides (4).

CONCLUSION

We have measured the characteristic mechanical parameters and analysed the fracture features of the metallic carbides TiC and WC. Fracture always occurs in the elastic region. In TiC, the most brittle of the materials studied, the type of rupture depends strongly on the mean of stressing. In WC the crystals are stronger than the crystal boundaries. Consequently the fracture is predominantly intergranular, apart from in compression where the carbide crystals are pressed against each other. The values of the critical stress intensity factor, K_{IC} , are low, confirming the brittle nature of these materials.

ACKNOWLEDGMENTS

This work was supported by the Délégation Générale à la Recherche Scientifique et Technique, Contract 76.7.0613.

REFERENCE

1. A.G. EVANS, in Fracture Mechanics of Ceramics, Edited by R.C. BRADT, D.P.H. HASSELMAN, F.F. LANGE, Plenum Press, Vol. 1, p. 17, (1974).
2. H.E. EXNER, A. WALTER, R. PABST, Mat. Sci. Eng., 16, 231, (1974).
3. J.L. CHERMANT, Bull. Soc. Franç. Ceram., 108, 16, (1975).
4. J.L. CHERMANT, F. OSTERSTOCK, J. Mat. Sci.,11, 1939, (1976).
5. F. OSTERSTOCK, Thèse d'Ingénieur-Docteur, Caen, Sept. 1975.
6. W.F. BROWN, J.E. STRAWLEY, A.S.T.M.-S.T.P., 410, (1967).
7. L.A. SIMPSON, J. Am. Ceram. Soc., 56 , 7, (1973).

8. P.L. GUTSHALL, G.E. GROSS, Eng.Fract. Mech., 1, 463, (1969).
9. R. PABST, Z. Werkstofftechnik, 6, 17, (1975).
10. W.S. WILLIAMS, J. Appl. Phys., 32, 552, (1961).
11. F.W. VAHLDIEK, J. Less Common Metals, 12, 429, (1967).
12. J.L. HENSCHALL, D.J. ROWCLIFFE, J.W. EDINGTON, J. Mater. Sci.,
       9, 1559, (1974).
13. V.A. IVENSEN, D.N. EIDUK, L.Kh. PIVOVAROV, Sov. Powd. Met.
       Metal. Ceram., 4, 300, (1964).

MECHANICAL PROPERTIES OF Al_2O_3-HfO_2 EUTECTIC MICROSTRUCTURES

C. O. Hulse

United Technologies Research Center

East Hartford, CT 06108

INTRODUCTION

The concepts associated with fracture mechanics suggest that ceramic eutectic microstructures could offer useful advantages in increasing the strength and perhaps reducing the brittle behavior of ceramic materials. For example, the presence of a strong whisker phase could function to limit the size of the critical micro-flaws present in the material or it could act to divert moving cracks so that they are required to produce more surface and travel in directions which are oblique to the applied tensile stresses. Also, the whisker phase might contract more than the matrix upon cooling from the temperature of fabrication and so place the continuous matrix phase in compression. A further possibility is that at elevated temperatures, the whisker phase may remain strong and reinforce the more ductile matrix phase in a classic composite manner.

An important advantage of directionally solidified eutectics for high temperature applications is that their microstructures are often extremely stable, practically to their melting point.[1] This stability results from the fact that their microstructures are produced directly from the molten state under conditions of thermodynamic equilibrium. If grain boundaries are present in these microstructures, they are relatively few in number and generally parallel to the axis of primary reinforcement.

A particular objective of this study was to examine the possibility that the high tensile stresses associated with a moving crack
might be sufficient to cause a transformation of the HfO_2 whisker
phase into the monoclinic structure if the hafnia were barely stabilized in the cubic or tetragonal structure by small additions of
Y_2O_3. In this event, the stress induced transformation might result
in an apparent plastic zone associated with the crack tip and some
noticeable improvement in mechanical properties.

EXPERIMENTAL PROCEDURE

Directionally solidified ingots of the Al_2O_3-HfO_2 eutectic were
prepared using the melting facility shown in Figure 1.   The sintered
charge rod inside a 3/8 in. I.D. tungsten tube was traversed down-
ward through a small furnace which consisted of a carbon ring susceptor surrounded by graphite and ZrO_2 fibrous insulation and a
three turn R.F. coil operating at 550 K.C.   Three passes were made
in argon, one just below the melting point and two molten zone passes
about 50°C above the eutectic melting temperature.

(VERTICAL—CENTER SECTION VIEW)

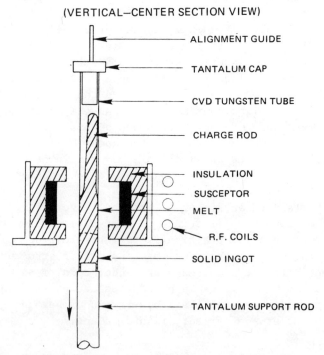

Fig. 1. Furnace for Directional Solidification of Ceramic Eutectics

The charge rods were prepared from -325 mesh powders of HfO_2, Al_2O_3, and Y_2O_3 supplied by Wah Chang, Adolf Meller and Molycorp, respectively, with all purities reported as 99.99 or better, neglecting ZrO_2. The batch powders were hand mixed in air and isostatically pressed into bars at 10,000 psi without the use of a binder. The final preparation step was hand sanding into triangular cross-section bars about 4 in. long after firing in argon to about 1400°C.

Work-to-fracture samples were prepared from unannealed ingots to be typically 0.25 in. x 0.25 in. x 1.5 in. with the direction of solidification parallel to the longest dimension. Two diagonal cuts were made with a 0.011 in. thick diamond saw to remove all material at the center of the 1.5 in. dimension except for a triangular web whose apex was inside the notch at the center of the sample. The sample configuration was similar to that used by Coppola, et al[2,3] and Simpson.[4] Three point flexure samples were diamond ground to nominal dimensions of 0.25 in. x 0.066 in. x 1.0 in. and tested on a 0.75 in. span after polishing on a glass plate through Linde A.

Work-to-fracture samples were broken with relatively stable fracture propagation using a 4-screw, 60,000 lb Tinius Olsen universal testing machine with the crosshead moving at .0025 in./min. Elevated temperature testing was done in air using alumina fixtures and solid alumina loading rams 2 in. in diameter. The bend testing was done in a separate facility in argon using molybdenum loading rams and fixtures.

RESULTS AND DISCUSSION

The Al_2O_3-HfO_2 eutectic composition was determined to be 44.5 Al_2O_3 and 55.5 HfO_2. The volume fraction of HfO_2 whiskers was calculated as 0.339 based on this composition and X-ray densities. A melting temperature of 1840°C was measured by directly viewing with an optical pyrometer the melting of a eutectic fragment held in an open tungsten wire basket inside an R.F. susceptor.

This eutectic forms a colony microstructure[5] when solidified at speeds greater than about 2 cm/hr. The colonies run parallel to the direction of solidification for distances up to a maximum of approximately 4 cm. Colony diameters were nominally about 150μ, fairly independent of solidification speed up to 30 cm/hr. The C-axis of the alumina matrix was parallel to the direction of solidification.

Within the colonies, the axes of the whiskers generally ran parallel
to the direction of solidification. Particularly when excess Al_2O_3
is present, however, the colonies contain a centered, threefold
spine of Al_2O_3 which divides the colony into three separate sections
which run parallel to axis of the colony. In each colony section,
the whiskers all tend to be tilted at various appreciable angles to
the direction of solidification.

The binary trough of the Al_2O_3–HfO_2 eutectic with Y_2O_3 additions
up to at least 10 weight percent was found to be at 44.5 weight percent Al_2O_3. X-ray patterns of powders prepared from eutectic ingots
without Y_2O_3 additions indicated that all the HfO_2 was monoclinic.
·The conversion of the HfO_2 to all tetragonal with no unexplained
X-ray lines was complete at 5 weight percent Y_2O_3 additions. A
series of ingots for further testing was then prepared with and without 5 weight percent additions of Y_2O_3 at solidification speeds of
10 cm/hr.

Thermal expansion data for the Al_2O_3–HfO_2 eutectic parallel to
the direction of solidification with and without 5 weight percent
Y_2O_3 additions compared with that for Al_2O_3 is presented in Figure 2.
Estimates of the thermal stresses present in the eutectic can be
estimated from the equation[6]

$$\sigma_1 = \frac{(\alpha_1-\alpha_2)\Delta TE_1}{1 + \dfrac{E_1V_1}{E_2V_2}}$$

σ_1 = stress in phase 1
α = thermal expansion coefficient
ΔT = temperature change
E = elastic modulus
V = volume fraction.

In the following calculation, Poissons ratio effects have been neglected and it was assumed that all stresses above 1525°C were
thermally relieved. Substitution of elastic modulus data for monoclinic and tetragonal ZrO_2[7], because no data for HfO_2 was available,

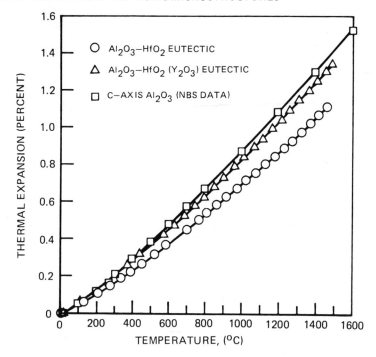

Fig. 2. Thermal Expansions of C-Axis Al_2O_3 and Al_2O_3-HfO_2 Eutectic With and Without 5 Weight Percent Y_2O_3 Additions

results in estimated tensile stresses of 13.6 and 103.5 ksi in the Al_2O_3 matrix with and without 5 weight percent Y_2O_3 additions, respectively. The corresponding room temperature compressive stresses in the HfO_2 are 26.7 and 202.0 ksi.

A tendency for the unstabilized ingots to contain cracks was observed in agreement with the above calculation. The Y_2O_3 stabilized ingots did not crack but grew with more poorly aligned structures and were difficult to polish. These ingots tended to have pull outs both at colony walls and within the regular eutectic structure. This was less of a problem in ingots solidified at speeds greater or less than 10 cm/hr.

A summary of work-to-fracture data for the eutectic with and without Y_2O_3 additions is shown in Figure 3. Energies of about 34 x 10^4 ergs/cm^2 (350 J/m^2) were measured for the eutectic solidified at 10 and 30 cm/hr. Much higher values of 105 and 138 x 10^4 ergs/cm^2 were measured for the eutectic solidified at 2 cm/hr and with excess Al_2O_3, respectively. SEM examinations of these fracture faces indicated that the fracture surface always contained the axis of the HfO_2 whiskers and that a very rough, splintered surface was formed.

Above temperatures of about 1000°C, large increases in the work-to-fracture energies were observed. The highest value was 380 x 10^4 ergs/cm^2 (3800 J/m^2) at 1500°C for the eutectic with no Y_2O_3 additions. A fracture surface produced at 1450°C is shown in Figure 4. Apparently, the alumina matrix is beginning to show some ductility and is being reinforced by the HfO_2 phase. Figure 5 shows a fracture surface produced at 1525°C. The whiskers now extend above the Al_2O_3 surface and apparently at one time were connected between the opened faces of the crack. Figure 6 presents diagrammatically the sequence of steps which may have occurred. Note that the deformed

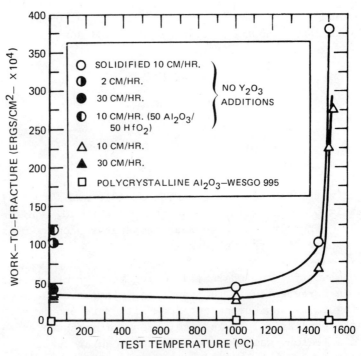

Fig. 3. Work-to-Fracture of Al_2O_3-HfO_2 Eutectic With and Without 5 Wt % Y_2O_3 Additions and Polycrystalline Al_2O_3 vs Temperature

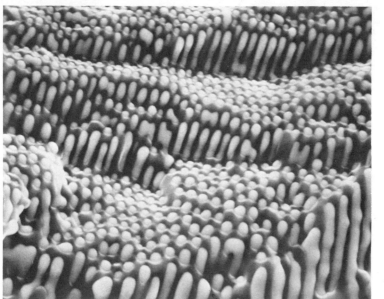

Fig. 4. Fracture Surface Formed at 1450°C in Directionally Solidified Al_2O_3-HfO_2 Eutectic with 5 Wt % Y_2O_3

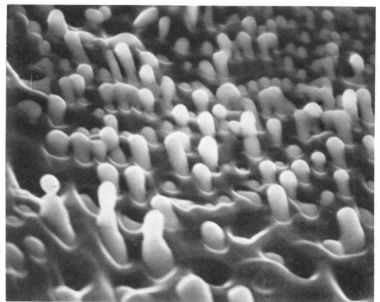

Fig. 5. Fracture Surface Formed at 1525°C in Directionally Solidified Al_2O_3-HfO_2 Eutectic with 5 Wt % Y_2O_3

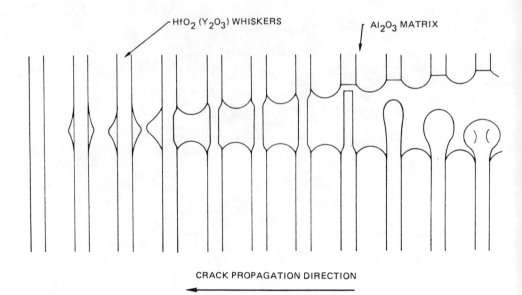

Fig. 6. Suggested Fracture Mode for Al_2O_3–HfO_2 (Y_2O_3) Eutectic at 1500°C

whiskers usually break at the point of attachment to the matrix so that only about one-half of the whiskers are pulled out in any one area. The HfO_2 whiskers are apparently not stable as whiskers at these temperatures when removed from the Al_2O_3 matrix. Thus, the stubs tend to coarsen and either form balls on their ends or become shorter and thicker by diffusional processes. Large amounts of energy must be needed to propagate a fracture through this material as both the whiskers and the alumina reinforced by the whiskers must have been extensively deformed.

Whisker stubs produced at 1500° and 1525°C were removed from the fracture surfaces by replica techniques and examined in the electron microscope. Electron diffraction experiments revealed only tetragonal crystal structures. The whiskers were fairly thick for good transmission and thus the volumes actually examined were those which had diffused down the stub and thus may have been converted back to the tetragonal form. Tilting experiments did not reveal the presence of any internal interfaces between the coarsened sheath and the core of the whisker.

Some preliminary three-point bend strength data are presented in Table I. The average bend strength of the eutectic parallel to the direction of solidification was 20 ksi at room temperatures and shows an appreciable increase above 1000°C where much higher values of work-to-fracture were measured. Attempts to produce better surface finishes by chemical polishing techniques were not successful.

SUMMARY AND CONCLUSIONS

The propagation of slow transverse cracks through the Al_2O_3-HfO_2 eutectic was studied using the work-to-fracture technique. This eutectic is rod-like with Al_2O_3 as the matrix phase and forms with a regular colony structure when directionally solidified at 10 cm/hr. Work-to-fracture energies at room temperature were typically 35 x 10^4 ergs/cm^2 (350 J/m^2). Above about 1000°C, there was a large increase in toughness with average energies of 293 x 10^4 ergs/cm^2 (2930 J/m^2) obtained at 1500° to 1525°C. This increase was attributed primarily to the extra energy required to deform the whisker phase which was still connected to the two fracture faces after the matrix had failed.

Work-to-fracture energies of 101 and 138 x 10^4 ergs/cm^2 were obtained at room temperatures for samples which were solidified at 2 cm/hr or contained excess Al_2O_3, respectively. These higher energies were associated with splintering of the structure parallel to the axes of the whiskers.

Table I

Three-Point Flexural Strengths of Al_2O_3-HfO_2
Eutectics Solidified at 10 cm/hr

Sample #	Additive (Wt %)	Temp. (°C)	Strength (ksi)
77-1A	–	24	20.1
77-1B	–	24	18.73
77-7	–	24	22.3
77-12	–	1100	70.2
77-44A	–	1450	59.3
77-43B	10 Y_2O_3	24	18.9

The average bend strength of the eutectic parallel to the direction of solidification was 20 ksi at room temperature and this increased by a factor of three at temperatures of 1100°-1500°C.

No direct evidence was found for the transformation of partially stabilized HfO_2 to the monoclinic structure by the high tensile stresses associated with the fracture process.

ACKNOWLEDGEMENTS

The author would like to thank Mr. N. Chamberlain for making the ingots, Mr. L. Jackman and Mr. G. McCarthy for SEM and TEM measurements, respectively, and Mr. R. Bailey and Mr. S. Kustra for making the mechanical tests. The author would also like to thank Dr. Earl Thompson for his continued interest and support.

REFERENCES

1. B. Bayles, J. Ford and M. Salkind, Trans. AIME, 239, 844 (1967).
2. J. Coppola and R. Bradt, J. Am. Ceram. Soc., 55, 455 (1972).
3. J. Coppola and R. Bradt, Ceram. Bull., 51, 847 (1972).
4. L. Simpson, J. Am. Ceram. Soc., 56, 7 (1973).
5. B. Chalmers, Principles of Solidification, J. Wiley & Sons (1964).
6. R. Novak and M. DeCrescente, J. Engrg. for Power, Trans. AIME, 379 (1970).
7. J. Lynch, C. Ruderer and W. Duckworth, Engineering Properties of Selected Ceramic Materials, Am. Ceram. Soc. (1966).

FRACTURE RESISTANCE AND TEMPERATURE IN METAL-INFILTRATED

POROUS CERAMICS

R. A. Queeney and N. Rupert

Department of Engineering Science and Mechanics

The Pennsylvania State University

INTRODUCTION

Specialized mechanical design applications may involve environmental and loading conditions too demanding for the application
of conventional structural metallic alloys. Ceramic materials may
prove viable design options under circumstances involving elevated
temperatures and loading particulars that give rise to excessive
wear of contacting parts. The current popularity of developmental
efforts to produce fully-dense ceramic bodies, as reported in the
ceramic literature, attests to the promise of these solids. At
least two difficulties will remain even though such bodies are successfully fabricated. First, ceramic materials exhibit a level of
fracture resistance, or notch sensitivity, that is an order of magnitude less than that found in most structural metals. Second, the
very properties that are superior to metallics--e.g. hardness--
render them difficult to machine into complex structural shapes and
machine members.

Porous, crystalline ceramic materials exhibit more facile
machining and mechanical finishing characteristics than fully dense
bodies, and the processes by which they are formed are often
economically more attractive, thus ameliorating the second of the
two difficulties mentioned above. However, the fracture resistance
of a porous body is significantly less than if the body were made
fully dense, suggesting that porous ceramics have relatively little
mechanical application potential. However, it might be possible
to fill all, or most, of the porosity with a suitable infiltrant to
raise fracture resistance levels while maintaining machineability.
Metallic alloys of relatively low melting point have been successfully

infiltrated into porous ceramics, and structural shapes have been
successfully fabricated (1). Successful infiltration of a porous
ceramic demands that the metallic infiltrant must wet the ceramic
surface; moreover, it has been demonstrated that the most successful infiltrants form a reaction layer of intermetallic composition
that adheres to both the pore wall and the filler metal, constraining the infiltrant to strain compatibility with the ceramic body at
their mutual interfaces. Although lowered melting point of the
metallic phase is desireable to facilitate the infiltration process,
it is clear that the metallic melting point will predetermine the
upper temperature limit of mechanical utilization of the composite.

Finally, although successful infiltration implies an inter-
connecting pore structure, some porosity is bound to be inaccess-
ible to the molten infiltrant by reason of being free-standing.
Residual porosities of from 5% to 10% have been characterized in
well-infiltrated solids (2). These residual localized stress risers,
coupled with cycling stresses--from either thermal constraint
stress fields or boundary fractions--will lead to stable crack development in mechanical applications likely to be encountered.
Fracture resistance, then, is at least one of the most important
mechanical response parameters to be determined in designing these
ceramic-metal composites.

STRESS ANALYSIS OF THE COMPOSITE

The exact stress fields in and around the filled pores in
the composite body is not attempted here, due to the extreme complexity of the geometries involved. Instead, an approximate
analysis of the micro-stress state has been attempted to provide
expectations and predictions of the fracture resistance of the solid
as a function of temperature. If the porosity is idealized to a
distribution of spherical inclusions, the thermally generated
stress fields are already known (3), provided that the interpore
separation λ is an order of magnitude greater than the pore diameter, or radius a. These tresselated stresses, generated by the
mismatch in thermal strain coefficients $\Delta\alpha$ between metal "i" and
ceramic matrix "m" will lower the strain energy density locally,
and, from (3), the strain energy release rate G will be reduced by
ΔG, relative to the fully dense ceramic case, as

$$\Delta G = \left\{\frac{1}{\lambda}\right\}^2 \frac{p^2 (4a)^3 \pi^2 (1+\nu_m)}{E_m} \tag{1}$$

and:

$$P = \frac{\Delta\alpha \; \Delta T}{\dfrac{1+\nu_m}{2E_m} + \dfrac{1-2\nu_i}{E_i}} \qquad (2)$$

In equations (1) and (2), ΔT is the change in temperature over which the mechanical response is determined, E is the elastic modulus and ν the poisson's ratio of the materials: subscripts "m" and "i" refer to the ceramic matrix and metallic infiltrant, respectively. When equation (1) is inserted into an energy balance expression, such as due to Griffith[4], to ascertain crack stability, a positive increase in crack extension stress, linearly proportional to positive temperature changes, is obtained.

The strength-enhancing tresselated stress fields derivable from equation (1) cannot be indefinitely supported by the metallic infiltrant phase, however, as one can expect the yield strength σ_0 of the alloy to degrade with temperature.

Plasticity aspects of fracture mechanics, seldom a concern in the fracturing of pure ceramics or ceramic/ceramic composites, will limit the strength enhancement due to elastic tresselated stress fields developed above. The extent of plastic flow in the infiltrant can be approximated by the plane strain plastic zone radius r_p ahead of the crack in the crack plane, given by (5):

$$r_p \cong 2 \frac{1}{6\pi} \left(\frac{K_I}{\sigma_o} \right)^2 \qquad (3)$$

Here, K_I is the stress intensity factor for the structural configuration in question and has an upper limit of K_{IC}, the fracture toughness. Expression (3), derived for a homogeneous solid, can only be considered approximate for the composites of concern here. Two plastic zone sizes should strongly effect the changes in fracture resistance as temperature changes occur. The first significant deviation from linearly increasing (with temperature) fracture resistance will be encountered when the plastic zone is equivalent to an average pore diameter ahead of the crack tip. Those pores closest to the crack tip will then be unable to carry further tresselated stress (neglecting alloy strain hardening) and the improvement in fracture resistance predicted by equation (1) will be lessened. The second significant change in the fracture resistance with temperature will occur when all pores within one average interpore spacing are saturated with stress levels equivalent to the yield stress. Figure (1) qualitatively illustrates these expectations.

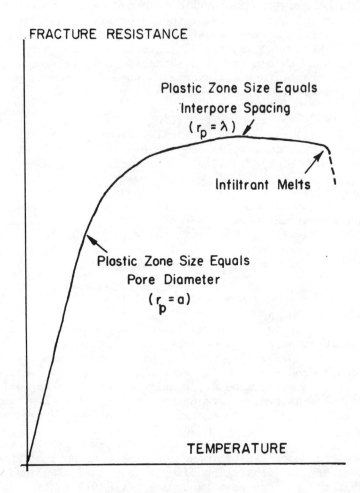

Figure 1. Expected Fracture Resistance (K_{IC}) Response of a
Metal-Infiltrated Porous Ceramic vs. Temperature

EXPERIMENTAL STUDIES

Two composites were available for experimental study. One was
a commercial graphite (Grade 2020, Stackpole Carbon, St. Marys, PA)
and the other was an experimental grade of reaction-sintered Si_3N_4.
Both porous solids were infiltrated with 2024 Aluminum alloy, using
a pressure infiltration technique, and both were prepared by Amos J.
Shaler Associates of State College, PA. The resulting microstructures are shown in Figure 2 below. Pore diameters and interpore

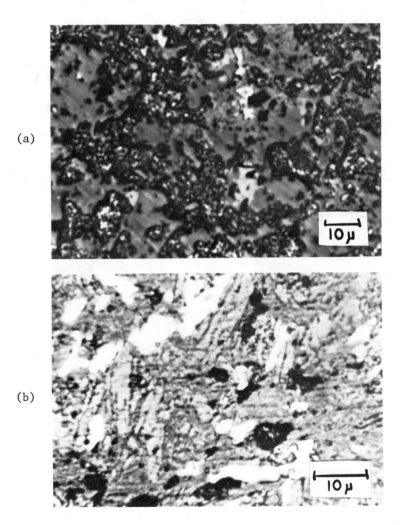

(a)

(b)

Figure 2. Microstructures of 2024 Aluminum-Infiltrated
a. Graphite and b. Si_3N_4

spacings were taken from suitably prepared plane sections as in
Figure 2. It is obvious, from these photomicrographs, that the
porosity is anything but spherical; hence, characterizing the composite in this manner must be viewed as only an unrefined mode of
analysis. The aluminum alloy infiltrant was in an overaged, or T7
heat treatment state. The bulk properties of this alloy form are
tabulated below (6).

Yield Strength (MN/m^2)	323	309	248	130	62	42	27
Temperature, °C	24	100	150	205	260	315	370

Table 1. Yield Strength Levels and Temperature for 2024-T7
          Aluminum Alloy

Fracture resistance testing was performed on beam specimens
with square cross-sections 0.63 cm on a side and a span length of
5.0 cm. The beams were notched to a depth of 0.21 cm with notch
root radii kept to less than 1.2×10^{-3} cm. The beams were loaded
in four point bending with a center span of 2.5 cm at a rate of
2.5×10^{-2} cm/minute in air maintained at the desired testing temperature by thermal jacketing. The measured maximum load was
inserted in the appropriate stress intensity formula for this configuration (7) to calculate the fracture toughness as a measure of
fracture resistance.

RESULTS AND DISCUSSION

The measured fracture resistance of the aluminum-infiltrated
graphite and Si_3N_4 are shown in Figure 3 below. As can be seen
from the graphs, the fracture resistance does increase linearly,
then nonlinearly, and, finally, achieve and hold a constant value,
as predicted in Figure 1. Examining the mechanical response data
for the aluminum alloy in Table 1, it can be seen that strength
considerations of the alloy alone clearly do not determine the
composite's ability to resist crack extension, since the strength
of the alloy degrades markedly above 200°C whereas the composite's
fracture toughness is constant.

The interplay of the plastic zones developed in the infiltrant
ahead of the crack can be seen in Table 2 below. The microstructural features of the two solids are also tabulated there for comparison. The predictions made in the previous discussion relating
to plasticity effects are closely supported by the experimental
results for the Si_3N_4 but not as well correlated for the graphite.
Examining the photomicrographs of Figure 2, it is apparent that
the Si_3N_4 has a more homogeneous distribution of porosity, whereas

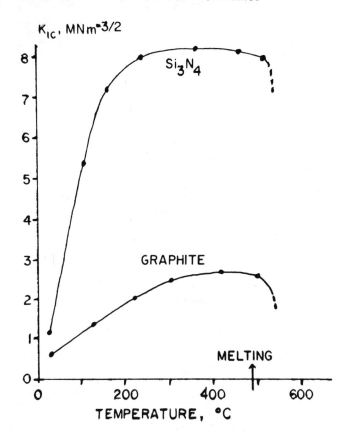

Figure 3. Fracture Resistance K_{IC} of 2024-T7 Aluminum Infiltrated Graphite and Si_3N_4 vs. Temperature

	Average Pore Diameter (microns)	Average Interpore Spacing (microns)	Plastic Zone Size at:	
			K_{IC}, nonlinear (microns)	K_{IC}, max
Si_3N_4	10.2	55	3.5	50
Graphite	70	105	20	83

Table 2. Plastic Zone Sizes and Porosity Distribution in Infiltrated Composite.

the graphite pores vary greatly in size and uniformity of distribution. Despite the relatively poor correlation in the case of
the graphite, the order of magnitude of the plasticity measures
and the microstructure is correct.

SUMMARY AND CONCLUSIONS

The fracture resistance of a metal-infiltrated porous ceramic
depends upon the thermoelastic properties of both ceramic and
metal. The maximum values of fracture resistance that can be
obtained, however, are limited by the thermoplastic mechanical
response (yield strength at temperature) of the metal and the size
and spacing of the filled pores. Pore spacing seems to determine
the maximum fracture resistance of the composite, at least whenever
the ceramic itself displays little or no plastic deformation in
the temperature range of interest. The present study indicates
that porous ceramics, when infiltrated by even a weak metallic
alloy, can achieve fracture resistance levels equal to those
associated with fully dense bodies of the ceramic (7). A more
careful choice of infiltrant will surely allow the useful range of
application temperature to be considerably extended.

REFERENCES

1. J. C. Conway and A. J. Shaler, Am. Ceram. Soc. Bull., 50, 636
    (1971).
2. R. A. Queeney and R. L. Turner in Fracture Mechanics of
    Ceramics, 2, 807 (1974), Plenum Publishing Co., New York.
3. J. Selsing, J. Am. Ceram. Soc., 44, 419 (1961).
4. A. A. Griffith, Phil. Trans. Roy. Soc., 221, 163 (1920).
5. G. R. Irwin in Proceedings of Seventh Sagamore Army Materials
    Research Conference, 4 (1960), Syracuse University Press,
    Syracuse, New York.
6. Aluminum Standards and Data, 31, (1969) The Aluminum Associa-
    tion, New York.
7. F. F. Lange, J. Am. Ceram. Soc., 57, 84 (1974).

REACTION SINTERED Si_3N_4 : DEVELOPMENT OF MECHANICAL PROPERTIES RELATIVE TO MICROSTRUCTURE AND THE NITRIDING ENVIRONMENT

M W Lindley, K C Pitman and B F Jones

Admiralty Materials Laboratory
Poole, Dorset, BH16 6JU, UK

ABSTRACT

Fracture mechanics measurements have enabled the critical defects to be assessed in silicon compacts as they are progressively nitrided. From these data and microstructural studies, a structural model has been developed to explain the development of strength in reaction sintered Si_3N_4. The nitriding technique and gas composition are shown to have a significant influence on fracture properties and can be interpreted in terms of this model.

INTRODUCTION

This paper describes the use of linear elastic fracture mechanics in a fundamental study of the factors which control the strength of reaction sintered silicon nitride, Si_3N_4. Whilst the more usual role of fracture mechanics is to provide design data for engineering purposes, this work indicates its value as an investigative tool for the materials scientist. In the present case it provides a means of estimating the size of critical defects under circumstances where the direct identification and measurement of such features have been impossible. The difficulty of identifying critical defects in reaction sintered silicon nitride is well known[1] and although in certain cases special flaws can be identified[2,3] such cases are not typical in our experience. However, measurement of the critical stress intensity factor K_{IC} of the material provides a means of estimating the critical defect size 2a from the equation[4]

$$K_{IC} = \sigma Y \sqrt{(2a)} \qquad (1)$$

921

where σ is the stress to propagate the defect and Y is a geometric factor.

The need to identify the size of critical defects when they are not directly visible arises because there are linear relationships between strength and nitrided density for most silicon powders which are independent of the green density of the silicon compact, but dependent on the particular silicon powder and nitriding technique[3,6]. The implication of the existence of these linear relationships is best illustrated by considering two green silicon compacts X and Y prepared from the same silicon powder but compacted to different green densities ρ_x and ρ_y such that $\rho_x > \rho_y$. If these materials are nitrided to different degrees of conversion such that their nitrided densities are the same, then they will have equivalent strengths[3,5,6]. Their structures in terms of the proportions of silicon, silicon nitride and porosity will be very different but since $\sigma_x = \sigma_y$ and for a flaw of length 2a at the surface

$$\sigma = \frac{K_{IC}}{Y\sqrt{2a}} \tag{2}$$

then

$$\frac{K_{IC_x}^2}{a_x} = \frac{K_{IC_y}^2}{a_y} \tag{3}$$

Because of the structural differences (Table 1) it was thought unlikely that $K_{IC_x} = K_{IC_y}$ and $a_x = a_y$, so it was concluded that there must be a subtle relationship between these parameters to produce the observed strength equivalence. To define this relationship it is necessary to determine a, and K_{IC} as they change during the nitriding process and in the absence of a means of directly measuring '2a' a fracture mechanics approach was adopted.

DENSITY Mg m^{-3}	% CONVERSION TO SILICON NITRIDE		PROPORTIONS OF PHASES BY VOLUME (%)					
			SILICON NITRIDE		SILICON		POROSITY	
	lgd	hgd	lgd	hgd	lgd	hgd	lgd	hgd
1.80	33.6	18.8	26.1	15.9	41.9	55.8	32.0	28.3
2.40	94.9	75.1	73.7	63.5	3.2	17.1	23.1	19.4

TABLE 1. Structures of low green density 'lgd' (1.47Mgm^{-3}) and high green density 'hgd' (1.60Mgm^{-3}) silicon compacts nitrided to equivalent densities.

MATERIAL SELECTION AND EXPERIMENTAL PROCEDURE

There appear to be at least four potential sources of defects in reaction sintered silicon nitride but the ability of any one type to influence fracture will depend on its relative size and distribution:

(a) Mechanical damage of surfaces during machining[2].
(b) Impurity particles or voids generated by impurities[3].
(c) Porosity developed from voids in the original compact[3,5,7].
(d) Large silicon particles which are retained (unconverted) in the final material or holes which develop due to the melting of large silicon particles during nitriding[7].

Silicon nitrides, produced from various silicon powders have been studied at AML and the results from these experiments allowed the selection of a material in which defects of the type described in (a), (b) and (d) could be eliminated as the source of failure and where it was known that the strengths of bars of this material were controlled by surface breaking features (c) of the bulk microstructure[3].

Compact Preparation and Nitriding

The silicon powder (median particle size 13 μm) was isostatically pressed at 31 or 185 MNm^{-2} and argon sintered for 4 h at 1175°C to produce compacts with green densities of 1.47 Mgm^{-3} (lgd) and 1.60 Mgm^{-3} (hgd) respectively. Strength bars 4.57 x 4.57 x 30.00 mm^3, modulus bars 5.00 x 5.00 x 50.00 mm^3 and K$_{IC}$ double torsion plates 25.4 x 25.4 x 2.00 mm^3 were cut from each billet. Specimens were reacted in a 'static'[8] nitriding system at 1 bar to various levels of conversion in a closed-end mullite furnace tube (capacity ~3l) at temperatures in the range 1200 to 1330°C.

Measurement of Strength, Moduli of Elasticity and K$_{IC}$

Strengths were determined in the as-nitrided condition in three-point bend with a span of 19.05 mm. Rigidity modulus (G), Young's modulus (E) and Poisson's ratio (ν) were determined by resonant frequency methods[9].

A short double torsion[10] specimen was selected for the determination of K$_{IC}$ because of limited furnace space and the requirement to nitride strength and modulus bars and K$_{IC}$ plates at the same time. It was noted in an earlier investigation[3] that the load to propagate a crack became lower as the crack approached the end of the plate, thus casting some doubt on the values of K$_{IC}$ measured when the crack was in this end region of the sample. In

a short specimen of the type used here there is an increased
chance that measurements may be made in the region where the end
effect occurs and it was therefore necessary to design a testing
technique which allowed us to identify and exclude data of this
type. Natural cracks were initiated in the specimens at a crosshead speed of 0.05 mm min⁻¹ and the load removed. The load
required to propagate this crack at the same cross-head speed was
then determined, the load being removed as soon as significant
crack movement was detected by a drop in load on the recording
chart. The load was then reapplied and the procedure repeated
so that a number of values (up to 5) for the load to extend the
crack were obtained. Only when the first two or three load
values for any one specimen were similar was it concluded that
the initial crack was far enough from the end of the plate to
provide accurate values of K_{IC}. This procedure resulted in the
rejection of approximately 25% of the plates tested. When these
conditions applied the first load recorded to propagate the crack
was used to calculate K_{IC}.[10] Confidence in the K_{IC} data obtained
by this procedure was confirmed by testing 25.4 and 100 mm long
plates in both the green and fully nitrided condition when the
values obtained from both geometries were in excellent agreement[3].

MATERIALS NITRIDED UNDER 'STATIC' CONDITIONS

The variation of strength and Young's modulus, Poisson's ratio
and K_{IC} with density were measured and the strength and K_{IC} data
are shown in Figs. 1 and 2.

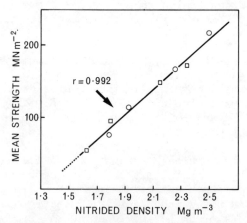

Fig. 1.   Strength versus nitrided density for 'static' nitriding.
         o hgd compacts; □ lgd compacts; r = correlation coefficient.

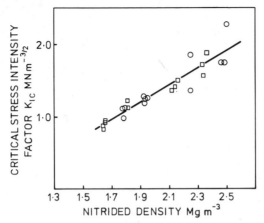

Fig. 2.   Critical stress intensity factor versus nitrided density
          for lgd and hgd compacts nitrided under 'static' conditions.
          Symbols as in Fig. 1.

Assuming that the flaws approximate to a sharp penny shaped
crack over the range of densities studied [e.g.7] then

$$Y = \left(\frac{4}{\pi}\right)^{\frac{1}{2}} \qquad (4)$$

and the critical defect size was calculated at various densities
(see Fig. 3) from

$$2a = \frac{\pi K_{IC}^2}{4\sigma^2} \qquad (5)$$

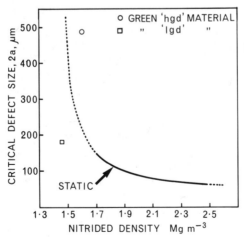

Fig. 3.   Critical defect size versus nitrided density for 'static'
          nitriding.

Confidence in the 2a curve is high for values above 1.7 Mgm^{-3}. At
lower densities the curve is less satisfactory due to uncertainties in the extrapolation of the σ and K_{IC} lines. The evidence is
conclusive, however, that in the density range 1.7 to 2.6 Mgm^{-3},
the critical defect size decreases with increasing density. σ, ν
E, K_{IC} and 2a were all independent of the green density (lgd or
hgd).

MICROSTRUCTURAL MODEL FOR REACTION SINTERING OF SILICON COMPACTS

After consideration of the data a microstructural model for
reaction sintered Si_3N_4 compacts has been formulated: it is presented in detail elsewhere[3]. When the isostatically pressed powders
have been sintered in argon, particles are joined by necks of silicon, or silica derived from the oxide surface layer present on the
original particles. The strengths of bars machined from this
material are controlled by either interparticle voids or by larger
defects which must result from the breaking of interparticle necks
in the regions close to the surface. The evidence suggests that
during the early stages of nitridation the interparticle necks are
destroyed resulting in a decrease in strength and modulus. Evidence for this includes the fact that strengths and elastic moduli
lower than the measured green properties have been recorded for
materials nitrided to low weight gains, and the critical defect
size of green compacts probably increases during the early stages
of nitriding (Fig. 3). Additionally, under the same nitriding conditions the same strength/density relationship is obtained for
compacts of a particular powder independent of the critical defect
size in the green silicon compact.

As the necks between silicon particles are gradually eliminated, silicon nitride begins to form in the void space between sili-

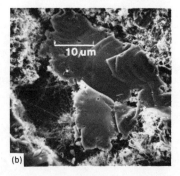

Fig. 4.   Scanning electron micrographs of hgd compacts reacted in
'static' nitrogen (circa 20% conversion to Si_3N_4). In (a)
silicon particle is marked A, developing silicon nitride B
and the void C between these features is noted.

con particles. The improvement in strength and stiffness following the initial decline must correspond to the development of a continuous skeletal network of whisker-like silicon nitride. It is the densification of this network on further nitridation which gives rise to the linear portions of the strength/density and modulus/density relationships. Once the silicon nitride network has formed, the major voids in the material will be holes which contain the reacting silicon particles. As the reaction proceeds, the silicon particles reduce in size providing additional silicon nitride for the thickening and extension of the silicon nitride network. The mechanism for these processes is in agreement with the theories proposed for the formation of silicon nitride involving vapour-phase transport of silicon monoxide formed from silica and silicon[8,11,12]. The mechanism is supported by the microstructural evidence of Fig. 4 a and b which show the interface between a silicon particle and the silicon nitride network to be a gap in a compact converted under "static" conditions. Consideration of the mismatch in thermal contraction coefficients of silicon and silicon nitride suggests that it is unlikely that the silicon will contribute to the mechanical properties of the structure at room temperature.

The stage of the process at which the silicon nitride network becomes continuous will depend on both the geometry of the structure (i.e. the size and distribution of the silicon particles and the voids between them) and the proportion of the silicon nitride formed which contributes to the network. The properties of the network at the stage when it becomes continuous will depend on the

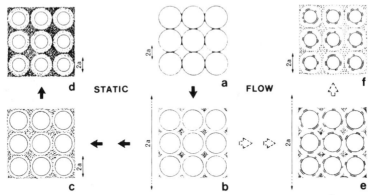

Fig. 5. Schematic model of development of structure of reaction sintered silicon nitride. (a) GREEN COMPACT- silicon particles bonded by silicon or silica necks; (b) EARLY STAGES, STATIC or FLOW (weight gain < 5%)- necks destroyed, nitride network not continuous; (c) STATIC (weight gain ~15%)- nitride network continuous; (d) STATIC (weight gain ~25%)- nitride network thickening; (e) FLOW (weight gain ~15%)- nitride layer on particles, network not continuous; (f) FLOW (weight gain ~25%)- continuous nitride network just established.

volume of the silicon nitride contributing to the network and the
detailed structure of the network and whisker dimensions. A simplified schematic representation of the early stages of conversion is
shown in Fig. 5 (a–d). The contrast between "static" and "flow"[8]
situations is discussed in detail in a later section. The main
experimental observations can now be discussed in terms of the
proposed mechanism of formation of reaction sintered Si_3N_4.

THE EXISTENCE OF STRENGTH/DENSITY RELATIONSHIPS

Strength/density relationships are independent of green
density because K_{IC}, ν, E and 2a are the same for lgd and hgd
compacts at any particular density despite the fact that lgd
material will contain a higher proportion of silicon nitride and
less silicon than hgd material (see Table I). The particle sizes
in the green silicon compacts are identical, but the particles
will be further apart and the void spaces larger in lgd compared
to hgd compacts. It follows, therefore, that a higher degree of
conversion from silicon to silicon nitride will be required to
form a continuous network in the more open lgd structure. It must
also be remembered that whilst all the silicon nitride formed
contributes to the measured density of the material, it will not
all contribute to the continuous network, i.e. a proportion of the
silicon nitride will be "redundant" as far as mechanical properties are concerned. A higher proportion of the silicon nitride
formed in the lgd material will be "redundant" compared to that in
hgd material. Whilst a quantitative analysis of this situation is
impossible because the amount of "redundant" silicon nitride is
unknown, it seems plausible that the proportion of silicon nitride
contributing to K_{IC} and Young's modulus at a given density for lgd
and hgd material could be similar, as a result of greater "redun-
dancy" in lgd. It also seems plausible that K_{IC} and 2a could be
similar in the two materials as a result of the combination of
the geometric relationship of the green structures and the degrees
of conversion to silicon nitride required to produce lgd and hgd
materials of the same density.

EFFECT OF NITROGEN GAS FLOW

The influence of nitriding in a flowing rather than a static
nitrogen atmosphere on the properties of reaction sintered silicon
nitride has been reported[8]. Whilst linear strength/density
relationships were observed it was noted that nitrided strengths
at any given density were 50–75 MNm^{-2} lower for specimens nitrided
in flowing nitrogen compared to those of specimens nitrided in
static nitrogen. To investigate this effect the experiments
described above for static nitriding conditions were paralleled
using a gas flow rate of 100 ml min^{-1}. Fig. 6 shows strength

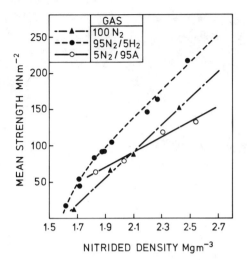

Fig. 6. Mean strength versus nitrided density of hgd compacts for
different nitriding conditions: gas flow rate 100 ml min^{-1}.

versus nitrided density and Fig. 7 compares the critical defect
sizes produced under static conditions with those produced in
flowing nitrogen. In the density range 1.7 - 2.5 Mgm^{-3} larger
defect sizes occur in specimens nitrided in flowing nitrogen.

During the early stages of nitridation the necks which join
the silicon particles in the green compacts are gradually elimin-

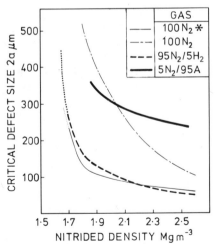

Fig. 7. Critical defect size versus nitrided density of hgd compacts
for different nitriding conditions: gas flow rate 100 ml min^{-1}
(*static nitriding).

 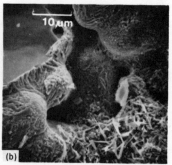

Fig. 8.   Scanning electron micrographs of hgd compacts reacted in
          'flowing' nitrogen (circa 20% conversion to Si_3N_4). The
          smoother edges of the reacting silicon particles are evi-
          dent and the void space C is less clearly defined. A dis-
          continuous surface layer can be seen along the line L-L in
          (a). The increased surface layer formation on hgd compacts
          reacted in very high purity flowing nitrogen can be seen in
          (b).

ated under both static and flow conditions. This process is
likely to be more efficient under flow conditions due to the
removal of silicon monoxide from the furnace atmosphere in the gas
stream[11,12]. It is difficult to imagine, however, that after elimination of the necks, the silicon particle sizes, spacings or the
void spaces are very different for the compacts nitrided under
"static" or "flow" conditions (see Fig. 5, stage b). It is probably the mechanism of network growth which is the most important
difference between the two situations (see Fig. 5, stages cd and
ef). It is well established that differences in reaction rate
exist between "static" and "flow" systems[8,12] and the examination
of microstructures has indicated the tendency for the formation of
surface layers on the silicon particles nitrided in "flow" (Fig.
8a and b) whereas no such layers are observed after "static" nitriding (Fig. 4a and b). Layers of this nature will not contribute
at the early stages to the continuous network of silicon nitride
which governs mechanical properties and as such can be considered
as part of the "redundant" silicon nitride. Thus for any particular compact the use of "flow" rather than "static" nitrogen
conditions will result in a greater proportion of "redundant"
silicon nitride in the structure at any particular nitrided density as illustrated in Fig. 5. This will be reflected in a lower
value of Young's modulus and a much larger defect size, and since
K_{IC} is not greatly changed[3], a lower strength results.

EFFECT OF NITRIDING GAS COMPOSITION

Other important observations have been made concerning the

influence of the nitriding gas composition on the strength of
reaction sintered silicon nitride. For example the presence of
hydrogen in a flowing nitriding gas has been shown to eliminate
the strength reducing effect of gas flow and produce strengths
equivalent to those obtained under static nitriding conditions[13,14],
whereas diluting a flowing nitriding gas with argon (5 x 10^{-2} bar
nitrogen partial pressure) results in strengths[15] only 50% of those
achieved in flowing nitrogen/hydrogen. Strength data for hgd silicon compacts reacted in flowing nitrogen, 95% nitrogen/5% hydrogen[13,14], and 5% nitrogen/95% argon[15] are compared in Fig. 6. It is
clear that the nitriding gas composition can have a very large
influence on the strength of reaction sintered silicon nitride.
However whilst differences in the detailed structure of the nitride network between these materials could be observed, critical
defects could not be conclusively identified. Fracture mechanics
data therefore were obtained over a range of nitrided densities
for the above nitriding gases using the same experimentation, as for
earlier work with nitrogen. The critical defect sizes and their
variation with nitrided density are shown in Fig. 7.

    Clearly the presence of hydrogen results in a reduction of
the critical defect size in specimens nitrided in a flowing gas,
which is thought to be associated with the greatly enhanced reaction rates observed [14]. When hydrogen was added to the flowing
nitriding gas K_{IC} values were lower (circa 1.75 MNm$^{-3/2}$ c.f.
2.0 MNm$^{-3/2}$ for nitrogen alone at a density of 2.5 Mgm^{-3}), E
values were higher, the structure of the nitride network was finer
and silicon nitride surface layers were absent. This implies a
reduction in the proportion of redundant silicon nitride which
would in turn explain the higher strengths. Water vapour (in
place of hydrogen) additions to a flowing nitriding gas can have
a similar influence on the properties of the resulting nitride[14].

    Microstructural examination of samples reacted in flowing
5% N$_2$/95%A exhibited a very coarse structure in the nitride network and enhanced surface layer formation compared to samples
reacted in flowing nitrogen: K_{IC} values were marginally higher at
equivalent densities but the most obvious effect was the much
larger critical defect size explaining why strengths are low.
These observations used in conjunction with the microstructural
model suggest an increased proportion of redundant silicon
nitride. A further consideration for these partially nitrided
materials is the very slow reaction rates observed with flowing
5%N$_2$/95%A which resulted in loss of volatile reacting species
from the compact and their deposition elsewhere in the furnace as
reflected by a significant difference between the measured weight
gain and residual silicon content and the lower fully nitrided
density.

CONCLUSIONS

A fracture mechanics approach has been employed to determine the critical defect size in a material where direct identification of critical defects proved impossible. The information obtained has allowed the development of a model to explain the formation of reaction sintered silicon nitride which has important implications in relation to the optimisation of the material. Application of this model and linear elastic fracture mechanics have shown that compared to flowing nitrogen alone:-

a) hydrogen in the nitriding gas reduces the critical defect size resulting in increased strengths.

b) a low nitrogen partial pressure increases the critical defect size resulting in decreased strengths.

ACKNOWLEDGEMENTS

The authors acknowledge experimental assistance from colleagues at AML. This article is published by permission of the Controller HMSO, holder of Crown Copyright ©

REFERENCES

1. D J Godfrey and K C Pitman, "Ceramics for High Performance Applications", ed. J J Burke, et. al. (Brook Hill, Chestnut Hill, Mass. 1974), 425.
2. R W Rice, J. Mater. Sci. 12 (1977) 627.
3. B F Jones, K C Pitman and M W Lindley, J. Mater. Sci. 12 (1977), 563.
4. P C Paris and G C Sih, ASTM Spec. Tech. Publ. No. 381 (1965).
5. B F Jones and M W Lindley, J. Mater. Sci. 10 (1975), 967.
6.   idem   , Powder Met. Int. 8 (1976), 32.
7. A G Evans and R W Davidge, J. Mater. Sci. 5 (1970), 314.
8. B F Jones and M W Lindley, J. Mater. Sci. 11 (1976), 1288.
9. W R Davis, Trans. Brit. Ceram. Soc. 67 (1968), 515.
10. A G Evans, J. Mater. Sci. 7 (1972) 1137.
11. D P Elias and M W Lindley, J. Mater. Sci. 11 (1976) 1278.
12. D P Elias, B F Jones and M W Lindley, Powder Met. Int. 8 (1976), 162.
13. B F Jones and M W Lindley, J. Mater. Sci. 11 (1976), 1969.
14. M W Lindley, D P Elias, B F Jones and K C Pitman, to be published.
15. D P Elias, K C Pitman and M W Lindley, to be published.

MECHANICAL PROPERTIES OF POROUS PNZT POLYCRYSTALLINE CERAMICS

Dipak R. Biswas and Richard M. Fulrath

Materials and Molecular Research Division, Lawrence
Berkeley Laboratory and Department of Materials Science
and Mineral Engineering, University of California
Berkeley, CA 94720

Niobium-doped lead zirconate-titanate (PNZT) was used to
investigate the effect of porosity on the mechanical properties of
a polycrystalline ceramic. Spherical pores (110-150μm diameter)
were introduced by using organic materials in the initial specimen
fabrication. The matrix grain size (2-5μm) was kept constant.
Small pores (2-3μm diameter) of the order of the grain size were
formed by varying the sintering conditions. The effect of porosity
on strength was predicted quite well by Weibull's probabilistic
approach. The Young's modulus showed a linear relationship with
increase in porosity. A decrease in fracture toughness with
increase in porosity was also observed. It was found that at equivalent porosities, small pore specimens gave a higher strength,
Young's modulus and fracture toughness compared to specimens containing large pores. Fracture surface analysis, by scanning electron microscopy, showed fracture originated either at the tensile
surface or, at the edge of the specimen.

I.   INTRODUCTION

Most polycrystalline ceramic materials contain some porosity
after firing. This porosity reduces the mechanical strength considerably and is extremely important where ceramics are used as
structural materials. During the past twenty years, a number of
studies have been performed to elucidate the effect of porosity on
strength[1-8] and Young's modulus.[7] General relationships are beginning to emerge. Most of these investigations considered only the
total porosity and neglected the size and shape of the pores. The
volume and size effect of porosity on strength was observed by
Hasselman and Fulrath[9] and Bertolotti and Fulrath[10] for a sodium
borosilicate glass with artificially introduced spherical pores of

933

various sizes and volume fractions. Bertolotti and Fulrath[10] found
that the strength was dependent on both the pore size and the volume
fraction of porosity. In the case of porous polycrystalline ceramics, inherent flaws are present as well as pores. The effective
flaw size for failure can be considered as the combination of inherent flaws on two sides of the pore plus the pore size. Thus,
the effective flaw length will be larger for large pore specimens
compared to small pore specimens assuming that the inherent flaws
are the same for both cases. Therefore, the large pore specimen
will fail at a lower stress level than the small pore specimen. In
this study, an attempt has been made to perform a systematic experimental study of the mechanical properties of a well-character-
ized polycrystalline ceramic containing controlled porosity with
variations in the volume fraction of pores and pore size.

II. EXPERIMENTAL PROCEDURE

Preparation of Powders and Fabrication of Specimens

Niobium-doped lead zirconate-titanate (PNZT) ceramic was used
in the present study. Its composition was $Pb_{0.99} \square_{0.01}$
$(Zr_{0.52}Ti_{0.46}Nb_{0.02})O_3$, where $\square$ is a lead vacancy. The powders of
PbO, ZrO_2, TiO_2 and Nb_2O_5 were milled in a vibratory energy mill
for four hours using isopropyl alcohol as a liquid medium, then
dried and calcined at 850°C for four hours. The calcined powders
were then mixed with 5.5 w/o excess PbO, milled for four hours using
isopropyl alcohol and polyvinyl alcohol (as a binder) in water, and
then air dried. Two types of porosity were considered in this study.
Large spherical pores (110–150μm diameter) of controlled amounts
were introduced by mixing with organic materials, cold pressing or
isostatic pressing, and then decomposing at 250°C for twelve hours
prior to sintering. Sintering was carried out at 1200°C for sixteen
hours in one atmosphere pressure of oxygen. Small pores (2–3μm)
were generated by varying the sintering conditions; namely, green
density, and sintering time and temperature (1100°C to 1150°C for
one minute to one hour). In both cases, a packing powder ($PbZrO_3$ +
ZrO_2) technique was used to control PbO loss from the specimen during
sintering.

Density Measurements

The apparent density of large pore specimens was determined
from dry weight (W_D) and suspended weight (W_S) in isopropyl alcohol
relative to a nickel metal standard. Calculation of the density of
the specimen was made by using the equation

$$\rho_{specimen} = \left[\frac{W_D}{W_D - W_S}\right]_{specimen} \times \left[\frac{W_D - W_S}{W_D}\right]_{standard} \times \rho_{standard} \quad (1)$$

where $\rho_{standard}$ is the density of nickel, 8.91 gms/cc. The theoretical density of PNZT ceramic was assumed to be 8.00 gms/cc. The bulk density of small pore specimens was calculated from the dry weight and the dimensions.

Mechanical Property Measurements

Strength of the ceramics at room temperature was measured by using a 4-point bending machine with a 0.75 in. overall span and a 0.25 in. inner span. The specimen dimensions were approximately 1.0 in. x 0.3 in. x 0.05 in.

Young's modulus (E) at room temperature was determined by a sonic resonance technique[11] using rectangular specimens with dimensions of 3.0 in. x 0.25 in. x 0.20 in. Once the exact resonant frequency was obtained, Young's modulus was calculated by using the resonant frequency, dimensions, and mass of the specimen. The resonant frequency used to calculate Young's modulus is theoretically correct only for materials without internal damping. To determine the internal damping, the damping capacity[12] of PNZT ceramics was measured from the width at the half-maximum value of the resonance curve. The value was obtained by finding the resonant frequency at the maximum amplitude and then locating the two frequencies above and below the resonant frequency at which the amplitude decreased to half of the resonant amplitude. Damping capacity δ was calculated from the differences of the two frequencies Δf and the resonant frequency f as

$$\delta = 1.8136 \, \frac{\Delta f}{f} \tag{2}$$

Fracture toughness (K_{IC}) of the ceramics was determined by the double torsion method.[13]

After fracturing the specimen, the fracture surfaces were observed by using scanning electron microscopy. In some cases, the "river pattern" (markings on the fracture surface) were traced back to determine the fracture origin.

III.   RESULTS AND DISCUSSION

A typical microstructure of PNZT ceramics is shown in Fig. 1. The average grain size for large pore (110–150μm) specimens was 2–5μm; and for small pore (2–3μm) specimens, 1–3μm.

Strength

The strength of the ceramics containing large spherical pores and small pores is shown in Fig. 2. In the case of large pores,

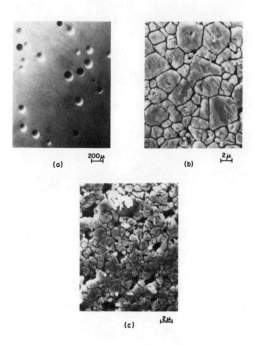

Fig. 1.   Microstructure of PNZT ceramics (with 5.5 w/o excess PbO
in starting mixture showing (a) spherical porosity, 4.9 v/o and
110-150μm pore size, (b) grain size distribution in nearly theoretically dense specimen, and (c) grain and pore distributions in
PNZT-small pore (4.8 v/o) specimen.

strength decreases rapidly with initial increase in porosity and
then levels off at about 8 v/o porosity. By using Weibull's
probabilistic approach to brittle strength, the experimental results
were analyzed and it was found[14] that the strength depends only on
the total porosity. However, in the case of small pores, Weibull's
approach is not applicable. The experimental results follow a
simple model of decreasing cross-sectional area with increasing
porosity up to about 5 v/o. The strength values up to 8 v/o of
small pores shows a higher value compared to specimens containing
large pores. The bending strength is directly related to the load
at the point of fracture. For small pore specimens, failure occurs
at a higher stress and the load at fracture is higher compared to
large pore specimens. This behavior is attributed to the postulate
that the small pore specimens contain smaller effective critical
flaws that cause failure than the flaws in the large pore specimens.

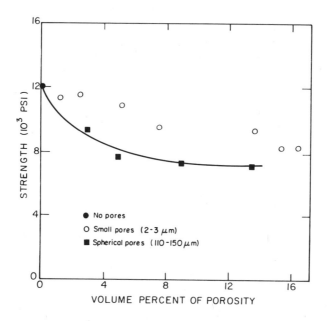

Fig. 2. Porosity dependence of strength of PNZT ceramics containing spherical pores and small pores versus the volume percent of porosity.

Young's Modulus

The experimental results for Young's modulus are shown in Fig. 3. The linear relationship

$$E/E_o = 1-KP \qquad\qquad (3)$$

describes the experimental data for large pore specimens. E_o was found to be 11.0×10^6 psi with a value of 2.5 for K. For the same pore content, Young's modulus for the small pore material appears to be slightly higher.

The plot of damping capacity against the volume percent of porosity is shown in Fig. 4. Damping capacity shows an increase with increase in porosity. A similar trend was observed by Marlowe and Wilder[15] in case of polycrystalline yttrium oxide. An interesting observation in Fig. 4 is that for small pores, the damping capacity is approximately one third that of the large pore specimens. It is even less than the extrapolated damping capacity value for zero porosity.

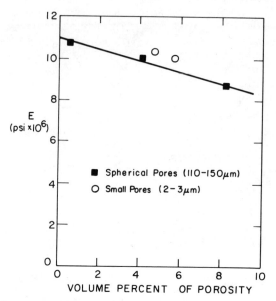

Fig. 3. Young's modulus of polycrystalline PNZT ceramics as a function of volume percent of porosity.

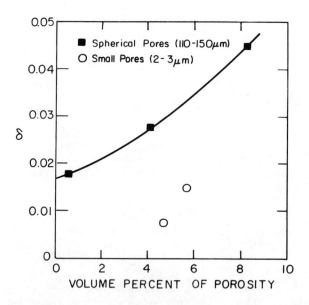

Fig. 4. Damping capacity of polycrystalline PNZT ceramics as a function of volume percent of porosity.

In fabricating the specimens, the starting powder (PNZT + 5.5 w/o excess PbO) was the same for all series, but in the case of small pore specimens, mainly the sintering temperature and time were varied to generate different amounts of porosity. The excess PbO forms a lead oxide rich liquid at about 890°C. In sintering large pore specimens at 1200°C for sixteen hours, it is expected that most of the excess PbO evaporates to the packing powder. In the case of small pore specimens which are fired at 1100°-1150°C, it is suspected that some PbO rich liquid remains and is converted during cooling to crystalline PbO and PNZT along the grain boundaries. Thus, the grain boundary structure for small pore specimens may be different resulting in a more continuous PNZT matrix. Therefore, the dissipation of energy will be less for the small pore specimens, as observed, which is reflected in a low damping capacity.

Fracture Toughness

The experimental results for fracture toughness are shown in Fig. 5. K_{IC} decreases with increase in porosity (110–150μm); and a small pore (2–3μm) specimen shows a slightly higher value. By definition, the strength, σ_f, is proportional to K_{IC}; the higher strength for small pore specimens would correspond to a higher fracture toughness, as observed in Fig. 5.

As mentioned, the critical flaw size for small pore specimens is smaller than that for large pore specimens. The flaw size "a" has been calculated from the experimental results using the Griffith-Irwin relationship:

$$a = f(Z/Y)(K_{IC}/\sigma_f)^2 \qquad (4)$$

where Z is the flaw shape parameter and Y is a geometrical constant. At an equivalent porosity (~4.8v/o), when the K_{IC} value is taken from Fig. 5 and the σ_f value is taken from Fig. 2, it is found that the critical flaw size of large pore specimens is about one and a half times larger than for small pore specimens assuming that f(Z/Y) is the same in both cases. Therefore, the strength of small pore specimens (4.8 v/o porosity) should be expected to be about 1.25 times higher than for large pore specimens. This agrees quite well with the experimental results as shown in Fig. 2. The flaw size calculated from K_{IC} and σ_f values is an appreciable fraction of the specimen thickness. Similar results have also been obtained for rock and chalk. Such a large flaw was not detected by conventional electron microscopy. Thus, there must be some uncertainties in interpreting the calculated values as the actual critical flaw size for brittle materials.

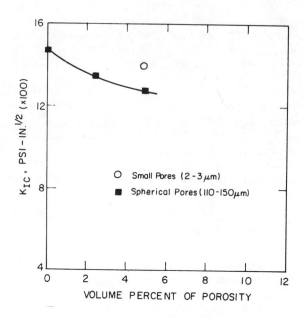

Fig. 5. Dependence of room temperature fracture toughness of polycrystalline PNZT on porosity.

Fractographic Analysis

Intergranular fracture was the primary fracture mode with some transgranular fracture observed at high magnification. By tracing back the river pattern, it was found that the fracture originates either at or near the tensile surface, or at the edges of the tensile surface of the specimen. To determine the edge effect on the strength, three highly dense specimens were fabricated by the same processing technique and then were cut into thin slices for strength measurement. In one-third of the specimens the edges were carefully rounded by using a rotating diamond wheel and polished by using 0.3μm Al_2O_3. For another third of the specimens, after rounding and polishing the edges, the tensile surface was polished. The final third of the specimens were tested in the as-cut condition. Insignificant differences in strength were observed between the three types of specimens. By examining the fracture surface, it was found that the edge failure could be eliminated by rounding off the edges, as shown in Fig. 6. Fracture from the tensile surface could also be minimized by polishing the tensile surface (Figs. 6 b and d).

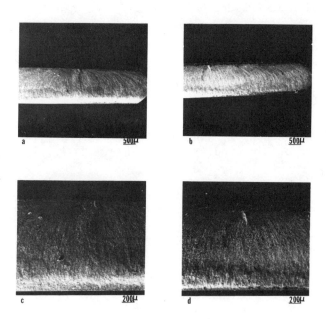

Fig. 6. Fracture surfaces of the PNZT specimens after rounding the edges showing the fracture origin (a and c) from the tensile surface, (b and d) from beneath the polished tensile surface.

IV. CONCLUSION

Polycrystalline niobium-doped lead zirconate-titanate ceramics, with small pores, exhibited a higher strength and fracture toughness and slightly higher Young's modulus, than the same material with large pores at equivalent porosity. The higher strength of the small pore specimens is attributed to the smaller effective flaw size that causes failure. Because the strength is proportional to the fracture toughness, the fracture toughness for small pore specimens is higher than for large pore specimens, as observed. Edge failures in bend specimens are eliminated by rounding off the edges.

ACKNOWLEDGEMENTS

The authors would like to acknowledge Professors D. P. H. Hasselman, J. A. Pask and I. Finnie for their helpful comments and suggestions. Discussions with J. Wallace is also acknowledged. This work was supported by the Energy Research and Development Administration.

REFERENCES

1. M. Y. Bal'shin, Powder Metallography (in Russian), Metallurgizdat, Moscow (1948).

2. E. Ryshkewitch, "Compression Strength of Porous Sintered Alumina and Zirconia," J. Am. Ceram. Soc., 36 [2], 65-68 (1953).

3. W. Duckworth, "Discussion of Ryshkewitch Paper," Ibid., [2] 68 (1953).

4. R. L. Coble and W. D. Kingery, "Effect of Porosity on Physical Properties of Sintered Alumina," Ibid., 39 (11) 377-85 (1956).

5. F. P. Knudsen, "Dependence of Mechanical Strength of Brittle Polycrystalline Specimens on Porosity and Grain Size," Ibid., 42 (8) 367-87 (1959).

6. E. M. Passmore, R. M. Spriggs and T. Vasilos, "Strength-Grain Size - Porosity Relations in Alumina," Ibid., 48 (1)1-7 (1965).

7. J. E. Bailey and N. A. Hill, "The Effect of Porosity and Micro-structure on the Mechanical Properties of Ceramics," Proc., Brit. Ceram. Soc., No. 15, 15-35 (1970).

8. S. C. Carniglia, "Working Model for Porosity Effects on the Uniaxial Strength of Ceramics," J. Am. Ceram. Soc., 55 (12) 610-18 (1972).

9. D. P. H. Hasselman and R. M. Fulrath, "Micromechanical Stress Concentrations in Two-Phase Brittle-Matrix Ceramic Composites," Ibid., 50 (8) 399-404 (1967).

10. R. L. Bertolotti and R. M. Fulrath, "Effect of Micromechanical Stress Concentrations on Strength of Porous Glass," Ibid., 50 (11) 558-562 (1967).

11. D. P. H. Hasselman, "Tables for the Computation of Shear Modulus and Young's Modulus of Elasticity from the Resonant Frequency of Rectangular Prisms," Applied Research Branch, Research and Development Div., The Carborundum Co., Niagara Falls, NY, (1961).

12. D. R. Biswas, "Influence of Porosity on the Mechanical Properties of Lead Zirconate-Titanate Ceramics," Ph.D. Thesis. University of California, Berkeley, CA 94720. (LBL-5479) (1976).

13. D. P. Williams and A. G. Evans, "A Simple Method for Studying Slow Crack Growth," J. Test Eval., 1 (4) 264-70 (1973).

14. O. Vardar, I. Finnie, D. R. Biswas and R. M. Fulrath, "Effect of Spherical Pores on the Strength of Polycrystalline Ceramics," Int. J. of Fracture, 13 (2) April (1977).

15. M. O. Marlowe and D. R. Wilder, "Elasticity and Internal Friction of Polycrystalline Yttrium Oxide," J. Am. Ceram. Soc., 48 (5) 227-233 (1965).

FRACTURE OF BRITTLE PARTICULATE COMPOSITES

D. J. Green* and P. S. Nicholson

McMaster University

Hamilton, Ontario, Canada

1. INTRODUCTION

In recent years, several mechanisms have been proposed for inc-reasing the fracture toughness of ceramic materials. The validity and applicability of a number of these mechanisms remains to be demon-strated both theoretically and experimentally. It does appear, how-ever, that moderate increases in toughness are feasible, so that provided the intrinsic strength of a material is not degraded and, that this strength can be retained, the use of ceramics in many new applications should be possible. This approach seems very promising, particularly in many energy-related situations, where the special properties of ceramics, such as high temperature stability and hard-ness can also be exploited.

It is the aim of this paper to consider one of these toughening mechanisms, i.e., the impedance of a crack by second-phase particles, an idea originally proposed by Lange [1]. It is proposed in this mechanism that when a crack intersects an array of obstacles, it is bent to some angle ϕ ($0 \leq \phi \leq \pi$) before it can move on, as shown in Fig. 1. In order for this mechanism to be useful in toughening brittle materials, the stress needed to bypass the obstacles must be greater than the fracture stress of the matrix, i.e., there must be an inc-rease in stored elastic energy associated with this bypassing mechan-ism. The angle ϕ is a measure of obstacle strength so that for strong obstacles $\phi \sim 0$, while for weak obstacles $\phi \simeq \pi$.

Clearly, there are several basic questions that must be answered for this process to be accepted and used for microstructural design. It is hoped in this paper to review some of the answers that have been

---

* Presently at CANMET, Energy, Mines and Resources, Ottawa, Canada

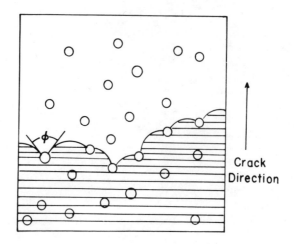

Figure 1.   Impedance of a crack front by a random array of spherical
            obstacles.

found and to point out areas where future research is needed in the
understanding of these crack-particle interactions.

2. OBSTACLE INTERSECTION

To take advantage of any local impedance to crack motion in
particulate composites, it is important to choose a system where the
the crack front must intersect the obstacles rather than stay within
the matrix. It is not immediately clear that the process of particle
avoidance would require less energy than particle intersection. From
surface area considerations, however, it seems unlikely that increases
in fracture surface area greater than a factor of two could be generated from surface roughness effects in these type of composites. It
will be shown later, that increases in toughness much greater than
this can occur when a crack front intersects an array of impenetrable
particles.

The behaviour of a crack front in the vicinity of a particle will
depend on the strain fields associated with the particles. Two different types of particle strain fields will be discussed in this section. The exact solution to the crack-path problem near second-phase
particles has not been solved but some simpler fracture mechanical and
stress-field problems have been analysed and these indicate the general nature of the interaction and the structural parameters that
control the process.

2.1 Self-Stresses

Even in the absence of an applied stress, particulate composites can contain stresses associated with the particles. These self-stresses can be caused by coherency between the particle and the matrix, by difference in coefficient of thermal expansion between the particle and the matrix or by a phase transformation during some earlier heat treatment. In the absence of a crack, the magnitude and spatial variation of the strain field can be calculated from elasticity theory [2,3]. It was found that, provided neither the shape of the particle nor its transformation strain depends on the size of the particle, the maximum stress in the neighbourhood of any particle is independent of its size [4]. For a simple 'mis-fitting' sphere model the maximum tensile stress occurs at the particle-matrix interface. For example, in a system where the matrix shrinks more than the particles, there is a tensile radial stress at the interface of a rigid particle of $\sim -4G_m\delta$, where G_m is the shear modulus of the matrix and δ is the mismatch strain. Even for small values of δ , these are large stresses that can interact with nearby defects to nucleate a crack and in some cases may approach the theoretical cleavage stress. In brittle composites, these stresses are not easily relaxed and are often large enough to fracture either the matrix, the particle or the interface during fabrication [5,6]. For particles less than a certain critical size, however, spontaneous cracking does not occur [5]. This critical size effect was proposed to be a result of the elastic stored energy of a misfitted particle being proportional to its volume, while the crack resistance force is related to its area [5]. This concept is somewhat analogous to the loss of coherency process in precipitation systems.

The stresses round a misfitted particle fall off rapidly with distance from the particle but in particulate composites where high volume fractions are often used, the average stresses in the system can still be large. For example, a volume fraction V_V of particles in a matrix cooled as before, results in an average tensile radial stress of roughly $-4G_mV_V\delta$. Taking $V_V = 0.25$ and $\delta = -0.02$, an average stress value of $G_m/50$ is obtained, which is still higher than the fracture stress of most brittle, particulate composites.

It is important to consider the interaction of a crack front with the self-stresses round a particle, with respect to crack trajectory. In terms of the prior stress field, high-expansion particles (positive δ) give rise to a tangential tensile stress field round the particle, while low-expansion particles give rise to a tensile radial stress field. From these considerations, it would be expected that the crack should avoid high-expansion inclusions and intersect low-expansion inclusions. This type of behaviour has been observed by Davidge and Green [5] and by Binns [6].

In order to evaluate the influence of these self-stresses on the local stress intensity factor, the effect of the stresses in loading the crack must be considered. For example, for a long Mode I crack situated along the equator of a well-bonded, spherical inclusion with a misfit strain δ (inset Fig. 2) can be simply treated using linear fracture mechanics[7]. The maximum interaction will occur on the x-y plane that passes through the centre of the sphere. The change in stress intensity factor (ΔK_s) is given by [7]

$$\Delta K_s = \frac{-3D\delta}{16}\left[\frac{r}{t}\right]^{5/2} (2\pi r)^{1/2} \tag{1}$$

where D is a constant that depends on the elastic properties of the matrix and the inclusion. The results of this calculation are shown in Figure 2 for a rigid inclusion ($D=4G_m$) with different values of $\delta r^{1/2}$. It can be seen that high values of ΔK_s can be attained even in the absence of any applied stress. For a given value of r/t, the effect of increasing δ or r can be significant, and in some cases exceed typical values of K_{IC} for brittle materials. For this crack configuration $K_{II} = 0$, so that the crack can be either attracted or repelled by the inclusion, depending on whether δ is negative or positive. This effect should lead to a local change in crack shape as the crack approaches the particle. For the case of an arbitrary-located crack, the situation is rather more complex. The value of K_{II} is no longer zero so that a shear stress component is now available to change the crack trajectory and the sign of K_I no longer depends on δ but also on crack location. The use of transformation-induced self-stresses as a means to toughen ceramic materials has recently discussed by Evans et al [8].

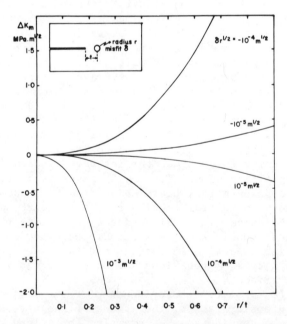

Figure 2. Change in stress intensity factor on equator of a misfitted, spherical rigid inclusion ($G_m = 30$ GPa).

2.2 Elastic Modulus Interactions

In the absence of self-stresses, once a body containing second-phase particles is stressed, a second kind of local stress field is set up that depends on the difference in elastic properties of the particle and the matrix[3,9,10]. Considering again a long Mode I crack approaching a well-bonded, spherical inclusion along its equator. The interaction between the two stress fields can be studied by considering the stressing of the particle by the crack front and the back-influence of the particle stress field on the crack[7]. The situation is similar to that considered in the last section. The change in stress intensity factor (ΔK_m) from the unperturbed value $(K_I{}^o)$ is given by [7]

$$\Delta K_m = K_{\bar{I}}{}^o \left\{ -0.375\left[\frac{r}{t}\right]^3 \left[A + 2\left(\frac{2\upsilon_m}{1 - 2\upsilon_m}\right)C\right] -0.82B\left[\frac{r}{t}\right]^5 \right\} \qquad (2)$$

where υ_m is the Poisson's Ratio of the matrix and A, B and C are constants that depend on the elastic properties of the matrix and the inclusion[7]. The results of the above calculation are shown in Figure 3 for the extreme cases of a rigid inclusion and a spherical hole and it can be seen that the maximum values of $\Delta K_m = K_I' - K_I{}^o$ are less than $\sim25\%$ $K_I{}^o$, even for these extreme cases. The results of equation (2) indicate that the crack will be attracted by soft obstacles and

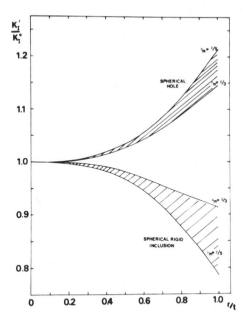

Figure 3. Elastic modulus interaction for a long, mode I crack on the equator of a spherical hole or rigid inclusion.

repelled by hard. This problem has been considered by several other authors[11-15] for the case of a cylindrical inclusion that is lying with its cylindrical axis parallel to the crack front. These approaches were mainly concerned with the effect of small cracks in the vicinity of the inclusion, which is useful in considerations of crack initiation. For this case much higher values of ΔK_m can be attained, particularly at the crack tip nearest the inclusion.

The problem again becomes more complex for an arbitrary-located crack front, where $K_{II} \neq 0$. This problem has been studied by Erdogan [15], who found numerical values of K_I and K_{II} for a crack near a cylindrical inclusion. Fortunately, the results of this work can be understood in terms of the prior stress distribution around the inclusion, i.e., the influence of the shear stress component will be such that the crack will tend to move towards the position of maximum stress concentration in the prior stress field. In other words, the crack has a tendency to avoid hard particles and be attracted to soft. This interaction should also be reflected as a local change in crack shape for spherical obstacles.

Experimental proof of local changes in crack trajectory in the vicinity of holes has been discussed by Hancock[16] for a degraded polystyrene and by Wagoner and Hirth[17] for an aluminum alloy. Indirect evidence for these elastic modulus interactions can often be obtained from fractographic observations. For example, in their study of glass composites, Stett and Fulrath[18] noted that for non-bonded particles or spherical holes, the crack intersected the inclusions, while for well-bonded, hard particles, the crack usually remained in the matrix. The local change in crack shape for a crack approaching a spherical hole, expected from these modulus strain field considerations, has been observed by Green et al[19].

2.3 Other Effects

In the discussion of the local strain-field interactions, it was assumed that the particles were well-bonded to the matrix. In many brittle, particulate systems, these interfaces may be weak and may fail either prior to or during the crack-bypassing process. In these cases the particles may completely debond and will then act more like spherical holes. The strength of the interface will be dependent on whether the particles are a result of a precipitation or crystallization process or were simply added as a filler prior to fabrication. For the former case, the interfacial strength will depend on the coherency of the interface, while for the latter it will depend on the nature and extent of the chemical reaction that occured at the interface during fabrication.

The change in crack trajectory considered in the last two sections assumed that the presence of a shear stress on the crack plane would be sufficient to change the crack direction. In most systems, this appears not to be so and a critical value of K_{II}/K_I is needed

before the trajectory will change[17,20]. For example, in systems
where the matrix has preferred cleavage dirctions, changes in crack
path may be difficult and in some cases may not happen at all, making
obstacle intersection inevitable. Clearly, more work is needed in the
understanding of this problem. Currently, these theoretical considerations indicate the nature of the problem but often only fractographic observation clearly indicate whether obstacle intersection
did or did not occur.

3. THE MICROSCOPIC CRACK BOWING STRESS

For crack-particle interactions to be a useful toughening mechanism, it is important to assess the magnitude of the local stress
needed to change the crack shape as it bypasses the obstacles on a
microscopic level. The unit step in this process will be the bowing
of the crack between a pair of obstacles. In practice, a segment of
crack front, bowing between one pair of obstacles will be influenced
by the neighbouring segments, so that this effect must also be included in the calculations. It will be assumed in this discussion that
the crack front intersects the particles. The maximum interaction will
occur in systems where the particles can be considered impenetrable
($\phi \approx 0$), in these cases the stress needed to circumvent the obstacles
will be independent of their character.

As the stress is increased on a brittle, particulate composite,
a well-developed crack will move forward and press against the particles that lie in the crack plane. For the obstacles to be impenetrable, the increase stored elastic energy for particle penetration must
outweigh the increase associated with the crack bowing round the
particles; the energy associated with particle penetration will be
related to the atomic force-displacement curves, and hence the fracture toughness, of the particles and the matrix. For tough particles,
an equivalent statement would be to say that the particle exerts a
strong, short-range repulsive force on the crack that prevents it
from penetrating the obstacle. Crack propagation will occur when a
stress is reached, at which the crack front can bow between the obstacles and bypass them. It should be noted, however, that when a
particle is circumvented by crack bowing, the obstacles will still
resist crack opening, as they are left as ligaments behind the crack
front. Unfortunately, there has only been two attempts at calculating
the stress needed to circumvent a pair of impenetrable obstacles, both
of which involve several assumptions.

3.1 The Line Tension Analogy

Lange[1] postulated that a crack front could have a line tension,
in analogy to dislocation theory. Based on further assumptions that
the particles were dimensionless points and that the crack shape at
breakaway is semi-circular ($\phi = 0$), Lange was able to derive the

the following expression for the composite fracture surface energy (Γ_c)

$$\Gamma_c = \Gamma_o + \frac{T}{2C} \tag{3}$$

where Γ_o is the fracture surface energy of the matrix, T is the crack line tension and 2C is the interobstacle spacing. An inverse relationship between Γ_c and C has been observed in several fracture studies of brittle, particulate composites, especially at low volume fractions, (for a list of these studies, see references 21 and 22). Unfortuneately, it was found that T was not a constant but depended on particle size and indeed the concept of a crack line tension is rather tenuous. For example, a crack front does not have the same radial stress distribution as a dislocation and for a crack, this cannot be integrated to produce a unique line tension.

3.2 Motion of Secondary Cracks

Evans[23] has discussed the bowing of a crack front between a pair of impenetrable obstacles from a fracture mechanical viewpoint, that compares the stress to move semi-circular or semi-elliptical cracks σ_E between two obstacles to the stress to move a straight primary crack (σ_c). For systems where the primary crack length is much larger than 2C, values of $\sigma_E/\sigma_c > 1$ were obtained, indicating that for these systems increases in Γ_c could be obtained as

$$\left[\frac{\sigma_E}{\sigma_c}\right]^2 = m\left[\frac{\Gamma_c}{\Gamma_o}\right] = \left[\frac{K_{IC}{}^c}{K_{IC}{}^o}\right]^2 \tag{4}$$

where m is the ratio of the Young's Moduli of the composite and the matrix (m = E_c/E_o). Using a combination of numerical and analytical solutions, Evans was able to show that the ratio of (σ_E/σ_c) is a complex function of r_o/C, where $2r_o$ is the length of the obstacle in the crack direction. It should be noted that the value of m influences the Γ_c/Γ_o ratio. The strength of the obstacles in this model was measured by a parameter r, rather than φ, where r is the distance between the primary crack front and the origin of the secondary crack, as shown in Figure 4. For the obstacles to be impenetrable, Evans assumed that the secondary crack will break away from the obstacles when r = $2r_o$ and that 2C does not vary during the bypassing process. Figure 4 also shows the (σ_E/σ_c) ratio from a recent numerical recalculation of the Evans' approach[22]. This recalculation differs from Evans, in that higher values of σ_E/σ_c were obtained and for r_o/C = 0, it was shown that $\sigma_E/\sigma_c \simeq 1$; a result that would be expected intuitively.

A major drawback of this approach was that the secondary cracks were assumed to have a particular shape. For the case of semi-

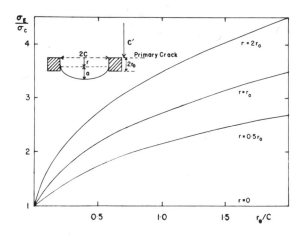

Figure 4. Magnitude of σ_E/σ_c as a function of r_0/C for impenetrable
obstacles ($C' \gg C$)

elliptical, secondary cracks, however, this approach did allow the
crack shape at breakaway to be predicted ($a/2C*$), where a and C are
the minor and major axes of the semi-ellipse. The values of $a/2C*$ are
shown in Figure 5 from a recent analysis[22]. This analysis differs
from that of Evans[23], in that at high volume fractions (large r_0/C),
the crack shape tends to be semi-circular, while at low r_0/C, the
crack front is predicted to remain straight, again behaviour that is
expected intuitively for the growth of a semi-elliptical crack.

3.3 Refinements to Evans' Model

In the last section, it was assumed that crack breakaway occurs
when $r = 2r_0$. In these crack-particle interactions, however, there
are two effects that generally reduce r to a lower value, even for
impenetrable obstacles, and hence reduce the critical stress. The
first effect is a result of the variation in interparticle spacing
during the bypassing process and the second effect is due to the
interaction between neighbouring crack segments.

The variation in interparticle spacing depends in detail on the
particle shape. For circular obstacles, the interparticle spacing
varies from a minimum value at $r = r_0$ to maximum values at $r = 0$ and
$r = 2r_0$. Figure 4 can be used to study the influence of this variation
on σ_E/σ_c and the results are shown in Figure 6 for $r_0/C = 1$. The maximum value occurs at a value of r that is less than $2r_0$ ($\sim 1.6r_0$),
which reduces the value of σ_E/σ_c by $\sim 15\%$. The effect of the variation
in 2C is large at large r_0/C and can reduce the critical value of r
to $\sim r_0$. At low r_0/C, the effect is less important and $r \approx 2r_0$.

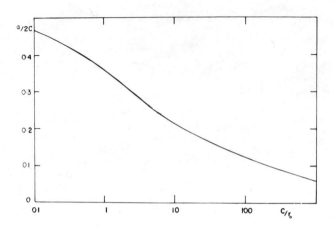

Figure 5.   Variation of crack shape at breakaway for impenetrable
            obstacles

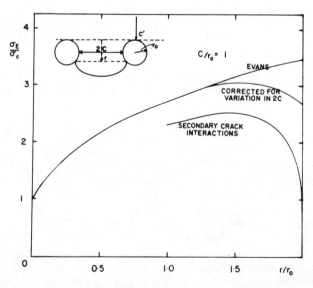

Figure 6. Effect of variation in interparticle spacing and crack in-
          teraction on the strength of circular impenetrable obstacles

The interaction between neighbouring crack segments has been considered by Evans[23], assuming that the obstacle width does not change during the bypassing process. For circular obstacles, however this width (2W) varies from 0 at $r = 0$ and $r = 2r_0$ to a maximum of $2W = 2r_0$ at $r = r_0$. The interaction between the neighbouring segments can be studied if it is assumed that they are coplanar. The value of σ_E is reduced by these interactions to σ_E', which can be determined from the solution for a linear array of equally-spaced, coplanar cracks (length 2C) [24], i.e.,

$$\frac{\sigma_E'}{\sigma_E} = \left[\frac{\pi}{2(1 + W/C)}\right]^{1/2} \left(\tan\left[\frac{\pi}{2(1 + W/C)}\right]\right)^{-1/2} \qquad (5)$$

The values of σ_E/σ_C' as a function of r for circular obstacles are also shown in Figure 6 for $r_0/C = 1$. The critical value of r is reduced even further, to roughly $r = 1.4r_0$. This second refinement appears to have a substantial effect on the critical stress especially at low values of r_0/C. It should be noted, however, that the assumption that the neighbouring segments are coplanar is open to question. Fracture surface steps are usually formed in these composites as the neighboring segments break away from the particles. This second refinement should, therefore be regarded as the maximum effect of these interactions. Finally, it should be noted that Figure 4 is also useful for determining the critical stress for penetrable obstacles, provided the critcal value of r is known or can be determined.

The Evans' model is an approximate approach to these crack-particle interactions, but the above analysis shows that it is a powerful advance in understanding the bypassing process until the complex, three-dimensional problem is solved more accurately.

3.4 Other Contributions to the Critical Stress

There are theoretical reasons for supposing that tough particles contribute to the fracture toughness of particulate composites in other ways than than that considered in the last two sections. First, there is the influence of the particle strain fields (Sections 2.1 and 2.2), that can influence the critical stress in two distinct ways. The first effect is a short-range interaction between the particle and the crack front. For situations where the interaction is repulsive, the crack will stand-off from the surface of the particle, increasing the effective value of r_0. The second effect is a long-range interaction, that may effect the local elastic energy of a bowing segment. From the discussion in Section 2, it would appear that self-stresses could be important in both these effects. The elastic modulus interaction may be important, however, once the crack front reaches the interface. For example, it has been shown by Erdogan and Gupta[25] that soft particles tend to reduce the crack opening displacement, while hard particles will increase this parameter.

Secondly, it must be remembered that other fracture mechanisms may be operating in these composites as well as crack bowing. For example, impenetrable obstacles will still be left behind the crack front even after the completion of crack bowing. A similar problem may also be possible with the fracture surface steps. The effect of these ligament zones has been estimated by Evans et al[8].

4. THE MACROSCOPIC CRACK BOWING STRESS

The calculations of Sections 3.2 and 3.3 were concerned with the local stress to bow a crack front between two obstacles but the value of the applied stress should be different for general bypassing to occur.

4.1 Stress Concentrations

Stress concentrations cause the local stress to differ from the applied stress. Even if it is assumed that holes, notches and other large inhomogeneities are absent, there are still the strain fields associated with the particles. For the case of self-stresses (Section 2.1), the average stress on a fracture plane could be quite different from the applied stress.

4.2 Particle Distribution

If all the particles had the same size, spacing, shape and strength, the composite would behave in the way calculated. In all real composites, however, a distribution of these parameters will exist so that the critical stress will be associated with some mean value. Consider a random distribution of equiaxed particles (uniform size $2r_o$), the problem is to calculate the effective obstacle spacing. This problem has been considered extensively for dislocation-particle interactions[4], but not for crack-particle interactions. The number of particles that a crack front touches will depend on the crack shape and the obstacle strength. In the dislocation problem, Kocks[26] has suggested that a useful approximation for strong obstacles is given by

$$2C + 2r_o = N_A^{-1/2} = \left(\frac{2\pi}{3V_v}\right)^{1/2} r_o \tag{6}$$

where N_A is the number of particles per unit area of slip plane. This equation, which may be useful for crack particle interactions, is really a volumetric particle centre-to-centre distance so that the value of r_o/C will depend on how the crack intersects the obstacles. For example, if the crack intersects the obstacles on a random plane, the correction for the finite size of the obstacle ($2r_o$) is sometimes replaced by $\pi r_0/2$. A similar correction should also be applied to 2C. In fracture studies of brittle particulate composites, workers have often used the mean free path to describe the effective obstacle spacing, which is given by[27]

$$\frac{C}{r_0} = \frac{2(1 - V_v)}{3V_v} \tag{7}$$

This equation can only apply if the crack front remains straight (as r_0/C tends to zero). The situation would become even more complex if the variation in particle size, shape and strength were also considered and there is clearly a need for some detailed calculations in this area. Deviations from randomness will also be important in determining these distribution parameters. For clustered distributions, the particle size and spacing are increased, while for uniform distributions only the spacing will be affected.

5. EXPERIMENTAL ASPECTS

Experimental evidence is important in substantiating the predictions of the previous theoretical calculations. Fractographic observations clearly indicate whether the crack intersects the particles or not, so that in well-defined systems, the influence of the particles on crack trajectory may be assessed. The calculations in Sections 3.2 and 3.3 allowed the value of the critical stress to be predicted as a function of r_0/C, which can then be related to the increase in composite toughness (Γ_c/Γ_0). The value of Γ_c can be measured experimentally and then compared to this theory. This has been done for several model particulate systems but the results of these studies are often ambiguous. For example, Figure 7 shows the value of σ_E/σ_c' predicted by a refinement of the Evans approach for spherical obstacles[28], in comparison with the experimental data of Lange[29]. If the effective obstacle spacing is taken as the mean free path, the comparison is good. If equation (6) is used, there is poor agreement. It is often not clear from these experiments whether the particles failed or not during crack bowing. The situation is futher complicated when it is realized that other fracture mechanisms may be operating in these composites. For example, the particles may act as preferential sites for matrix microcracking or the particles may themselves crack or debond. It is important in studying model composite systems to be able to distinguish the major energy-absorbing mechanisms, some of which have been reviewed recently[8,30].

Many of these experimental problems may, however, be resolved using the technique of ultrasonic fractography. This technique has been described elsewhere[19,22,31], but it basically allows successive positions of the crack front to be imprinted on the fracture surface at microsecond intervals. Ultrasonic fractography allows quantitative measurements to be made of crack shape and obstacle strength, which can then be compared with the theoretical predictions. Local changes

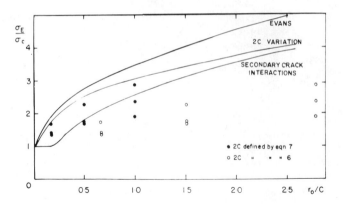

Figure 7 Comparison of experimental data with a refined version of
         the Evans' model for different definitions of the effective
         obstacle spacing

in crack velocity can also be determined by this technique. In this
way, ultrasonic fractography could be a powerful tool for studying
different particulate systems. Figure 8 shows an example of an ultrasonic fractograph for a crack front interacting with a pore in glass
[19]. The attraction to the hole during crack approach and the subsequent impedance to crack motion are evident.

     This paper has not discussed in detail the origin of particle
impenetrability, which is an important area. In this respect, Evans
and Graham have shown that as $a/2C$ increases, the stress intensity
factor at the obstacle also increases[32]. In terms of the earlier calculations this implies that particle penetration is more likely to
occur as r_0/C increases (high V_V). The nature of the interface will
also be important, though it has been shown that a crack interacting
with an array of non-bonded obstacles or pores can still increase
toughness due to local crack blunting[22]. This effect is not as large
as the interaction with impenetrable obstacles. Finally, it should be
noted that Lange[21] has reviewed the fracture properties of brittle,
particulate composites in more general terms.

6. CONCLUSIONS

Brittle materials can be made more resistant to fracture by dispersing
particles of a second phase in them. In these particulate composites,
the particles clearly disturb the fracture process and can lead to
local changes in the crack shape. The impedance to crack motion by
particles has been proposed as a means to toughen ceramics. The microscopic stress to bend a crack between two obstacles has been calculated

Figure 8. Ultrasonic fractograph of a crack front interacting with
         a spherical pore in glass

approximately in terms of the particle size and spacing and refined
to account for particle shape and crack interactions. Moderate increases in toughness could be derived by this mechanism, especially
in situations where the particles can be considered impenetrable.
The spatial distribution of the particles and their strain fields will
influence the macroscopic stress needed for crack propagation. The
geometry of these crack-particle interactions is similar to its dislocation analogue, while there is a difference in the physics. Experimentally, the technique of ultrasonic fractography will be
extremely useful in detailed observations of crack shape and particle
strength and further work could improve the understanding of this
crack-bowing mechanism.

7. REFERENCES

1. F. F. Lange, Phil. Mag., 22, 983 (1970).
2. N. F. Mott and F. R. N. Nabarro, Proc. Phys. Soc., 52, 86 (1940).
3. J. D. Eshelby, Progress in Solid Mechanics, (eds. I. N. Sneddon
    and R. Hill), Volume 2, Chapter 3. Wiley, New York 1961.

4. L. M. Brown and R. K. Ham, Strengthening Methods in Crystals, (eds. A. Kelly and R. B. Nicholson, Chapter 2. Wiley, New York 1971.
5. R. W. Davidge and T. J. Green, J. Mater. Sci., $\underline{3}$, 620 (1968).
6. D. B. Binns, Science of Ceramics, (ed. G. H. Stewart), p315. Academic Press, London 1962.
7. A. K. Khaund, V. D. Kristic and P. S. Nicholson, Submitted to J. Mater. Sci., February 1977.
8. A. G. Evans, A. H. Heuer and D. L. Porter, Proc. 4th Intl. Conf. on Fracture, (ed. D. M. R. Taplin), Volume 1, p529. Univ. of Waterloo Press, Waterloo 1977.
9. J. N. Goodier, J. Appl. Mech., $\underline{1}$, 39 (1933).
10. R. V. Southwell and H. J. Gough, Phil. Mag., $\underline{1}$, 71 (1926).
11. O. Tamate, Intl. J. Frac. Mech., $\underline{4}$, 257 (1968).
12. C. Atkinson., Scripta Met., $\underline{5}$, 643 (1971).
13. D. J. Cartwright and D. P. Rooke., Eng. Frac. Mech., $\underline{6}$, 563 (1974).
14. J. Tirosh and A. S. Tetelman, Intl. J. Frac. Mech., $\underline{12}$, 187 (1976).
15. F. Erdogan., Fracture Mechanics of Ceramics, (eds. R. C. Bradt, D. P. H. Hasselman and F. F. Lange), Volume 1, p245. Plenum Press, New York 1974.
16. J. Hancock, J. Mater. Sci., $\underline{10}$, 1230 (1975).
17. R. H. Wagoner and J. P. Hirth, Metall. Trans., $\underline{7A}$, 661 (1976).
18. M. A. Stett and R. M. Fulrath, J. Am. Ceram. Soc., $\underline{53}$, 5 (1970).
19. D. J. Green, J. D. Embury and P. S. Nicholson, J. Mater. Sci., $\underline{12}$, 987 (1977).
20. E. Sommer, Eng. Frac. Mech., $\underline{1}$, 539 (1969).
21. F. F. Lange, Fracture and Fatigue of Composites (eds. L. J. Broutman and R. H. Krock), Chapter 1. Academic Press, New York 1973.
22. D. J. Green, Crack-Particle Interactions in Brittle Composites, Ph. D. Thesis, McMaster University 1977.
23. A. G. Evans, Phil. Mag., $\underline{26}$, 1327 (1972).
24. P. Paris and G. C. Sih., ASTM Spec. Tech. Publ. 381, 30 (1965).
25. F. Erdogan and G. D. Gupta, Intl. J. Frac., $\underline{11}$, 13 (1975).
26. U. F. Kocks, Acta Met., $\underline{14}$, 1629 (1966).
27. R. L. Fullman, Trans. AIME, $\underline{197}$, 447 (1953).
28. D. J. Green, Unpublished Results.
29. F. F. Lange, J. Am. Ceram. Soc., $\underline{54}$, 614 (1971).
30. A. G. Evans and T. G. Langdon, Progress in Mater. Sci., $\underline{19}$, (1976).
31. D. J. Green, J. D. Embury and P. S. Nicholson, Ceramic Microstructures '76, (eds. R. M. Fulrath and J. A. Pask), In Press 1977.
32. A. G. Evans and L. J. Graham, Acta Met., $\underline{23}$, 1303 (1975).

SOME EFFECTS OF DISPERSED PHASES ON THE FRACTURE BEHAVIOR OF GLASS

J. S. Nadeau, R.C. Bennett

Department of Metallurgy
University of British Columbia
2075 Wesbrook Place
Vancouver, B.C. Canada
V6T 1W5

I. INTRODUCTION

The strength of glass can be increased or decreased by dispersed phases[1]. Since many ceramic bodies consist of a glass matrix in which crystalline phases are embedded, it is important to know the mechanism by which these phases influence the strength. There are basically two mechanisms: 1) by altering the inherent crack size, 2) by impeding or aiding crack propagation. These follow directly from the Griffith-Irwin equation which has in it a term for crack size and another for "fracture energy". Because these two ways of altering the strength of glass exist, two theories for the effects of dispersions including porosity, on glass strength have developed.

The first, due to Hasselman and Fulrath[2], is flaw-oriented. It suggests that the spacing between particles of the dispersed phase controls the inherent flaw size. The second theory, proposed by Lange[3,4] and elaborated by Evans[5], attributes the strengthening of some ceramics by a dispersed phase to the resistance to crack growth resulting from direct interaction between the crack and the particles. There is experimental evidence in support of both theories and of course it is quite possible that both are correct in different circumstances. There are a number of experimental results that are better explained by one theory than the other and a few results that cause difficulties for both.

The principal objection to the Hasselman-Fulrath theory is that it allows for no effect of fracture energy even though there is considerable evidence that it is changed by dispersions in glass.

961

Lange[4] observed an increase in the fracture energy by more than a
factor of three in a 25% crystal-glass composite. Hing and
McMillan[6] and later, A.S. Rao[7], observed a similar increase in the
fracture energy of a glass-ceramic over that of the parent glass.
Rao, observed that the strength and fracture energy increased in
parallel as the proportion of crystalline phase increased. The
flaw size remained constant.

Recently, Borom[8] compared the temperature dependence of the
strength of two glass ceramics with that of the parent glass. At
room temperature, one glass-ceramic was about 2.3 times as strong
as the matrix glass. The strength of the glass-ceramic decreased
rapidly with temperature while that of the matrix-glass increased
slightly. At 400°C they had approximately the same strength. It
is difficult to account for this behavior in terms of the
Hasselman-Fulrath theory. If the dispersed phase increases the
strength by limiting the flaw size at room temperature, why is it
less able to do so at elevated temperatures? Borom argued that
part of the strengthening effect of the dispersed phases was due to
thermal expansion mismatch with the matrix. In other words, the
fracture energy was increased by internal stresses.

The Lange-Evans theory has difficulty accounting for the
numerous examples of weakening by dispersed phases in glass. In
many cases the weakening is simply a manifestation of cracking due
to unequal thermal expansion between matrix and dispersed phase,
thus to an increase in flaw size. However Frey and McKenzie[9] observe
a lower strength in a well bonded 25% Z_rO_2-glass composite than for
the matrix-glass. Hasselman and Fulrath[2] observed weakening of
glass by up to 25 percent volume of Al_2O_3 particles and Davidge
and Green[10] observed weakening of glass by well bonded thoria spheres. It is possible that these observations can be explained by
the presence of stresses around the particles which aid crack
growth and in effect, lower the fracture energy.

In the present work, the interaction between a single, growing
crack and pores and second phase particles was examined. The
strength of the interaction was indicated directly by the force
needed to drive the crack. The results help to clarify the mechanisms by which dispersed phases influence the strength of glass.

II. EXPERIMENTAL PROCEDURE

The tests were done on commercial soda-lime float-glass plate.
Specimens were 22.8 cm x 7.6 cm x 0.3 cm. An artificial microstructure was produced on one surface by cutting with a diamond saw
or drilling shallow holes. The plate was tested in the double
torsion configuration as shown in Fig. 1. An advantage of this

test method was that it was only necessary to prepare one surface
of the specimen. The crack emerged at right angles to the surface
with the microstructure and then curved back along the bottom surface as described elsewhere.

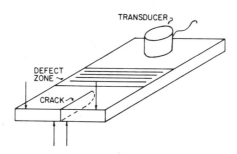

Fig. 1 Double torsion specimen. Grooves and holes were machined in
        the defect zone. Acoustic emission transducer was used in
        all experiments but results will be reported separately.

The artificial microstructures consisted of an array of grooves
or holes. The grooves were 0.15 or 0.3 mm wide and ranged in depth
from 0.05 mm to 1.25 mm. The holes were 0.5 mm diameter and 0.5 mm
deep. They were drilled on a staggered pattern to give an area
fraction of about 20%.

Low melting glasses were fused into the grooves or holes of
some specimens. This gave a well-bonded second phase having a
thermal contraction coefficient different from that of the matrix.
Thus, at room temperature there were residual stresses in these
specimens.

The two low-melting glasses were "sealant glasses" produced by
Owens-Illinois. Grades 00130-A and 00158-A having thermal con-
traction coefficients of $60 \times 10^{-7}°C^{-1}$ and $102 \times 10^{-7}°C^{-1}$ respectively were used. They were both fired at 610°C while the plate
containing them rested on a bed of fine alumina powder in a thick
aluminum box. Two fillings and firings were needed to fill the
grooves or holes. The plate had a thermal contraction coefficient
of about $85 \times 10^{-7}°C^{-1}$.

The specimens were pre-cracked and then loaded in an Instron
machine. The crack was driven through the microstructure using a
constant cross-head speed. In a smooth glass plate this type of
test causes the crack to grow at a constant velocity[12]. In the
present tests however, the crack was frequently arrested by the
microstructure. The crack velocity was visually recorded and was

typically about 10^{-4} m s^{-1}.

III. RESULTS

Unfilled Grooves

The results of tests on unfilled grooves will be described first. Most of these grooves were 0.3 mm wide. Typical loading curves for grooves of this width and various depths are shown in Fig. 2. For the deep grooves the crack was arrested and then the load increased until breakaway occurred causing a load drop. So much energy was stored during the arrest period for a deep groove that it jumped quickly to the next arrest point and a saw-tooth loading curve was produced as in Fig. 2a. For shallower grooves, the breakaway was not so violent and the crack slowed to a normal velocity after leaving the groove. As it approached the next groove, its velocity increased and the load dropped. This is seen in the two lower curves of Fig. 2. The arrest points are marked by arrows. The arrest position was clearly delineated by a hackle mark on the fracture surface as shown in Fig. 3. In this position the crack front had curved well out of the original crack plane. The crack had in effect, been blunted. Successive positions of the crack as it approached the arrest position are revealed by minor hackle marks in Fig. 3b.

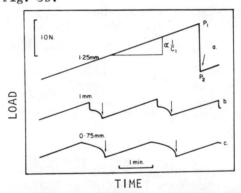

Fig. 2   Typical loading curves for grooved specimens. Crack arrest points are indicated by arrows. Breakaway occurs at highest load. Figures on curves indicate depth of grooves.

The effectiveness of a groove in arresting a crack increased with the depth of the groove. However the apparent stress intensity at breakaway K_{ba}, for grooves < 1 mm deep, was about the same as the equilibrium stress intensity for an ungrooved specimen being loaded at the same cross-head speed. Thus these specimens were neither strengthened nor weakened (in the spirit of this particular test) by the presence of the grooves. In a phenomenological sense, the drop in load as the crack approached a groove was compensated by a rise in load during the period of arrest.

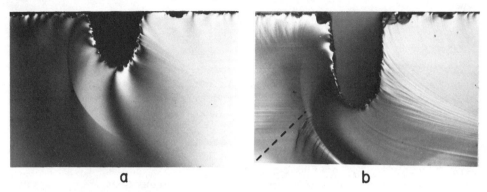

a   b

Fig. 3   The arrest position on two grooves of different depths.
         a) 0.75 mm deep b) 1.25 mm deep. Broken lines indicate
         position of breakaway crack.

For grooves deeper than about 1 mm, the breakaway event was more complex and more violent. The stress intensity at breakaway was higher than the equilibrium stress intensity for an ungrooved specimen as shown in Fig. 4. Breakaway was preceded by the formation of a wedge-shaped crack at the base of the plate as indicated by broken lines in Fig. 3b. Interaction between this new crack and the main crack in its arrested position, produced chevron shaped marks on the fracture surface, (Fig. 3b). For very deep grooves considerable shattering was produced around the arrest position.

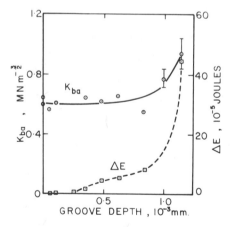

Fig. 4   Breakaway stress intensity and energy, versus depth of
         unfilled single grooves.

The arrest and breakaway sequence was accompanied by the storage and release of elastic energy in the specimen. The energy, ΔE, associated with one event was computed from the loading curve using the relation:

$$\Delta E = \frac{1}{2} P_1 C_1 (P_1 - P_2)$$ (1)

where P is the load and C is the specimen compliance as shown on Fig. 2a. The value of this energy is not a fundamental characteristic of the breakaway event. It depends very much on the spacing of the grooves as shown in Fig. 5. However, when the groove spacing is kept constant and the depth is varied, the breakaway energy reveals the threshold value of groove depth to produce crack arrest. This is shown in Fig. 4. Furthermore the breakaway energy is a sensitive indicator of the change in breakaway mode at large groove depths, which is also shown in Fig. 4. The breakaway energy appears to be correlated with the amount of acoustic emission at breakaway and this will be reported in a separate paper.

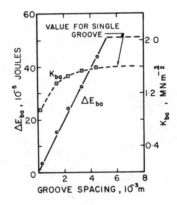

Fig. 5   Breakaway stress intensity and energy, versus spacing of 1.2 mm deep grooves.

The effectiveness of a groove in arresting a crack was not a unique function of groove depth. In specimens of different thickness, having the same groove depth, the energy stored during arrest was greater the thinner the specimen. The strength of the interaction depended upon the ratio of groove depth, D to remaining thickness λ, as shown in Fig. 6. This reveals a critical value of the ratio D/λ, above which the strength of the plate (again, in the spirit of this test) is increased by the presence of the groove.

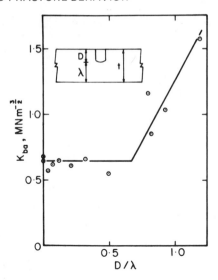

Fig. 6   Breakaway stress intensity versus D/λ   where λ = t − D.

Filled Grooves

The two sealant glasses used to produce specimens with internal stress, were chosen to have thermal contraction coefficients
lower (A-glass) and higher (B-glass) than that of the glass plate.
That this was achieved, was revealed by the birefringence colors
observed around the filled grooves.   They were similar to each
other in spatial distribution but the colors were exactly reversed
as indicated in Fig. 7.   A simple bending experiment revealed that
the first order color for compression was orange and for tension
was blue.   Thus the stress distributions are qualitatively the
expected ones for the intended thermal expansion mismatch of the
composite specimens.

As expected, the difference in stress distribution around the
two types of filled grooves, caused very different interactions
with the advancing crack.   The grooves containing "A-glass" (lower
contraction coefficient than matrix) arrested the crack and required
a very high stress intensity for breakaway.   The grooves containing
"B-glass" attracted the crack and no significant crack arrest
occurred.

The strength of the interactions with a series of grooves spaced
11 mm apart is shown in Fig. 8.   The "A-glass" increased the
strength of the plate by about a factor of two.   With the "B-glass"
the plate was weaker in the vicinity of the grooves than elsewhere.

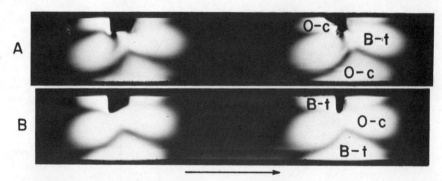

Fig. 7   Patterns of birefringence around grooves containing A-glass
         (lower contraction coefficient than matrix) and B-glass,
         (higher contraction coefficient than matrix). Letters on
         figures indicate colors in polariscope, O-orange, b-blue,
         and stresses, T - tensile and c - compressive.

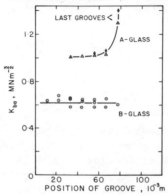

Fig. 8   Breakaway stress intensity in grooves containing A-glass
         (lower contraction coefficient than matrix) and B-glass
         (higher contraction coefficient than matrix. Grooves
         spaced 11 mm apart.

     The details of the interactions are shown in Fig. 9. In the
case of the "A-glass" the crack passed beneath the filled groove
aided by the tensile stress in that region. However it could not
penetrate into the filled groove because of the compressive stress
there. Eventually the crack passed completely around the filled
groove without breaking it. At that point the tensile component
of stress on the groove was greatly increased because it had been
nearly severed from the matrix. It then fractured, with the crack
passing through it in the direction opposite to the direction of
propagation of the general crack front.

     The grooves containing "B-glass" offered no resistance to crack
advance and the crack easily severed the filled groove due to the

tensile stress there. This is shown in Fig. 9b. The compressive
stress beneath the filament did not retard the crack because it
was relaxed by the severing of the filled groove.

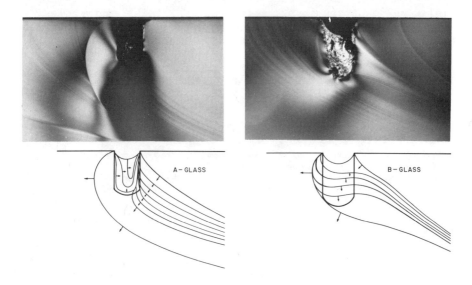

Fig. 9   Optical micrographs and schematic drawings of interactions
         of crack with grooves containing A-glass and B-glass (lower
         and higher contraction coefficient than matrix respectively)
         Arrows indicate direction of crack motion.

The Interaction with Holes

A reasonable objection to the experiments on grooved specimens
is that the structure resembles that of fiber reinforced composites
but not of dispersion strengthened ceramics. The crack cannot
avoid going through the grooves. However specimens with an array
of shallow holes drilled in the surface are a good two-dimensional
simulation of a conventional microstructure. These specimens are
more difficult than grooved ones to produce and thus only a limited
amount of testing has been done on them. A detailed quantitative
report on the behaviour of the drilled specimens must await further
testing. However the results to date indicate that the interactions
between shallow drilled holes and the advancing crack are qualitatively the same as for the grooves. Holes filled with "A-glass"
interact strongly and strengthen the specimen. Empty holes and
holes filled with "B-Glass" attract the crack and offer little
resistance to its progress. The crack path as it passes through
an array of holes in each type of drilled specimen is shown in
Fig. 10. The holes filled with "A-glass" attract the crack along

a radial direction due to the circumferential tensile stress in the
matrix, (Fig. 10a). (The holes are not all properly filled with
glass and therefore do not all react in the same way.) The holes
filled with "B-glass" attract the crack to the interface,where the
radial tensile stress is highest, (Fig.10b). The empty holes
attract the crack because of the concentrated stress on the diametral plane that is coplanar with the crack, (Fig. 10c).

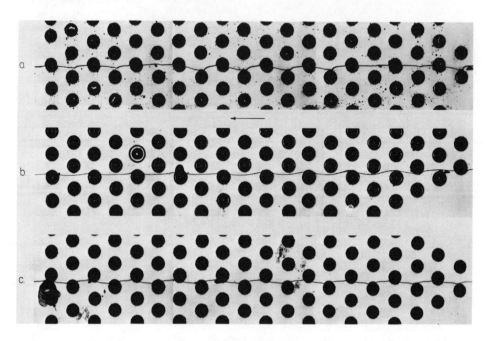

Fig. 10 Crack path through arrays of holes containing a) A-glass
         (lower contraction coefficient than matrix), b) B-glass
         (higher contraction than matrix), c) empty holes. All
         holes 0.5 mm deep and diameter.

IV. DISCUSSION AND CONCLUSIONS

The experiments on empty grooves and holes have helped to
reveal some of the phenomena that could influence the strength of
porous ceramics. They have shown that pores can arrest cracks.
The arrest has two causes both of which would presumably arise in
a real material:

1) The crack is lured forward by the stress concentration
    around the pore. However breakaway is not aided by the
    stress concentration and the crack must wait until the
    load rises again to the breakaway value.

2) The crack curves slightly out of the main crack plane due
to the stress field around the pore. Extra energy is
expended in reorienting the crack and in forming steps on
the fracture surface.

The phenomena reported above have been observed in porous
ceramics by Lange[3] and have recently been examined by the technique
of ultrasonic fractography by Green Nicholson and Embury[13].

There is a critical relation between the pore size and the
length of the pinned crack as revealed by Fig. 6. This suggests
that the spacing of pores along the crack front should be made as
small as possible to make the breakaway stress intensity as high as
possible. Unfortunately the data of Fig. 5 show that the next pore
ahead of the crack must be as far away as possible to avoid aiding
the breakaway event. These two rather incompatible requirements
mean that in a uniform dispersion of pores there is an optimum spacing. For the pores (grooves) used in the present experiments the
best spacing was about equal to the pore size. According to the
Fullman equation this occurs at a volume fraction of porosity of 0.4.
These experiments do not reveal how the size of the pores would
affect the results.

The experiments in which the grooves or holes were filled with
glass of different thermal expansion than the matrix showed that
slowly growing cracks interact strongly with internal stresses. The
stress in the particle appears to be more important than the stress
in the matrix because it is distributed over a much smaller volume
and therefore has higher intensity. Thus strengthening is obtained
when the particle is under compression and weakening when it is
under tension. So far these observations are supported by the
experiments with drilled holes. However there is considerable
evidence in the literature to conflict with this result and further
study is needed to rationalize the differences.

REFERENCES

1. D.B. Binns, Science of Ceramics, Vol. 1 315, G.H. Stewart Ed.
Academic Press, (1962).

2. D.P.H. Hasselman and R.M. Fulrath, J. Amer. Ceram. Soc. 49
68, (1966).

3. F.F. Lange, Phil. Mag., 22 983, (1970).

4. F.F. Lange, J. Amer. Ceram. Soc. 54, 614 (1971).

5.   A.G. Evans, Phil. Mag. 26 1327, (1972).

6.   P. Hing and P.W. McMillan, J. Mater. Sci. 8 1041, (1973).

7.   A.S. Rao, Ph. D. Thesis, University of British Columbia, 1977.

8.   M.P. Borom, J. Amer. Ceram. Soc. 60, 17(1977).

9.   W.J. Frey and J.D. MacKenzie, J. Mater. Sci. 2 124, (1967).

10.   R.W. Davidge and T.J. Green, J. Mater. Sci. 3 629, (1968).

11.   J.S. Nadeau, J. Amer. Ceram. Soc. 57 303 (1974).

12.   A.G. Evans, J. Mater. Sci. 7 1137 (1972).

13.   D.J. Green, P.S. Nicholson and J.D. Embury, Fracture 1977,
      Intl. Cong. on Fracture 4, Waterloo, Canada, 1977, Vol. 3 941.

FRACTURE TOUGHNESS OF REINFORCED GLASSES*

J. C. Swearengen, E. K. Beauchamp, R. J. Eagan

Sandia Laboratories

Albuquerque, New Mexico 87115

ABSTRACT

Fracture toughness (K_{Ic}) values of hot pressed borosilicate glasses containing dispersed Al_2O_3 spheres have been obtained using notched and thermally precracked specimens in three-point-bend tests. We found that K_{Ic} increases linearly with volume fraction of inclusions to a value three times that of the base glass at 40 volume percent alumina. The toughness increases directly with composite elastic modulus; however, the increase is greater than expected from a simple concept of matrix stress reduction due to high-modulus inclusions. Fractography reveals substantial differences in the interaction of the crack front with inclusions, depending on the sign of the thermal expansion mismatch, $\Delta\alpha$, between the crystal and the glass. Nevertheless, the effect of $\Delta\alpha$ on K_{Ic} is small compared with that of the modulus.

INTRODUCTION

In recent years a number of studies have been conducted with the goal of discerning the basic mechanisms of strengthening in glass/particle composites. Some of these efforts concentrated on various glass-ceramics [1-5] while others were based upon "model" systems wherein the properties and relative amounts of each phase can be selected [5-10]. Most of the studies have focused on the relationship between microstructure and fracture strength, but a few addressed fracture energy variations as well [5,9,10].

*This work supported by the United States Energy Research and Development Administration.

Much diversity is to be found in the interpretations given to the foregoing studies, but some consensus does exist. For example, it is clear that composite mechanical properties are affected by the number, size, and shape of the inclusions, as well as by the relative mechanical and physical properties of the inclusion and matrix [6,9, 10]. Stresses induced by the difference in thermal expansion between matrix and inclusion can alter the fracture morphology, and apparently induce microflaws if the difference is large [7-8]. Finally, the strength-controlling flaws are to be found in the glass phase, not within the crystalline phase or the inclusions [1,4,6,8].

The stress necessary for propagation of preexisting flaws in a brittle infinite plate containing an edge crack of length a can be given in terms of fracture surface energy, γ, and Young's modulus, E, by

$$\sigma_f = \left[\frac{2E\gamma}{\pi a} \right]^{\frac{1}{2}} . \tag{1}$$

Alternatively, by considering only the stress concentration at a crack tip, one can predict strength in terms of the fracture toughness, K_{Ic}, [11]. Equivalence of the two formulations yields

$$K_{Ic} = \left[\frac{2E\gamma}{(1 - \nu^2)} \right]^{\frac{1}{2}} \tag{2}$$

where the factor $(1 - \nu^2)$ accounts for plane strain conditions (ν is Poisson's ratio). Thus K_{Ic} can be thought of as depending upon the surface energy and elastic constants of the material, i.e., K_{Ic} is a true material constant.

In this study we attempt to gain further understanding of the mechanisms whereby glasses can be reinforced by a dispersion of alumina particles. We chose borosilicate glasses for the matrix because their thermal expansion coefficient can be readily changed by adjusting the B_2O_3/Na_2O ratio, without inducing major changes in K_{Ic} or E; furthermore, their relatively low viscosity facilitates fabrication of composites by hot pressing. The experimental results are presented in terms of K_{Ic} and E for the composites. Mechanisms of reinforcement are discerned by evaluating the measured variables in light of evidence gained from fracture surface analyses.

EXPERIMENTAL PROCEDURE AND RESULTS

Material Preparation and Characterization. Four sodium borosilicate glasses were formulated to have thermal expansion coefficients approximately equal to, greater, and less than the expansion of Al_2O_3. The glasses contained 75 mole percent SiO_2 with B_2O_3 and Na_2O in molar ratios of 0.2, 0.67, 1.3 and 3.0 B_2O_3/Na_2O. In subsequent discussion, these glasses will be referred to as 75-0.2

glass, 75-0.67 glass, etc. 1 Kg batches of reagent grade powders
were fused in platinum crucibles in an electric furnace, with continuous stirring to ensure homogeneity. The melts were air quenched,
then ground (dry) in an alumina ball mill to -200 mesh.

Spherical polycrystalline Al_2O_3 particles[*], sized to 25 ± 7 μm,
were blended with the powdered glasses to form mixtures containing
nominally 0, 2, 10, 15, 20, 25, and 40 volume percent Al_2O_3. The
powder mixture was poured into graphite dies lined with mica paper,
then hot pressed in a vacuum of ~10 Pa. The mixture was pre-pressed
(6.9 MPa) at room temperature; the pressure was released and the
mold was heated to 550°C, at which time the pressure was reapplied.
The temperature was then increased to 730 ± 10°C and held for 10
minutes after ram travel due to compaction ceased. Pressure was
released after cooling to 675°C. The composites were annealed at
550°C for one hour, then cooled slowly to room temperature. In
each case the glass darkened during processing; the darkening may be
associated with limited reduction during vacuum hot pressing. For
the purpose of this study, the discoloration is unimportant because
all glasses were treated the same way and composite properties were
determined relative to the hot pressed glass.

Final dimensions of the pressed blocks were ~ 57 x 26 x 13 mm.
Rectangular bars 53.9 x 1.27 x 9.5 mm were diamond-sawed from the
blocks for fracture toughness, moduli, density and thermal expansion
measurements. Young's moduli, shear moduli, and Poisson's ratio were
measured ultrasonically, using the pulse-overlap technique. Expansion coefficients (α) of the unreinforced glasses were measured from
20°C to the set point. The difference between α for each glass and
that of Al_2O_3 (78×10^{-7}°C^{-1}) is listed under the glass system
designation in Table I.

Polished cross-sections from typical composites are displayed
in Figure 1. Since final porosity was nearly zero, as evident from
the figure, actual volume percent Al_2O_3 (φ) was calculated from
measured densities of glass and composite using a tabulated value
for Al_2O_3 of 3.987 g/cc.

K_{Ic} Measurements. V-shaped edge notches were cut at the mid-
point of one 53.9 mm edge of the rectangular bars with a 0.20 mm
wide diamond wheel. Sharp cracks were induced by heating a region
just ahead of the notch with a miniature acetylene torch, while
viewing the region through a low-power microscope. With some practice using this technique one can consistently produce planar cracks
having nearly straight fronts. The precracked samples were stored
for about one day in a dessicator until time of test; fracture toughness tests were conducted under three-point bending in an atmosphere

---

[*]Linde Crystal Products Dept., Union Carbide Corp., San Diego, Calif.

Table I
Compilation of Data for Reinforced Glasses

Glass System (α glass – α inclusion x $10^{-7}°C^{-1}$)	φ Volume Percent Al_2O_3	K_{Ic} (MN – m$^{-3/2}$) Mean	K_{Ic} (MN – m$^{-3/2}$) Std Dev	E (GPa)	ν
75 – 0.2 (+27)	0	0.84	0.03	68.0	0.217
	1.82	1.00	0.09	68.6	0.22
	8.54	1.29	0.14		
	14.1	1.46	0.19	86.0	0.22
	16.2	1.65	0.065		
	17.8	1.42	0.10	91.2	0.22
	20.6	1.48	0.24	91.8	0.23
	21.1	1.66	0.115		
	34.1	2.42	0.21		
	38.9	2.49	0.25	136.15	
75 – 0.67 (+4)	0	0.88	0.06	77.35	0.203
	2.15	1.01	0.11	79.75	0.204
	6.70	1.25	0.15	89.80	0.20
	14.2	1.48	0.107	96.83	0.21
	17.9	1.65	0.116		
	21.3	1.91	0.20	109.0	0.225
	21.7	1.84	0.175	106.7	0.237
	35.0	2.39	0.13	144.0	0.22
75 – 1.3 (-16)	0	0.83	0.025	80.2	0.205
	2.3	0.93	0.03	80.5	0.20
	9.20	1.23	0.06		
	18.0	1.44	0.09	103.3	0.21
	20.4	1.59	0.059		
	28.6	1.86	0.09	122.5	0.21
	34.5	2.22	0.11	135.5	0.21
75 – 3.0 (-37)	0	0.83	0.03	68.9	0.20
	1.9	1.09	0.16	69.6	0.20
	11.9	1.17	0.075	79.0	0.20
	15.6	1.31	0.085		
	20.6	1.80	0.116		
	22.5	1.49	0.06	95.3	0.22
	37.3	2.55	0.29	126.6	0.22

controlled to 50% relative humidity and 24°C. The span was 45.8 mm and the crosshead rate was 0.0085 mm/sec. The precrack length was determined by post-fracture observation of the fracture surface; location was aided by a pencil mark applied at the crack tip prior to fracture.

(a) (b)

Figure 1. Polished cross sections of hot pressed composites:
(a) 75-0.67 glass, ϕ = 2%; (b) 75-1.3 glass, ϕ = 35%.

Fracture toughness (K_{Ic}) was determined from specimen dimensions and fracture load by use of the equations of Brown and Srawley
[11] and assuming linear elastic behavior. Ten to twelve samples
were fractured for each composite, and the mean values and standard
deviations computed for each set. In all, about four hundred specimens were fractured for this study: the results are summarized
along with other pertinent data in Table I. The mean values for
K_{Ic} are plotted in subsequent figures.

Fractographic Analysis. A technique for marking the fracture
surface to reveal effects of particles on crack propagation has been
suggested by the work of Kerkhof and Richter [12] and recently employed by Khaund et. al [13]. The technique consists essentially of
applying a superposed oscillatory stress which will cause small local
perturbations in the crack front as it propagates. The ripples produced on the fracture surface mark the position of the crack front as
a function of time and can be used to infer local fracture direction
and velocity (since the frequency of oscillation is known).

For the crack/particle interaction studies, slow fracture velocities were obtained by notching bars (remnants from the K_{Ic} tests),
then initiating and running the crack with thermal stresses. (A 400

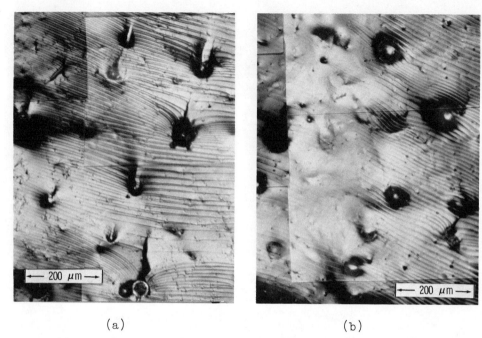

(a)   (b)

Figure 2.   (a) Fracture surface of 75-0.2 glass/crystal composite (ϕ=2 v/o).   (b) Fracture surface of 75-3.0 glass/crystal
composite (ϕ=2 v/o).   Fracture propagated from bottom to top in
both cases.

watt soldering gun provided the localized heating.)   Fracture velocities of 1-10 mm/sec were typical.   The modulated stress was obtained by coupling a vibratory engraver (60 Hz) to the bar while the
tip of the soldering gun was moved across the bar.   Most of the
fractographic examinations were made on samples containing only 2
v/o alumina spheres so that interaction between spheres was minimized.

A typical area of the fracture surface of the high expansion
(75 - .0.2) glass containing 2 v/o alumina spheres is shown in Figure
2a (this is part of a montage originally photographed at 220X).   Two
features are prominent: 1) strong curvature and pile-up of the markings at the inclusions indicating retardation of the crack front and
2) long "tails"* trailing many of the spheres.   Generally, the fracture seems to have passed the spheres very close to the interface.
At higher magnification it appears that a thin layer of glass remains

*Tails have the same genesis as "hackle."   Both are produced when a
crack separates into two or more non-coplanar sections as it advances.   Completion of the fracture requires lateral growth of one
of the sections to intersect the other [14].

on the surface of the spheres. Fracture rarely propagates through
the inclusions. The straightness of the markings as the crack
front approached the inclusions indicates there was no acceleration
or deceleration as the crack approached the inclusions. Except at
the inclusions, the fracture surface is relatively flat. However,
it appears that the crack front deviates from its plane as it approaches the sphere so that it tends toward the midplane of the
sphere as if attracted by local stresses. The velocity increases
rapidly once the fracture breaks away from an inclusion. Very small
inclusions (<1 μm) resulting from grinding the glass with alumina
balls are scattered through the glass. Crack front interaction with
these inclusions has the same general character but on a smaller
scale compared to the interaction with the spheres.

The appearance of the fracture surface in the lowest-expansion
(75 - 3.0) glass with 2 v/o alumina spheres (Figure 2b) contrasts
markedly with that of the high-expansion glass. In the 3.0 glass,
direct encounters resulting in either slowing of the fracture or
production of a marked tail, were very rare. Generally the surface
appears wavy with many rounded bumps or depressions corresponding to
near misses with spheres. The general impression is that the crack
front turns away from the spheres. In many instances, the spheres
can be seen clearly under the bumps. The uniformity of separation
of the markings produced by the oscillatory stress indicates very
little velocity change except for a slight reduction where tails are
formed. There is also a slight increase in velocity in many instances as the crack front passes over the back half of a sphere.

The fracture surfaces of composites containing intermediate expansion glasses (75-0.67 glass and 75-1.3 glass) have an appearance
intermediate between those of the 75-0.2 and 75-3.0 glass composites.
The 75-1.3 glass composite shows more direct encounters between the
crack front and the inclusions than in the 75-3.0 glass but not as
many as in the 75-0.67 glass.

The changes in the character of the fracture surface as the
glass is varied from the 0.2 B_2O_3/Na_2O ratio to the 3.0 ratio are
summarized in Table II. This is a tabulation of observations on
sections of fracture surface of 1.3 x 1.9 mm area. (In the case of
the 75-3.0 glass, observations were made on an area 0.8 x 2.2 mm and
adjusted by an appropriate factor.) Since the events where the crystal fractured or where marked bowing occurred have been counted already in the preceding categories, these events are indicated in
parentheses in Table II.

The trend in the fracture behavior is evident. For composites
where the thermal expansion of the glass is greater than that of the
crystal (75-0.2 glass), the local tangential tensile stresses produced by the differential expansion tend to turn the crack front

Table II
Compilation of Interactions Between Fracture Front and Crystal
Inclusions (1.3 x 1.9 mm area)

B_2O_3/Na_2O Ratio	0.2	0.675	1.3	3.0*
Bump in Crack Plane (but no crack splitting)	9	18	45	69
Slight Hackle	4	17	14	5
Tail Produced	83	55	34	17
Crystal Fractured[†]	(7)	(3)	(0)	(3)
Marked Bowing of Crack Front in Passing Inclusion[†]	(58)	(31)	(3)	(0)

*
Counted in a 0.8 x 2.2 mm area and adjusted by a factor of 2.47/1.76.

[†]
These numbers are shown in parentheses to indicate that the inclusion has already been counted in the first three categories.

---

toward the sphere (as shown schematically in Fig. 3a), so that the crack is forced to intersect the sphere. As the expansion mismatch is reduced, the tendency for intersection is reduced. When the mismatch changes sign, the local thermal stresses (radial tensile) begin to repel the crack front (as shown in Fig. 3b) and direct encounters are minimized. Examination of fracture surfaces of composites with higher volume fractions of alumina show that this difference in the character of fracture for different expansion mismatch persists. However, except in the 2.0 v/o samples, interactions between particles make a meaningful quantitative comparison difficult.

Some additional insight into the nature of the interaction between the fracture front and inclusions is obtained from scanning

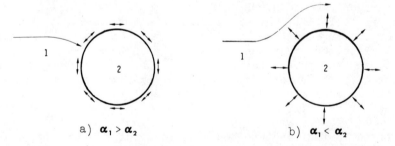

a) $\alpha_1 > \alpha_2$ b) $\alpha_1 < \alpha_2$

Figure 3. Schematic of fracture front interactions with spherical inclusions. The plane of the approaching crack is perpendicular to the page.

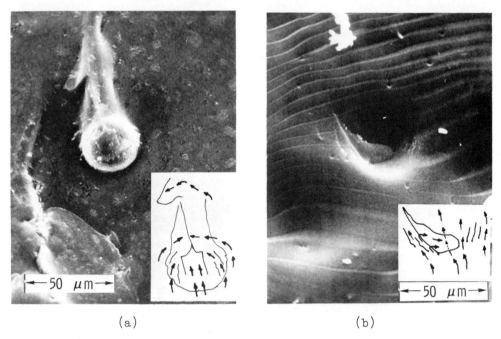

(a)   (b)

Figure 4.   Features on fracture surface of composites resulting
from interaction with inclusions: a) 75-0.2 glass b) 75-3.0 glass.
Fracture propagated from bottom to top in both cases.

electron microscopy. Typical behavior for the high-α (75-0.2) glass
is shown in Figure 4a. (This is one photo from a stereo pair.) The
spherical particle can be seen protruding from the fracture surface.
The crack front turned as it approached the sphere so that it intersected the midplane. From observations on the stereo pair and at
higher magnification, it is apparent that the fracture propagated as
indicated in the sketch. A central portion of the front moved directly over the sphere, leaving a very thin layer of glass on the
sphere. On either side, the fracture split as it encountered the
sphere and the resultant segments turned and intersected behind the
sphere to form a tail.

     The important difference between the behavior of the high-α and
the low-α glass is the way the crack front turns as it approaches the
sphere. In the low-α glass, the fracture turns to miss the sphere
instead of forcing the encounter as in the high-α glass; typical behavior for the low-α glass is shown in Figure 4b. The center feature
was produced when the fracture ran under a sphere. Fracture splitting
(hackle) is apparent on either side of the indentation. On the left
side a short tail has been produced. The L-shaped segment is an overlapping chip resulting from lateral fracture along a fracture split.
A more typical feature of the low-α glass is seen below and to the

left of the central feature. This is simply a shallow indentation
produced as the fracture ran beneath a sphere.

DISCUSSION

Fracture toughness (K_{Ic}) is plotted vs. volume percent of 25 μm
Al_2O_3 particles (ϕ) in Figure 5. It is evident that K_{Ic} increases
linearly with ϕ for composites in each glass system, according to the
relation

$$K_{Ic} = K_{Ic}^g + m \cdot \phi \qquad\qquad (3)$$

where the superscript g refers to the glass, and m is the slope of the
data in the figure. An increase in K_{Ic} with addition of higher-
modulus particles to a lower-modulus matrix (infinite in extent) has
been predicted by Khaund et. al. [15] and Evans [16]. They predict a
small effect, influenced more by modulus differences between inclusions and matrix than by thermal expansion mismatch, in accord with
the trends in Figure 5. The increase in crack propagation resistance
arises from the reduction of tensile stresses in the matrix in the
vicinity of high-modulus inclusions. Young's modulus and Poisson's
ratio for the composites are plotted versus volume fraction in Figure
6. It is apparent that the trend for E is concave upward, as observed
by Hasselman and Fulrath [17]. The differing residual thermal stresses in the four glass systems have no apparent effect on the trend.

When the data of Figures 5 and 6 are combined as in Figure 7,
the evidence shows that K_{Ic} increases with E, essentially without regard to the differences in thermal expansion coefficients. The
fractographic results of the preceding section showed that the stresses produced by thermal expansion mismatch can change the mechanics
of crack-particle interaction and force the development of additional
fracture surface. However, when fractographic observations are
evaluated in light of Figures 5-7, it becomes evident that increase in
composite modulus provides the primary toughening effect, and thermal
expansion effects are secondary: they serve only to order-rank the
composites in the sequence 0.67 -0.2 -1.3 -3.0 without changing the
basic strengthening mechanism embodied in Eq. (3). The effects of
occasional penetration of obstacles, as observed fractographically, is
not manifest per se in the K_{Ic} data, either because the fracture surface energy of the particles is similar to that of the glass, or pene-
trated obstacles are too few in number to contribute to the results.

With respect to Figures 5 and 7, the present data are not in
agreement with the flaw-size-limiting concept of strengthening proposed by Hasselman and Fulrath [6]. Their model requires that K_{Ic} be
a property of the glass phase only, and thus independent of the volume
fraction of reinforcement.

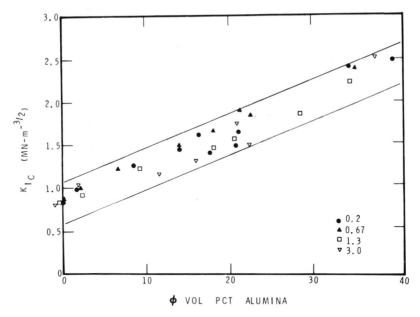

Figure 5. Fracture toughness (K_{Ic}) vs. volume percent alumina (ϕ) for reinforced glasses.

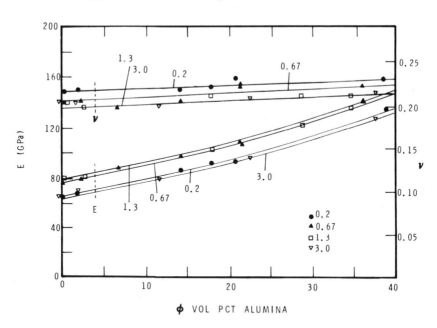

Figure 6. Young's modulus (E) and Poisson's ratio (ν) for reinforced glasses.

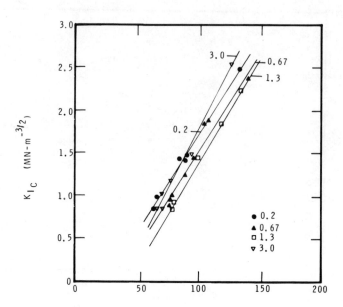

Figure 7. K_{IC} vs. E for reinforced glasses.

Lange [9] and Evans [18] have developed a model for strengthening of brittle composites in which the strength increase results from
an increase in effective surface energy (γ) due to crack bowing between obstacles. In order to compare our data with the predictions
of that model, we calculated γ values from K_{Ic}, E and ν according to
Eq. (2) and Table I. The results are plotted vs. $1/\lambda$ in Figure 8,
where

$$\lambda = \frac{2D(1 - \phi)}{3\phi} \qquad\qquad (4)$$

is the mean free path between obstacles and D is the particle diameter. Lange's data are also plotted for comparison. Our data, with
$D \simeq 25$ μm, fall into the particle diameter sequence observed by
Lange, and the trends are approximately linear. The intercept values
of γ for $1/\lambda = 0$ ($\phi = 0$) are those for the glass, in accord with
Lange's model. However, his data do not manifest a unique intercept.

The theoretical prediction for line tension effects developed by
Evans [18] is also plotted in Figure 8. The model is nearly sufficient quantitatively to explain our measurements, but predicts
curvature at small ϕ which we do not observe. Furthermore, the
model [9,18] was developed for the case of zero thermal residual
stress. By contrast, although we observe crack bowing in the 75-0.2
glass, we see none in the 3.0 glass. Nevertheless, both composites
have essentially the same dependence of γ on $1/\lambda$. When these results

are considered, the line-tension strengthening model does not appear to explain our observations satisfactorily.

In most of the studies cited earlier, wherein composite strength was determined as a function of amount of dispersed obstacles, it was found that strength does not rise monotonically with the number of obstacles. Instead, there exists a region at lower volume fractions where the strength is independent of volume fraction, or actually declines slightly with increasing ϕ [4,6,8,20]. The monotonically increasing value of K_{Ic} with ϕ, as embodied in Eq. (3), must be reconciled with this behavior observed for σ_f.

By combining Eqs. (1), (2), and (3) we obtain

$$\sigma_f = \frac{K_{Ic}^g + m \cdot \phi}{\sqrt{\pi a}} \quad . \tag{5}$$

To produce the observed trends in σ_f at small ϕ, the severity of strength-controlling flaws, as characterized by $\sqrt{a}$, must be increased by the addition of small volume fractions of particles. This concept is consistent with the model of Miyata and Jinno [19], wherein introduction of particles having matched expansion but higher modulus than

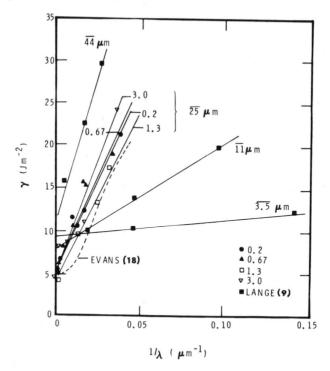

Figure 8. Fracture surface energy (γ) vs. (mean free path)$^{-1}$.

the matrix causes the nucleation of flaws larger than those originally present. At higher values of ϕ further crack propagation is
impeded by the particles, and strength increases with ϕ. Borom [20]
attributed the decrease of σ_f at low ϕ to the effects of thermal
mismatch at particles. We cannot absolutely rule out this possibility. However, the data of Frey and Mackenzie [8] indicate that
stresses due to $\Delta\alpha$ do not affect the relationship between σ_f and ϕ,
and our results (Figure 5) reveal a similar behavior for K_{Ic} vs. ϕ.

Borom [20] has proposed that strengthening in glass-ceramic composites results from load sharing, wherein the external load is ap-
portioned between matrix and particles according to their Young's
moduli. Under this scheme the stress in the matrix should be decreased, and the strength of the composite increased, in proportion
to composite modulus. This rationale is one-dimensional, in that it
does not account for stresses generated by Poisson effects. However,
Poisson's ratios for the borosilicate glasses of this study are
similar to those for Al_2O_3 [21], so that a one-dimensional simplified model may be sufficient.

The load sharing concept ought to apply to fracture toughness as well as to strength since, for constant a, K_{Ic} is proportional to σ_f. Figure 7 reveals that K_{Ic} does increase with E. However, the increase is greater than that which would be expected from
the simplified load-sharing model. The increase which we observe is
consistent with values which can be calculated from Lange's results
[9]. (Lange assumed Hasselman's modulus data [17] to be descriptive
of his material.) The difficulty in attempting to apply the load-
sharing model to fracture toughness measurements may be that load-
sharing is a global (i.e., continuum) concept, whereas the toughness,
as measured in this study, depends upon the localized interaction of
a large discrete crack with an array of dispersed obstacles. It is
possible that the short-range effects of modulus changes in the
vicinity of a crack front are greater than the global increases in
composite modulus measured by volume-averaging techniques. The analyses by Khaund [15] and Evans [16] present first steps in analyzing
these short-range interactions.

CONCLUSIONS

1. The addition of 25-μm-dia alumina particles to borosilicate
    glasses produces a linear increase in K_{Ic} with volume fraction,
    independent of the difference in thermal expansion coefficients
    of particles and glass.
2. The difference in thermal expansion between particle and matrix
    changes the nature of the interaction between cracks and par-
    ticles, but this change has a minor effect on composite
    toughness.

3. The fracture toughness of the composites increases directly
    with composite elastic modulus. However, the observed in-
    crease is greater than expected from a simple concept of
    matrix stress-reduction due to the presence of high-modulus
    inclusions.
4. To reconcile the trends in toughness with known trends in
    fracture strength, it is postulated that the introduction
    of small volume fractions of particles acts to introduce
    flaws larger than those originally present.

REFERENCES

1.   S. W. Freiman and L. L. Hench, J. Am. Cer. Soc. 55 (2) 86 (1972).
2.   M. P. Borom, A. M. Turkalo, and R. H. Doremus, J. Am. Cer. Soc.
     58 (9-10) 385 (1975).
3.   D. I. H. Atkinson and P. W. McMillan, J. Mater. Sci. 11 989,
     994 (1976).
4.   P. Hing and P. W. McMillan, J. Mater. Sci. 8 1041 (1973).
5.   J. C. Swearengen and R. J. Eagan, J. Mater. Sci. 11 1857 (1976).
6.   D. P. H. Hasselman and R. M. Fulrath, J. Am. Cer. Soc. 49 (2)
     68 (1966).
7.   R. W. Davidge and T. J. Green, J. Mater. Sci. 3 629 (1968).
8.   W. J. Frey and J. D. Mackenzie, J. Mater. Sci. 2 124 (1967).
9.   F. F. Lange, J. Am. Cer. Soc. 54 (12) 614 (1971).
10. D. R. Biswas and R. M. Fulrath, J. Am. Cer. Soc. 58 (11-12) 526
     (1975).
11. W. F. Brown, Jr. and J. E. Srawley, ASTM STP 410, Am. Soc. for
     Test. & Mater. (1966).
12. F. Kerkhof and H. Richter, Proc. 2nd International Conf. on
     Fracture, Ed. P. L. Pratt, Chapman & Hall, Ltd. London, (1969),
     p. 463.
13. A. K. Khaund, D. J. Green, P. S. Nicholson, and J. D. Embury,
     "Crack-Particle Interactions in Brittle Composites," Paper
     142-B-77, Am. Cer. Soc. 79th Annual Meeting, Chicago, Ill,
     April 1977.
14. V. D. Frechette, Proc. Br. Cer. Soc. 1965 (5) 97.
15. A. K. Khaund, V. D. Krstic and P. S. Nicholson, "The Influence
     of Elastic and Thermal Mismatch on the Local Crack Driving Force
     in Brittle Composites," to be published.
16. A. G. Evans, J. Mater. Sci. 9 1145 (1974).
17. D. P. H. Hasselman and R. M. Fulrath, J. Am. Cer. Soc. 48
     (4) 218 (1965).
18. A. G. Evans, Phil. Mag. 26 1327 (1972).
19. N. Miyata and H. Jinno, J. Mater. Sci. 7 973 (1972).
20. M. P. Borom, J. Am. Cer. Soc. 60 (1-2) 17 (1977).
21. R. L. Coble and W. D. Kingery, J. Am. Cer. Soc. 39 (11) 377
     (1956).

CONTRIBUTORS

Co-Chairmen

R. C. Bradt, Professor, Department of Materials Science and
Engineering, Ceramic Science and Engineering Section,
Pennsylvania State University, University Park,
Pennsylvania.

D. P. H. Hasselman, Whittemore Professor of Engineering,
Department of Materials Engineering, Virginia Polytechnic
Institute and State University, Blacksburg, Virginia.

F. F. Lange, Senior Research Scientists, Structural Ceramics
Group, Rockwell International/Science Center, Thousands
Oaks, California.

Conference Session Chairmen

E. M. Lenoe, AMMRC; Watertown, Massachusetts

R. W. Davidge, AERE-Harwell; England

V. D. Frechette, Alfred University; Alfred, New York

H. C. Palmour, III, North Carolina State University; Raleigh,
North Carolina

M. L. Williams, University of Pittsburgh; Pittsburgh,
Pennsylvania

K. R. Kinsman, E.P.R.I., Palo Alto, California

Conference Coordinator

Ronald Avillion, J. Orvis Keller Conference Center,
Pennsylvania State University; University Park,
Pennsylvania.

AUTHORS

M. Adams, Jet Propulsion Laboratory, Pasadena, California

W. F. Adler, Effects Technology Inc., Santa Barbara, California

C. G. Annis, Pratt and Whitney, West Palm Beach, Florida

K. Arin, General Electric Co., Schenectady, New York

M. Bakioglu, Lehigh University, Bethlehem, Pennsylvania

G. J. Bansal, Battelle Columbus Labs., Columbus, Ohio

L. M. Barker, Terra Tek, Salt Lake City, Utah

S. B. Batdorf, UCLA, Los Angeles, California

E. K. Beauchamp, Sandia Labs., Albuquerque, New Mexico

P. F. Becher, Naval Research Lab., Washington, District of
    Columbia

R. C. Bennet, University of British Columbia, Vancouver, British
    Columbia, Canada

D. R. Biwas, University of California, Berkeley, California

W. C. Bourne, The Pennsylvania State University, University
    Park, Pennsylvania

R. C. Bradt, The Pennsylvania State University, University Park,
    Pennsylvania

S. D. Brown, University of Illinois, Urbana, Illinois

J. G. Bruce, Honeywell, Minneapolis, Minnesota

F. E. Buresch, Kernforschungsanlage Julich, Julich, West Germany

J. S. Cargill, Pratt and Whitney, West Palm Beach, Florida

H. C. Chandan, Bell Laboratories, Norcross, Georgia

R. J. Charles, General Electric Co., Schenectady, New York

M. M. Chaudhri, Cavendish Laboratory, Cambridge, England

J. L. Chermant, Universite de Caen, Caen, Cedex, France

N. Claussen, Max Planck Institute, Stuttgart, West Germany

P. H. Conley, Koppers Co., Inc., Pittsburgh, Pennsylvania

R. W. Davidge, AERE-Harwell, Harwell, England

A. DesChanvres, University of Caen, Caen, Cedex, France

W. Duckworth, Battelle Columbus, Ohio

R. J. Eagen, Sandia Labs., Albuquerque, New Mexico

A. F. Emery, University of Washington, Seattle, Washington

F. F. Erdogan, Lehigh University, Bethlehem, Pennsylvania

A. G. Evans, Rockwell International, Thousand Oaks, California

H. Fessler, University of Nottingham, Nottingham, England

I. E. Field, University of Cambridge, Cambridge, England

V. D. Frechette, Alfred University, Alfred, New York

S. W. Freiman, U. S. Naval Research Laboratory, Washington,
      District of Columbia

E. R. Fuller, Jr., National Bureau of Standards, Washington,
      District of Columbia

R. M. Fullrath, University of California, Berkeley, California

A. Gaddipatti, University of Washington, Seattle, Washington

S. T. Gonczy, Northwestern University, Evanston, Illinois

R. S. Gordon, University of Utah, Salt Lake City, Utah

J. K. Gran, SRI, Palo Alto, California

D. J. Green, Canada Center for Min. & Energy Tech. Ottawa, Canada

R. M. Gruver, Ceramic Finishing Company, State College, Pennsylvania

T. K. Gupta, Westinghouse, Pittsburgh, Pennsylvania

J. T. Hagan, Cambridge University, Cambridge, England

S. A. Halen, USNRL, Washington, District of Columbia

D. P. H. Hasselman, Virginia Polytechnic Institute and State
University, Blacksburg, Virginia

N. Hattu, N. V. Philips Gloelampen Fabrieken, Eindhoven, Holland

B. J. Hockey, National Bureau of Standards, Washington, District
of Columbia

S. V. Hooker, Effects Technology, Santa Barbara, California

C. O. Hulse, United Technologies, East Hartford, Connecticut

C. A. Johnson, General Electric, Schenectady, New York

D. L. Johnson, Northwestern University, Evanston, Illinois

H. Johnson, National Bureau of Standards, Washington, District
of Columbia

B. F. Jones, Admiralty Materials Laboratory, Dorsett, England

R. L. Jones, SRI, Palo Alto, California

H. P. Kirchner, Ceramic Finishing Company, State College
Pennsylvania

A. S. Kobayashi, University of Washington, Seattle, Washington

B. G. Koepke, Honeywell International, Bloomington, Minnesota

F. F. Lange, Rockwell International, Thousand Oaks, California

J. Lankford, Southwest Research Institute, San Antonia, Texas

B. R. Lawn, University of New South Wales, Kensington, Australia

M. W. Lindley, Admiralty Materials Laboratory, Dorsett, England

W. J. Love, University of Washington, Seattle, Washington

B. D. Marshall, University of New S. Wales, Kensington, Australia

M. Matsui, NGK, Ltd., Nagoya, Japan

J. R. McClaren, AERE-Harwell, Harwell, England

C. L. McKinnis, Owens-Corning Fiberglas, Granville, Ohio

J. J. Mecholsky, Jr., USNRL, Washington, District of Columbia

T. Michalske, Alfred University, Alfred, New York

J. S. Nadeau, University of British Columbia, Vancouver, Canada

P. S. Nicholson, McMaster University, Hamilton, Ontario, Canada

H. A. Nied, General Electric, Schenectady, New York

W. L. Newell, USNRL, Washington, District of Columbia

I. Oda, NGK, Ltd., Nagoya, Japan

E. Olsen, IBM, San Jose, California

F. Osterstock, University of Caen, Caen, Cedex, France

R. F. Pabst Max-Planck Institute, Stuttgart, West Germany

K. C. Pitman, Admiralty Materials Labs., Dorsett, England

P. Pizzi, Fiat Research Center, Orbassano, Italy

B. J. Pletka, National Bureau of Standards, Wasington, District
   of Columbia

R. Pohanka, Office of Naval Research/NRL, Arlington, Virginia

R. A. Queeney, The Pennsylvania State University, University
   Park, Pennsylvania

R. W. Rice, U. S. Naval Research Laboratory, Washington,
   District of Columbia

J. E. Ritter, Jr., University of Massachusetts, Amherst,
   Massachusetts

E. M. Rockar, National Bureau of Standards, Washington, District
   of Columbia

D. J. Rowcliffe, SRI International, Menlo Park, California

N. L. Rupert, Naval Surface Weapons Center, Dahlgren, Virginia

W. D. Scott, University of Washington, Seattle, Washington

D. K. Shetty, University of Utah, Salt Lake City, Utah

G. Sines, UCLA, Los Angeles, California

A. D. Sivil, Lucas Ltd., Birmingham, England

V. Snijders, N. V. Philips Gloeilampen Fabrieken, Eindhoven, Holland

W. D. Snowden, General Electric Company, Nela Park, Ohio

T. Soma, NGK, Ltd., Nagoya, Japan

L. I. Staley, University of Washington, Seattle, Washington

P. Stanley, University of Nottingham, Nottingham, England

J. Steeb, Max-Planck Institute, Stuttgart, West Germany

M. V. Swain, Cambridge University, Cambridge, England

J. C. Swearengen, Sandia, Albuquerque, New Mexico

G. Tappin, AERE-Harwell, Harwell, England

B. K. Tariyal, Bell Laboratories, Norcross, Georgia

R. M. Thomson, National Bureau of Standards, Washington, District of Columbia

G. G. Trantina, General Electric Research & Development, Schenectady, New York

R. E. Tressler, The Pennsylvania State University, University Park, Pennsylvania

J. R. Varner, UCLA, Los Angeles, California

J. D. P. Veldkamp, N. V. Philips Gloeilampen Fabrieken, Eindhoven, Holland

A. V. Virkar, University of Utah, Salt Lake City, Utah

A. M. Walley, Cavendish Laboratory, Cambridge, England

S. M. Wiederhorn, National Bureau of Standards, Washington, District of Columbia

C. Cm. Wu, USNRL, Washington, District of Columbia

INDEX

Pages 1-506 are found in Volume 3, pages 507-987 in Volume 4.

Abrasive wear, 257
Acoustic emission, 246, 719
Alumina, 15, 22, 32, 186, 190,
        237, 245, 257, 299,
        379, 403, 451, 466,
        476, 483, 495, 568,
        607, 623, 651, 695,
        725, 762, 800, 826,
        838, 850, 908, 917
Alumina-boron nitride, 855
Alumina-hafnium oxide, 903
Alumina-titania, 800
Atomic lattice, 507

Barium titanate, 688, 857
Bend test, 183, 186-188, 191,
        198, 247
Benzene, 100
Beryllium oxide, 800, 821
Biaxial strength test, 36,
        419, 725, 754
Biaxial stress, 35, 82, 177,
        435, 449
Biological environment, 725
Boron, 99
Boron carbide, 99, 862, 885
Boron oxide, 151
Braze joints, 463
Brazil (tensile) test, 764

Calcium fluorite, 333, 865
Calcium oxide, 741
Carbon, 99
Cement, 761
Chlorine, 115
Cleavage, 337
Composites, 945, 961, 973

Compression
  microfracture, 245
  testing, 403, 761
Concrete, 762, 821
Cone cracks, 208, 308, 349, 367
Controlled flaws, 113, 211
Coplanar strain energy, 33
Corrosion
  molten salt, 113
  stress, 550, 581, 601, 652, 720
Crack
  blunting, 99
  branching, 237
  morphology, 254
  subsurface, 276
Crack density (bimodal), 8, 38,
        67, 125, 143, 174, 289,
        842
Crack extension
  criteria, 33, 55, 507
  viscous drag effects, 553
  See also Subcritical crack growth
Creep, 99
Cristobalite, 586, 717
Critical stress intensity tests
  ceramic braze, 463
  controlled flaw, 495
  double torsion, 190, 737
  dynamic impact, 495
  notched ring, 473
  short-rod, 483
  specimen for high temperature,
        451
  wedge-loaded double-cantilever,
        451
Cyclic fatigue, 670

995

Damage threshold
  impact, 309
Delayed failure
  See Subcritical crack growth
Design diagrams, 23, 676, 721,
    726
Double cantilever beam test,
    452, 641
Double torsion test, 190, 202,
    274, 291, 641, 687,
    711, 737, 753, 923,
    963
Dynamic fatigue, 168, 678, 726,
    761
Dynamic fracture, 23, 208, 237,
    495, 510

Eccentricity of loading, 408
Electrolytes, 556
  solid, 651
Epoxy-acrylate copolymer, 162
Erosion, 303, 328, 379
Ethylene vinyl acetate, 144,
    164
Eutectic microstructure, 903

Failure criterion, 12, 33,
    55, 67
Fatigue
  cyclic, 670
  dynamic, 168, 678, 726,
    761
  static, 21, 156, 171, 577,
    667, 726, 773, 787
  thermal, 93
Ferrites, 273
Fiber glass, 581
Fibers, 125, 143, 161, 581
Fibrous fracture, 811
Finite element method, 58, 473
Flaw density relations
  See Crack density
Fracture mirror, 187, 199, 238,
    239, 242
Fracture statistics, 1, 31, 57,
    61, 126, 183, 671
  See also Weibull
Fracture theories
  lattice theories, 507
  slow crack growth, 549

Gadolinium oxide, 800
Germanium, 381
Glass, 143, 231, 258, 273, 361,
    365, 395, 445, 468, 549,
    572, 602, 682, 695, 761,
    773, 787, 961, 973
  alumino silicate, 234, 558, 776
  barium silicate, 605
  borosilicate, 349, 558, 793, 973
  lead, 537, 570
  lead alkali, 558
  silicate, 123, 125, 144, 161,
    206, 557
  sodium alumino silicate, 776
  sodium borosilicate, 933
  sodium lime silicate, 22, 211,
    221, 234, 264, 287, 349,
    552, 602, 639, 667, 711,
    761, 787, 792, 855, 962
Glass ceramics, 190, 203, 745, 800
Grain size effects, 852, 863
Graphite, 16, 100, 436, 821, 917
Griffith theory, 6, 63, 115, 507,
    531, 663
Grinding damage, 273

Hafnium oxide, 801
Hardness, 573
Hertzian
  cracks, 208, 266
  elevated temperature fracture,
    369
  fracture, 208, 333
Hugoniot relations, 303, 333
Humidity, 549, 581, 585
Hydrogen fluoride, 117

Impact
  angle of impingement, 303, 387
  bending, 495
  damage, 303, 349, 365
  damage modes, 327
  damage threshold, 309
  elastic regime, 307
  elastic/plastic regime, 320
  elevated temperature studies,
    367, 381, 495
  energy losses, 373
  exploding wire tests, 333

Impact (cont'd)
  particle, 210
  photographic investigation,
     349
  plastic regime, 319, 361
  TEM study, 373
Inclusions, 945, 961
  as flaws, 865
Indenters, 205, 231
  blunt, 208
  flaw production, 444
  sharp, 212
Indentation fracture, 205, 231,
     245, 257, 303, 335, 349,
     365, 495
  compressive microfracture,
     245
  crack types, 214
  scratching, 257
  single flaw interaction, 303,
     444
  strength degradation, 205,
     231
  tests, 205, 231, 245
Internal stresses
  See Residual stresses
Iron, 511

Krebs-Ringer solution, 725

Lateral cracks, 213, 274, 308,
     354, 373
Lattice trapping, 507, 521
Lead titanate, 688, 802
Lead zirconate, 688
Lead zirconate titanate, 476,
     762, 858, 933
Line tension of cracks, 951

Machining, strength effects, 115
Magnesium fluorite, 324, 862
Magnesium oxide, 93, 186, 273,
     306, 333, 381, 407, 741
Magnesium oxide-silicon nitride,
     114
Magnesium oxide-titania, 800,
     852
Median cracks, 213, 257, 275
Microcracking, 245, 257, 799,
     821, 835, 854

Microstructure, 25, 68, 86, 101,
     113, 799, 849, 903, 921
Microvoids, 85
Molten salt corrosion, 113
Multiaxial stress
  See Polyaxial stress
Multiple flaw model, 70

Neutron scattering, 85
Niobium oxide, 800, 852
Nylon, 333
Nylon bead impact, 338

pH, effect on cracks, 553, 560
Phase transitions, 802, 813, 857
     877
Plastic
  flow, 384, 569, 857
  zone, 257, 307, 322, 335, 397,
     484, 915
Polyaxial stress, 4, 32, 177,
     419, 435, 754
Polyethylene, 135
Polyvinylidene fluoride, 144
Porcelain, 407, 605, 682, 711
Porosity, 70, 85, 849, 913, 933
Process zone, 551, 810, 821, 835
Proof testing 24, 166, 672, 704,
     722

Quantum tunnelling, 507
Quartz, 218, 343, 473, 802, 858
Quasistatic indentation, 257, 319,
     340, 380

Radial crack, 308, 354, 368
Refractory metals, 637
Residual stresses, 214, 268, 687,
     773, 787, 799, 863, 914,
     947, 963
Ring fracture, 342
Rock salt, 335

Sapphire, 252, 257, 299, 333,
     606, 885
  See also Alumina
Scratches, 257, 273
Second phases, 307, 803, 859,
     945, 961, 973
Silane, 164

Silicon, 100, 381, 436, 927
Silicon carbide, 99, 113, 178,
        381, 403, 435, 476, 495,
        623, 804, 858, 878
Silicon nitride, 32, 70, 85,
        113, 135, 177, 190, 318,
        379, 403, 466, 476, 495,
        623, 682, 737, 768, 811,
        828, 856, 878, 917, 921
Slip bands, 335
Slow crack growth, 507, 537,
        549, 597, 639, 651,
        659, 687, 711, 725,
        737, 745, 773, 787
Sodium, 115
Spinel, 324, 865, 885
Spin test, 177
Stainless steel, 630
Statistics
  See Fracture statistics
  See also Weibull
Strength degradation, 113, 143,
        150, 205, 219, 231,
        303, 325, 651
Strength-grain size relations,
        863
Stress corrosion, 550, 581,
        601, 652, 695, 711,
        720
Stress induced phase transforma-
        tion, 813, 877
Stress rupture, 99, 623
Subcritical crack growth
  See Slow crack growth
Sulfur, 115
Surface cracks (flaws), 17, 41,
        55, 115, 143, 195, 208,
        257, 273, 443

Tempered (tempering), 224, 231,
        237, 773, 787
Thermal expansion differences,
        803, 836
Thick films, 463
Threshold stress, 23, 55, 314
Titania, 800, 852
Titanium carbide, 891
Toluene, 186
Tungsten, 633

Tungsten carbide, 324, 333, 350,
        403, 489, 868, 891
Tungsten carbide-cobalt, 307, 853
Twinning, 245, 687, 853

Ultramicroscopy, laser, 581

Vapor transport crack blunting,
        93
Vitrified grinding wheels, 682
Volume flaws, 186

Wallner lines, 644, 945, 965
Weibull, 1, 31, 51, 125, 145,
        177, 189, 442, 671, 933

Yttrium oxide, 937

Zinc selenide, 855
Zinc sulfide, 318, 345, 473
Zircon, 804
Zirconia, 411, 466, 800, 828, 836,
        855, 877, 907, 962